EPILEPSY – HISTOLOGICAL, ELECTROENCEPHALOGRAPHIC AND PSYCHOLOGICAL ASPECTS

Edited by **Dejan Stevanovic**

Epilepsy – Histological, Electroencephalographic and Psychological Aspects
Edited by Dejan Stevanovic

Published by InTech
Janeza Trdine 9, 51000 Rijeka, Croatia

Publishing Process Manager Bojan Rafaj

Technical Editor Teodora Smiljanic

Cover Designer InTech Design Team

Additional hard copies can be obtained from orders@intechopen.com

Epilepsy – Histological, Electroencephalographic and Psychological Aspects,
Edited by Dejan Stevanovic
p. cm.
ISBN 978-953-51-0082-9

For EU product safety concerns:
IN TECH d.o.o., Prolaz Marije Krucifikse Kozulić 3, 51000 Rijeka, Croatia,
info@intechopen.com or visit our website at intechopen.com.

Meet the editor

Dejan Stevanovic, MD, was born in 1979 in Serbia. He obtained his medical education at the Medical School University of Belgrade, where he also finished child and adolescent psychiatry training and doctorial studies, but he was educated in research at various European centers. His main research interests are child neuropsychiatry, cognitive neurology, and pharmacoeconomics. Specifically, Dejan is working and researching children with epilepsy and neurobiological syndromes, but the patient-reported outcome concept as well. Currently, he is affiliated with the General Hospital Sombor in Serbia, serves editorial boards of several journals, and is a consultant for medical research.

Contents

Preface

As the very complex neurological condition primarily characterized by unexpected, episodic, and chronic nature of variety of seizures and neurophysiologic changes in the brain, epilepsy is also characterized by different neurodevelopmental, psychological, behavioral, educational, social, and many other difficulties. Therefore, many researchers all over the world investigating various aspects of epilepsy leave us with nearly 5000 new scientific documents every year. Reading all these documents is practically impossible, but unnecessary as well. Thus, periodically edited books about particular aspects of epilepsy are more than welcome.

With the InTech vision of including authors from different parts of the world, different educational backgrounds, and offering open-access to their published work, I my proud to present the latest edited book in epilepsy research, *Epilepsy: Histological, electroencephalographic, and psychological aspects.* Here are selected twelve interesting and inspiring chapters dealing with basic molecular and cellular mechanisms underlying epileptic seizures, electroencephalographic findings, and neuropsychological, psychological, and psychiatric aspects of epileptic seizures, but non-epileptic as well. Each chapter is organized in two parts. In the first, authors summarized the most relevant past research on the topic they had selected, while in the second, they gave their recent findings on the topic. All authors took responsibility to present only the most relevant information.

I am deeply convinced that the chapters will save our time reading many articles, but also contribute significantly to the growing nature of epilepsy research.

Dejan Stevanovic, MD
Department of Child Psychiatry
General Hospital Sombor
Serbia

Part 1

Histological Aspects

1

Cellular and Molecular Mechanisms Underlying Epilepsy: An Overview with Our Findings

Melda Yardimoglu, Gul Ilbay, Cannur Dalcik, Hakki Dalcik and Sibel Kokturk
Department of Histology & Embryology, Faculty of Medicine, Kocaeli University Turkey

1. Introduction

Epilepsy affects more than 50 million people worldwide. It is foreseen that around 50 million people in the world have epilepsy, or about 1% of the population. (http://www.epilepsyfoundation.org; http://epilepsy.med.nyu.edu/epilepsy/frequently-asked-questions: NYU Langone Medical Center, 2011). At the global level, it is estimated that there are nearly 50 million persons suffering from epilepsy of which three-fourths, i.e. 35 million, are in developing countries (http://www.searo.who.int.). It is the most common serious neurological condition. It can affect all age groups and it may be the result of an acute or chronic cerebral illness. Epileptic seizures begin simultaneously and several histopathological changes occur in both cerebral hemispheres. Epilepsy is a disorder of the central nervous system characterized by recurrent and sudden increase in electrical activity. Metabolic studies have shown that oxygen availability, glucose utilization, and blood flowall increase dramatically during epileptic seizures. It is also known that epileptic activity may induce some molecular and structural changes in the different brain regions (Ingvar & Siejo, 1983; Siesjo et al., 1986; Oztas et al., 2001).

Enolase is glyoclytic enzyme that converts 2-phosphoglycrate to phosphoenol pyruvate. It has three immunologically distinct subunits; α, β, and γ. The γ form is found primarily in the cytoplasm and process of neurons, which is referred to as neuron-specific enolase (NSE). NSE is a sensitive marker of neuronal damage in several central nervous system (CNS) diseases including epilepsy (Schmechel et al., 1978; Nogami et al., 1998a; Nogami et al., 1998b; Rodriquez-Nunez et al., 2000). Changes in membrane integrity as a result of neuronal injury can cause leakage of protein such as NSE from cytosol into extracellular space. Increased NSE in serum (sNSE) and in cerebrospinal fluid (cNSE) have been observed in animal model of traumatic and ischemic brain injury, cerebral hypoxia, and epileptic seizures (Hay et al., 1984; Persson et al., 1988; Hatfield & McKernan 1992; Barone et al., 1993; Brandel et al., 1999; Steinhoff et al., 1999). sNSE levels are also reported to increase in epileptic activities due to increased blood-brain barrier (BBB) permeability. Elevation of sNSE after SE correlated with overall histologic evidence for damage (Jacobi & Reiber, 1988; DeGiorgio et al., 1996; Sankar et al., 1997; Correale et al., 1998; B¨uttner et al., 1999; DeGiorgio et al., 1999; Schreiber et al., 1999). Although sNSE is not sensitive enough to detect neuronal damage, cNSE seems to be a reliable parameter for assessing neurological insult in patients (Lima et al., 2004). Although multiple reports have documented elevation in NSE levels following neuronal injury in various neurological disorders, little is known about the localization of NSE in different brain regions after chemically induced acute and chronic seizures. Therefore, the present work was designed to investigate changes in NSE immunoreactivity in different brain regions including the cerebral cortex, thalamus, hypothalamus, and hippocampus in the single- and repeated PTZ-induced generalized tonic-clonic seizures in rats.

2. Epileptic seizures

In simple terms, our nervous system is a communication network that controls every thought, emotion, impression, memory, movement, and upmost defining who we are. Nerves, throughout the body, function like telephone lines enabling the brain to communicate with every part of the body via electrical signals. In epilepsy, brain's electrical rhythms have a tendency to become imbalanced resulting in recurrent seizures (Schachter, 2006). Normally, the brain continuously generates tiny electrical impulses in an orderly pattern. These impulses travel along the network of nerve cells, called neurons, in the brain and throughout the whole body via chemical messengers called neurotransmitters. A seizure occurs when the brain's nerve cells misfire and generate a sudden, uncontrolled surge of an electrical activity in the brain. Another concept important to epilepsy is that different areas of the brain control different functions.

The International League Against Epilepsy (ILAE) Commission on Classification and Terminology has revised concepts, terminology, and approaches for classifying seizures and forms of epilepsy. Generalized and focal are redefined for seizures as occurring in and rapidly engaging bilaterally distributed networks (generalized) and within networks limited to one hemisphere and either discretely localized or more widely distributed (focal). Classification of generalized seizures is simplified. No natural classification for focal seizures exists; focal seizures should be described according to their manifestations (e.g., dyscognitive, focal motor). The concepts of generalized and focal do not apply to electroclinical syndromes. Genetic, structural–metabolic, and unknown represent modified concepts to replace idiopathic, symptomatic, and cryptogenic. Not all epilepsies are recognized as electroclinical syndromes. Organization of forms of epilepsy is first by specificity: electroclinical syndromes, nonsyndromic epilepsies with structural–metabolic causes, and epilepsies of unknown cause. Further organization within these divisions can be accomplished in a flexible manner depending on purpose. Natural classes (e.g., specific underlying cause, age at onset, associated seizure type), or pragmatic groupings (e.g., epileptic encephalopathies, self-limited electroclinical syndromes) may serve as the basis for organizing knowledge about recognized forms of epilepsy and facilitate identification of new forms (Berg, 2010).

Concepts and terminology for classifying seizures and epilepsies have, until recently, rested on ideas developed nearly a century ago. In order for clinical epilepsy and practice to benefit fully from the major technological and scientific advances of the last several years, advances that are revolutionizing our understanding and treatment of the epilepsies, it is necessary to break with the older vocabulary and approaches to classifying epilepsies and seizures. The Commission on Classification and Terminology made specific recommendations to move this process along and ensure that classification will reflect the best knowledge, will not be arbitrary, and will ultimately serve the purpose of improving clinical practice as well as research on many levels. The recommendations include new terms and concepts for etiology and seizure types as well as abandoning the 1989 classification structure and replacing it instead with a flexible multidimensional approach in which the most relevant features for a specific purpose can be emphasized. This is not a finished product and will take yet more time to achieve. Waiting any longer, however, would be a disservice to patient care and will continue the longstanding frustrations with the earlier system which, at this point in time, can be viewed as both antiquated and arbitrary (Berg et al., 2011). There are so many kinds of seizures that neurologists who specialize in epilepsy are still updating their thinking about how to classify them. Usually, they classify seizures into two types, primary generalized seizures and partial seizures. The difference between these types is in how they begin: Primary generalized seizures begin with a widespread electrical discharge that involves both sides of the brain at once. Hereditary factors are important in many of these seizures (Schachter, 2006; MedicineNet, Inc.).Partial seizures begin with an electrical discharge in one limited area of the brain. Some are related to head injury, brain infection, stroke, or tumor, but in most cases the cause is unknown (Steven C. Schachter, 2006; MedicineNet, Inc.). Identifying certain seizure types and other characteristics of a person's epilepsy like the age at which it begins, for instance, allows doctors to classify some cases into epilepsy syndromes. This kind of classification helps us to know how long the epilepsy will last and the best way to treat it.

Primary generalized seizures:Absence seizures are brief episodes of staring. During the seizure, awareness and responsiveness are impaired. People who have them usually do not realize when they have had one. There is no warning before a seizure, and the person is completely alert immediately afterwards (Schachter, 2006)

Simple absence seizures are just stares. Many absence seizures are considered *complex* absence seizures meaningthey include a change in muscle activity. The most common movements are eye blinkikgs. Other movements include slight tasting movements of the mouth, hand movements such as rubbing the fingers together, and contraction or relaxation of the muscles. Complex absence seizures are often more than 10 seconds long (Schachter, 2006), Atypical (a-TIP-i-kul) means unusual or not typical. The person will stare (as they would in any absence seizure) but often is somewhat responsive. Eye blinking or slight jerking movements of the lips may occur. This behavior can be hard to distinguish from the person's usual behavior, especially in those with cognitive impairment. Unlike other absence seizures, rapid breathing cannot produce them.

Myoclonic (MY-o-KLON-ik) seizures are brief, shock-like jerks of a muscle or a group of muscles. "Myo" means muscle and "clonus" (KLOH-nus) means rapidly alternating contraction and relaxation—jerking or twitching—of a muscle (Schachter, 2006). Even people without epilepsy can experience myoclonus in hiccups or in a sudden jerk that may wake you up as you are just falling asleep. These things are normal.

Muscle "tone" is the muscle's normal tension. "Atonic" (a-TON-ik) means "without tone," so in an atonic seizure, muscles suddenly lose strength. The eyelids may droop, the head may nod, and the person may drop things and often falls to the ground. These seizures are also called "drop attacks" or "drop seizures." The person usually remains conscious.

Muscle "tone" is the muscle's normal tension at rest. In a "tonic" seizure, the tone is greatly increased and the body, arms, or legs make sudden stiffening movements. Consciousness is usually preserved. Tonic seizures most often occur during sleep and usually involve all or most of the brain, affecting both sides of the body. If the person is standing when the seizure starts, he or she often will fall.

"Clonus" (KLOH-nus) means rapidly alternating contraction and relaxation of a muscle -- in other words, repeated jerking. The movements cannot be stopped by restraining or repositioning the arms or legs. Clonic (KLON-ik) seizures are rare, however. Much more common are tonic-clonic seizures, in which the jerking is preceded by stiffening (the "tonic" part). Sometimes tonic-clonic seizures start with jerking alone. These are called clonic-tonic-clonic seizures! This type is what most people think of when they hear the word "seizure." An older term for them is "grand mal." As implied by the name, they combine the characteristics of tonic seizures and clonic seizures. The tonic phase comes first: All the muscles stiffen. Air being forced past the vocal cords causes a cry or groan. The person loses consciousness and falls to the floor. The tongue or cheek may be bitten, so bloody saliva may come from the mouth. The person may turn a bit blue in the face. After the tonic phase comes the clonic phase: The arms and usually the legs begin to jerk rapidly and rhythmically, bending and relaxing at the elbows, hips, and knees. After a few minutes, the jerking slows and stops. Bladder or bowel control sometimes is lost as the body relaxes. Consciousness returns slowly, and the person may be drowsy, confused, agitated, or depressed.

2.1 Motor seizures

These cause a change in muscle activity. For example, a person may have abnormal movements such as jerking of a finger or stiffening of part of the body. These movements may spread, either staying on one side of the body (opposite the affected area of the brain) or extending to both sides. Other examples are weakness, which can even affect speech, and coordinated actions such as laughter or automatic hand movements. The person may or may not be aware of these movements (Schachter, 2006).

2.2 Sensory seizures

These cause changes in any one of the senses. People with sensory seizures may smell or taste things that are not there, may hear clicking, ringing, or a person's voice when there is no actual sound, or may feel a sensation of "pins and needles" or numbness. Seizures may even be painful for some patients. They may feel as if they are floating or spinning in space. They may have visual hallucinations, seeing things that are not there (a spot of light, a scene with people). They also may experience illusions—distortions of true sensations. For instance, they may believe that a parked car is moving farther away, or that a person's voice is muffled when it has actually clear (Schachter, 2006).

Autonomic seizures These cause changes in the part of the nervous system that automatically controls bodily functions. These common seizures may include strange or unpleasant sensations in the stomach, chest, or head; changes in the heart rate or breathing; sweating; or goose bumps.

2.3 Psychic seizures

These seizures change how people think, feel, or experience things. They may have problems with memory, garbled speech, ability to find the right word, or understanding spoken or written language. They may suddenly feel emotions like fear, depression, or happiness with no apparent reason. Some may feel as though they are outside their body or may have déja vu.These seizures usually start in a small area of the temporal lobe or frontal lobe of the brain. They quickly involve other areas of the brain that affect alertness and awareness. Thus, eventhough the person's eyes are open and they may move that seem to have a purpose, in reality "nobody's home." If the symptoms are subtle, other people may think the person is just daydreaming (Schachter, 2006). Some people can have seizures of this kind without realizing that anything has happened. Because the seizure can wipe out memories of events just before or after it, however, memory lapses can be a problem (Schachter, 2006).

Some of these seizures (usually ones beginning in the temporal lobe) start with a simple partial seizure. Also called an aura, this warning seizure often includes an odd feeling in the stomach. Then the person loses awareness and stares blankly. Most people move their mouth, pick at the air or their clothing, or perform other purposeless actions. These movements are called "automatisms" (aw-TOM-ah-TIZ-ums). Less often, people may repeat words or phrases, laugh, scream, or cry. Some people do things that can be dangerous or embarrassing, such as walking into traffic or taking their clothes off. These people need to take precautions in advance (Schachter, 2006). Complex partial seizures starting in the frontal lobe tend to be shorter than the ones from the temporal lobe. The seizures that start in the frontal lobe are also more likely to include automatisms like bicycling movements of the legs or pelvic thrusting (Schachter, 2006).

These seizures are called "secondarily generalized" because they only become generalized (spread to both sides of the brain) after the initial or "primary" event, a partial seizure, has already begun. They happen when a burst of electrical activity in a limited area (the partial seizure) spreads throughout the brain. Sometimes the person does not recall the first part of the seizure. These seizures occur in more than 30% of people with partial epilepsy (Schachter, 2006).

The concepts of generalized and focal when used to characterize seizures now explicitly reference networks, an increasingly accepted construct in neuroscience where networks are studied directly through the use of techniques such as functional magnetic resonance imaging (MRI). Berg and collagues (2011) explicitly acknowledged the group called "idiopathic generalized" epilepsies, although with a different name. For etiology, the terms idiopathic, symptomatic, and cryptogenic had become unworkable as descriptors of etiology and had, over time, taken on connotations of "good" and "bad" outcome. Epilepsies that later were recognized as monogenic syndromes such as autosomal dominant nocturnal frontal lobe epilepsy (ADNFLE) were classified as "cryptogenic" meaning "presumed symptomatic," as in secondary to a brain lesion. Current developments in molecular genetics and neuroimaging and other areas will, Berg and collagues (2011) predict, lead to a rational system for characterizing and classifying causes based on mechanisms. In moving forward to the next phase, Berg and collagues (2011) suggested the following terms and concepts:

Genetic: The epilepsy is a direct result of a genetic cause. Ideally, a gene and the mechanisms should be identified; however, this term would also apply to electroclinical syndromes for which twin or family segregation studies reproducibly show clinical evidence of a genetic basis (e.g., in the case of the genetic generalized epilepsies). At this time, channelopathies are the best example of genetic epilepsies (Berg et al., 2011).

Structural-Metabolic: The epilepsy is the secondary result of a separate structural or metabolic condition. Structural and metabolic were combined to separate the concept from genetic and also because the two are often inseparable (Berg et al., 2011).

Unknown: Plain and direct, this label simply and accurately indicates ignorance and that further investigation is needed to identify the cause of the epilepsy. Unlike cryptogenic (presumed symptomatic), it makes no presumptions and requires no explanation or reinterpretation (Berg et al., 2011).

2.4 Models of chemically induced epileptic seizures

A systemic administration of pentylenetetrazol (PTZ), an antagonist for the GABA (gamma-aminobutyric acid) receptor ion channel binding site was shown to cause generalized epilepsy in an animal model (Ahmed et al., 2005). Kindling is a model of epilepsy and epileptogenesis. Repeated application of subconvulsive doses of central nervous system (CNS) stimulants like PTZ (Corda et al.,1992) once every 24 to 48 hours over a period of time is also known to induce a permanent change in the epileptogenic sensitivity of the forebrain structures (Khanna et al., 2000). PTZ-induced seizure in rats, a relevant model of human absence and of generalized tonic-clonic epilepsy (ILE, 1989; Brevard et al., 2006).

In a single dose PTZ-treated group, rats were injected intraperitoneally (i.p.) with 55 mg/kg PTZ (Sigma Chemical Co) and observed for behavioral epileptic activity in our study. The animals in the repeated doses of PTZ-treated group were given 55 mg/kg PTZ i.p. on alternate days for six times and then the seizure activity was observed during each seizure period. After the last injection on the sixth day, similar procedure was applied as in the single dose PTZ-treated group. For the control group, saline solution was applied instead of PTZ. So, in our study, we also planned to examine hippocampal neurons in rat brain after the PTZ-induced epileptic seizures light and electron microscopically.

3. Histopathological changes of neurones in epilepsy

The extent that prolonged seizure activity, i.e. SE, and repeated, brief seizures affect neuronal structure and function in both the immature and mature brain has been the subject of increasing clinical and experimental research. The main emphasis is put on studies carried out in experimental animals, and the focus of interest is the hippocampus, the brain area of great vulnerability in epilepsy. Collectively, recent studies suggest that the deleterious effects of seizures may not solely be a consequence of neuronal damage and loss per se, but could be due to the fact that seizures interfere with the highly regulated developmental processes in the immature brain (Holopainen, 2008).

Holopainen (2008), provides not only up-to-date information of some of the processes involved in the complex reorganization cascade activated by seizures, but the aim is also to highlight the importance of the developing brain as a unique, dynamic structure within the field of neurochemistry and epilepsy research, and to awaken the interest for further new, innovative ways to approach this fascinating research field.

In epilepsy, several pathological changes typically occur in the brain, including neuronal loss, gliosis (Penfield, 1929; Steward et al., 1991), dendritic spine degeneration (Isokowa,1998, and abnormal synaptic reorganization (Babb et al., 1991; Mello et al., 1993;

Leite et al., 1996; Xiang-ming Zha et al., 2005). These changes lead to abnormally increased excitability and synchronization, and eventually to the occurrence of spontaneous seizures (Cavalheiro et al., 1991; Isokawa & Mello,1991; Bothwell et al., 2001).

It has been studied the effect of kainic acid (KA), a potent neuroexcitatory and neurotoxic analogue of glutamate, in the rat using a variety of light- and electron-microscopic techniques. The commonly affected areas include the olfactory cortex, amygdaloid complex, hippocampus, and related parts of the thalamus and neocortex (Schwob et al., 1980). Acute treatment with 30mg/kg KA did not produce major death of mouse hippocampal neurons, indicating that concentrations were not cytotoxic. Taken together, investigators' results provide new insights in the activation of several kinase-pathways implicated in cytoskeletal alterations that are a common feature of neurodegenerative diseases (Crespo-Biel et al., 2006).

Sankar et al. (2002) evaluated of the type of cell injury resulting from lithium-pilocarpine (LiPC) status epilepticus (SE) ultrasturucturally. Limbic system comprises of the brain which are important for memory, emotions, and cognitive functions (Wen et al., 1999).

Hippocampus is an important component of this system and it is widely accepted that it plays an essential role in memory. The hippocampus is a part of the brain located inside the temporal lobe. It forms a part of the limbic system and plays a part in memory and spatial navigation. It is known that the damage to the hippocampus can also result from oxygen starvation (anoxia) and encephalitis. Reductions in neuronal cell number were indicative of an abnormal development. The developmental structural abnormalities in the hippocampus may contribute to the cognitive impairments which result from isolation rearing in rats (Bianchi et al., 2006). However, our understanding of the cellular and molecular mechanisms underlying epilepsy remains incomplete.

4. Staining methods for neuronal damage localization of: Our findings

In our study, brainsections were stained with Cresyl Fast Violet (CFV) for Nissl staining and then these sections were examined under a light microscope (BX50F-3; Olympus, Tokyo, Japan). CFV binds very strongly to the RNA in the neuron's rough endoplasmic reticulum (Chan & Lowe, 2002).

NSE is a major neuronal protein that catalyzes the interconversion of 2 phosphoglycerate and phosphopyruvate. Immunocytochemistry was performed using the avidinbiotin-peroxidase method. The sections were incubated with anti-NSE primary antibodies (Zymed, Carlton Court, San Francisco) for 24 h at 4°C in a humidified chamber. Following washing in PBS-Tx, biotinylated anti IgG secondary antibodies were applied for 15 min at room temperature. Samples were washed with PBS-Tx and Streptavidin-peroxidase conjugate was applied to the sections for 15 min at room temperature. Following washing in Tris, 0.6% hydrogen peroxide and 0.02% diaminobenzidine (DAB) was applied 5 min at room temp. As control, the primary antibody was omitted and replaced with non-immune serum. Immunoreactivity of NSE was examined under a light microscope (BX50F-3; Olympus, Tokyo, Japan).

After perfusion, hippocampi were microdissected from each rat and were post-fixed in 2% osmium tetraoxide at 0.1 M, pH 7.4 phosphate buffer at 48C° for 1 hour, and stained with uranyl acetate during 2 hour. Later the sections were flatembedded in Durcupan. Semi-thin (1μm) sections were first stained with CFV and screened. Hippocampal regions were selected, and ultrathin sections were cut and placed on singlehole grids. After staining with uranyl acetate and lead citrate, the sections were examined by EM (Zeiss EM-9S).

The number of cells was quantified in 765x102 μm2 fields (counting frame) of hippocampal regions of rat brains with the X40 objective (Olympus) using a grid for determination of the sampling volume via the Cavalieri method (Michel RP & Cruz-Orive,1988). In the seven slices through hippocampus, number of neurons were examined among the acute-PTZ treated, chronic-PTZ treated and the control brains according to unbiased counting methods. The number of neurons were counted in CA1, CA3 and gyrus dentatus (GD) regions. The mean value and S.D. were calculated in the control and PTZ-induced groups. The data were statistically analyzed using the SPSS statistical software package. All groups were compared using ANOVA. Values were expressed as the mean ± standard error (SE).

Fifty-five mg/kg PTZ induced generalized tonic-clonic seizures in all animals. Following i.p. injections, generalized seizures started with the clonus of the facial and the forelimbmuscles, and continued with the neck and tail extensions, loss of straightening reflex with tonic flexion-extention, wild running and usually with extented clonic activities.

Different brain regions were examined for neuronal rER and NSE immunoreactivity in the control and PTZ-treated groups using light microscopy in our study. In the control brains, the observed morphology was as follow; nuclei of the neurons were huge in comparison with those of surrounding glial cells; DNA in the nucleus and nucleoli had similar staining properties; dispersed chromatin and prominent nucleoli reflect a high level of protein synthesis. The extensive cytoplasm was basophilic due to extensive rRNA damage. Nissl substance was stained with CFV to evaluate the morphology of neurons. Normal neuronal view was observed in the hippocampal regions from the control group by a light microscope; the nucleus was large with dispersed chromatin and prominent nucleoli and neuroplasm was basophilic due to extensive rRNA damage. CFV, for identifying the Nisll substance (GER) as dark blue material, revealed a granular appearance; nuclear DNA had a similar staining properties. Nissl method stained RNA, identifying the rER (Nissl substance) as purple blue (violet) material giving the neuronal cytoplasm a granular appearance (Figure 1. 1A, D, G, J). Aslight increase in Nissl stainingwas observed in the neurons of the cortex, thalamus, hypothalamus, and hippocampus of the single dose PTZ group rat brains comparing to the control group (Figure 1. 1B, E, H, K). However, slight decrease in the amount of nissl staining was noticed in III-VI layer of the cortex in the repeated dose PTZ-treated group (Figure 1. 1C). The NSE immunoreactivity was largely expressed in the brains of the control and seizing animals. This immunoreactivity was observed to be robust in the neuronal perikarya and dendrites. Representative coronal sections of NSE (+) cells depicting the cortex, thalamus, hypothalamus, and the hippocampus of the control, single and repeated dose PTZ-treated group are shown in Figure 1.2. The number of NSE (+) cells from the cortex, thalamus, hypothalamus, and hippocampus of all groups are shown in Table 1. In the cerebral cortex, no statistical significant difference was observed in the number of NSE (+) neurons in the single (B, E, H, K) and repeated (C, F, I, L) dose PTZ-treated groups compared to the control group (A, D, G, J), respectively. On the other hand, although a slight decrease in the NSE (+) immunoreactivity in the cortex of the repeated doses PTZ-treated group was noticed compared to the control group (Figure 1. 2A–C; Table 1.),

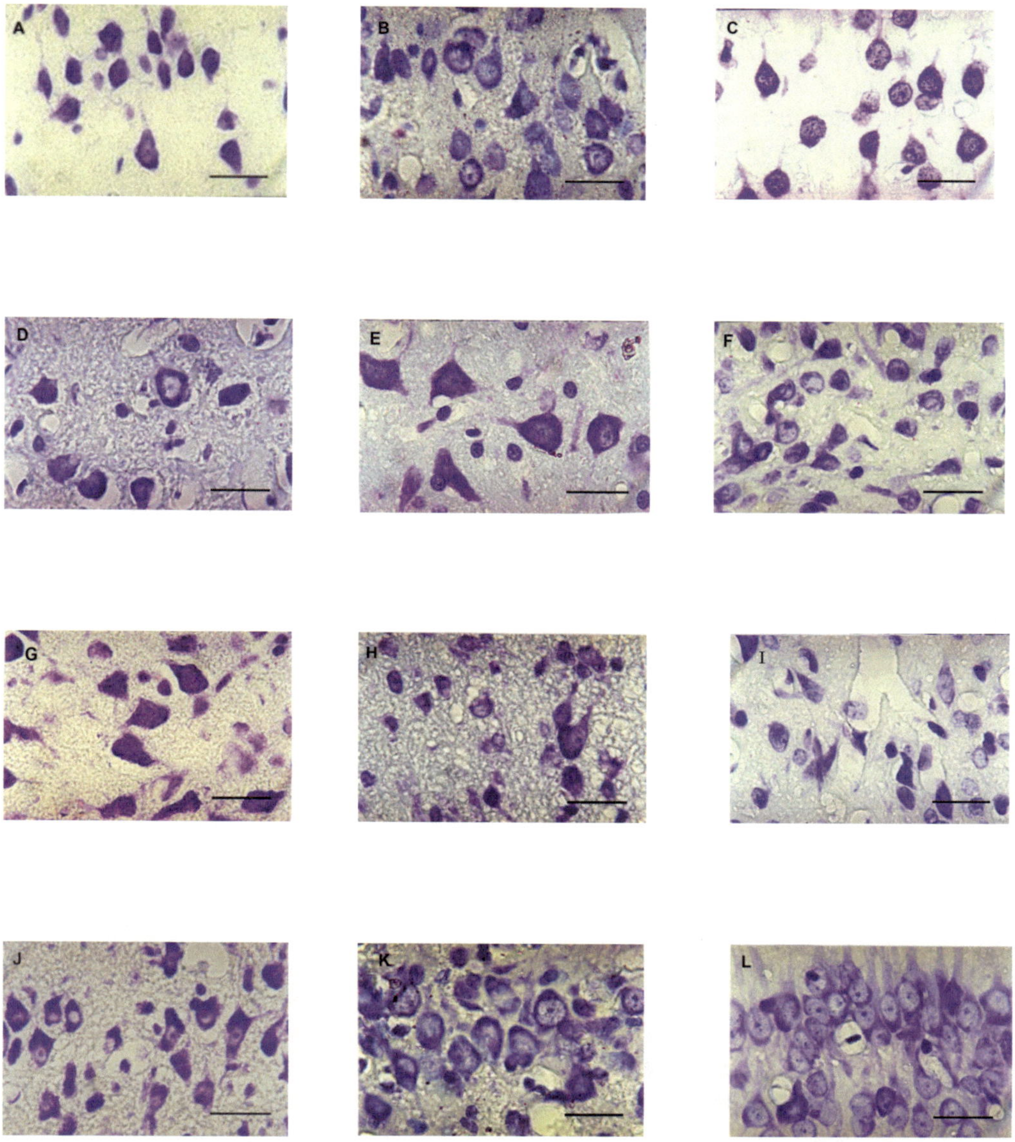

Fig. 1.1. Nissl staining of cerebral cortex (A, B, C), thalamus (D, E, F), hypothalamus (G, H, I), and hippocampus (J, K, L) of the control (A, D, G, J), single dose (B, E, H, K) and repeated doses (C, F, I, L) PTZ-treated group, respectively. Bar, 25 μm.Although there was a significant ($F = 13.05$; $df = 2$; $p < .001$) increase in the number of NSE (+) hypothalamic neurons of the single dose PTZ-treated group (Figure 1. 2H) compared to the control group (Figure 1. 2G), a significant ($p = .001$) decrease in the number of NSE (+) hypothalamic neurons was detected in the repeated doses PTZ-treated group compared to the single dose PTZ-treated group (Figure 1. 2H, I and Table 1.). In the hippocampus, no statistical significant difference was observed in the number of NSE (+) neurons in the single and repeated PTZ-treated groups. A slight increase of NSE immunoreactivity was seen in the hippocampus of the single dose PTZ-treated group (Figure 1. 2K) compared to the control (Figure 1. 2J) and repeated doses PTZ-treated group (Figure 1. 2L) rats.

a significant increase in the number of NSE (+) cortical neurons was observed in the single dose PTZ-treated group (Figure 1. 2B) compared to the repeated doses PTZ-treated group (Figure 1. 2C) ($F = 2.57$; $df = 2$; $p < .05$). In the thalamus, the number of the neuron showing NSE immunoreactivity was significantly ($F = 4.68$; $df = 2$; $p < .05$) increased in the repeated doses PTZ-treated group compared to the control group (Figure 1. 2 D, F and Table 1.).

Brain regions	Cortex[a]	Thalamus[b]	Hypothalamus[c]	Hippocampus[d]
Groups		Mean ± SE		
Control group	414.00± 51.41	224.00 ±14.86	124.50 ±6.91	231.30 ±23.14
Single dose PTZ-treated group	480.10±32.01 *	315.70±33.13	251.70±28.42**	263.30 ±10.97
Repeated doses PTZ-treated group	359.70±23.76	362.20±43.00**	144.66 ±14.52***	235.10 ±29.30

* p< .05 , compared with the repeted doses PTZ-treated group;
** p< .05 compared with the control group;
*** p< .05 compared with the single dose PTZ-treated group.

Table 1. Number of NSE (+) neurons in the cortex, thalamus, hypothalamus, and hippocampus in the control and epileptic brains following single dose and repeated doses PTZ-induced seizures

Neurons of CA1, CA3 and GD regions from the control group appeared to be normal (Fig 2 1a, b, c). A few necrotic neurons from the acute-PTZ group were seen in CA1 and CA3 regions (Fig. 2. 2a, b). There was not significant difference between the number of CA1 and CA3 neurons in the acute-PTZ group and control group (Table 2.). Necrotic neurons were seen extensively in the GD region of the acute-PTZ group (Fig. 2. 2c). There was significant difference between the number of GD neurons in the acute-PTZ group and that of control group (p<0.001; Table 2.). CFV showed a decreased Nissl of hippocampal neurons in the chronic-PTZ group compared to the control group. There was a characteristic view of neuronal damage in light microscopic analysis of hippocampus in the chronic-PTZ groups. In this group, both necrotic and apoptotic neurons were observed in the CA1 region (Fig. 2. 3a). Necrotic histological changes were as follows; perikaryal swelling, chromatolysis and decreasing of Nissl. Apoptotic histological changes were perikaryal shrinking and dark nucleus. There was significant difference between the numbers of CA1 neurons in the chronic-PTZ group and that of control group (p<0.001; Table 2.). Neuronal loss were observed with a resultant narrowing, sparse staining and a breach of continuity of staining in the CA1 region in the chronic-PTZ group (Fig. 2. 3b). In the CA3 region of the chronic-PTZ group, both few necrotic and apoptotic neurons were observed (Fig. 2. 3c). There was no significant difference between the numbers of CA3 neurons from the experimental groups and control group (Table 2.). In the chronic-PTZ group, both necrotic and apoptotic neurons were observed extensively in the GD region (Fig. 2. 3d). There was significant difference between the number of GD neurons in the chronic-PTZ group and control group (p<0.001; Table 2.). Hippocampal CA1 sections were examined to evaluate transmission EM in all groups. The ultrastructural appearance of the cytoplasmic organelles and nuclear components of CA1 neurons was normal in the control group (Fig. 2. 4a). Necrotic neurons were seen rarely in the CA1 region of the acute-PTZ group at a lower magnification. The necrotic degenerative changes were deformation of nuclear and perikaryal outlines, dilatation of the cistarnae of endoplasmic reticulum at a higher magnification (Fig. 2. 4b). In the chronic-PTZ group both necrotic and apoptotic neurons were observed in the CA1 region at lower magnification. EM revealed that dying neurons at the CA1 region showed an apoptotic cells with the regularly shaped, round clumps of condensed chromatin

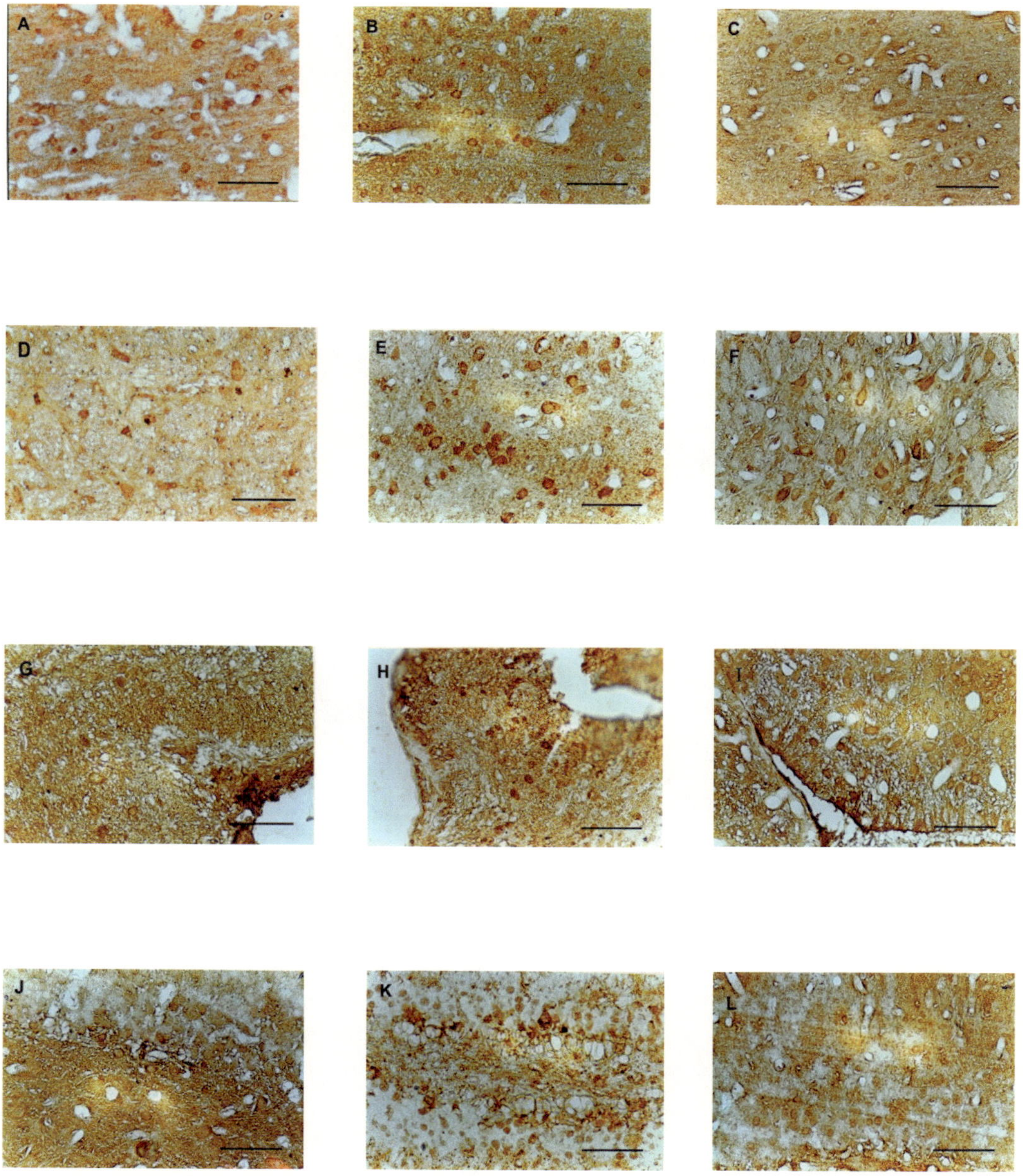

Fig. 2.2. NSE immunostaining of cerebral cortex (A, B, C), thalamus (D, E, F), hypothalamus (G, H, I), and hippocampus (J, K, L) of the control, single dose, and repeated doses PTZ-treated group, respectively. Bar, 100 μm.

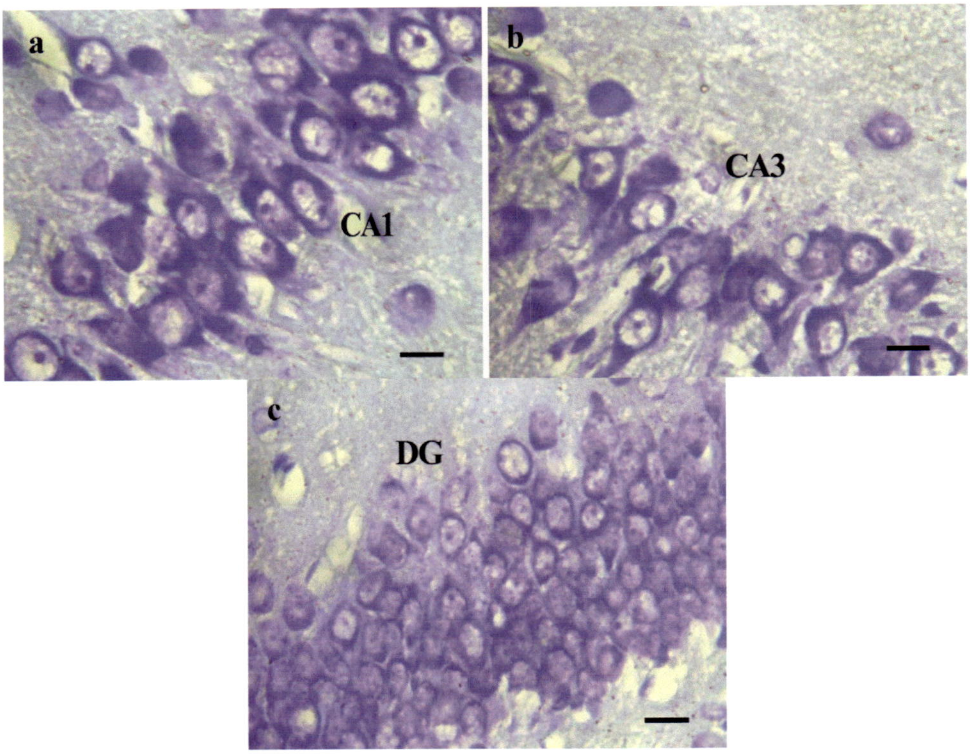
a
CA1
b
CA3
c
DG

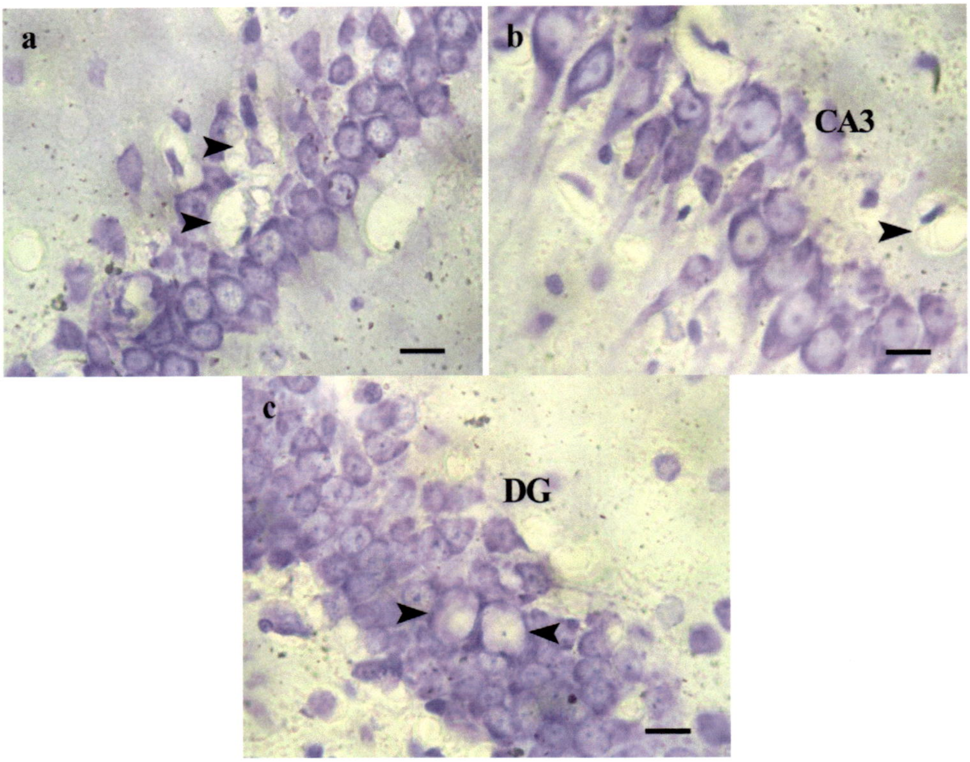
a
b
CA3
c
DG

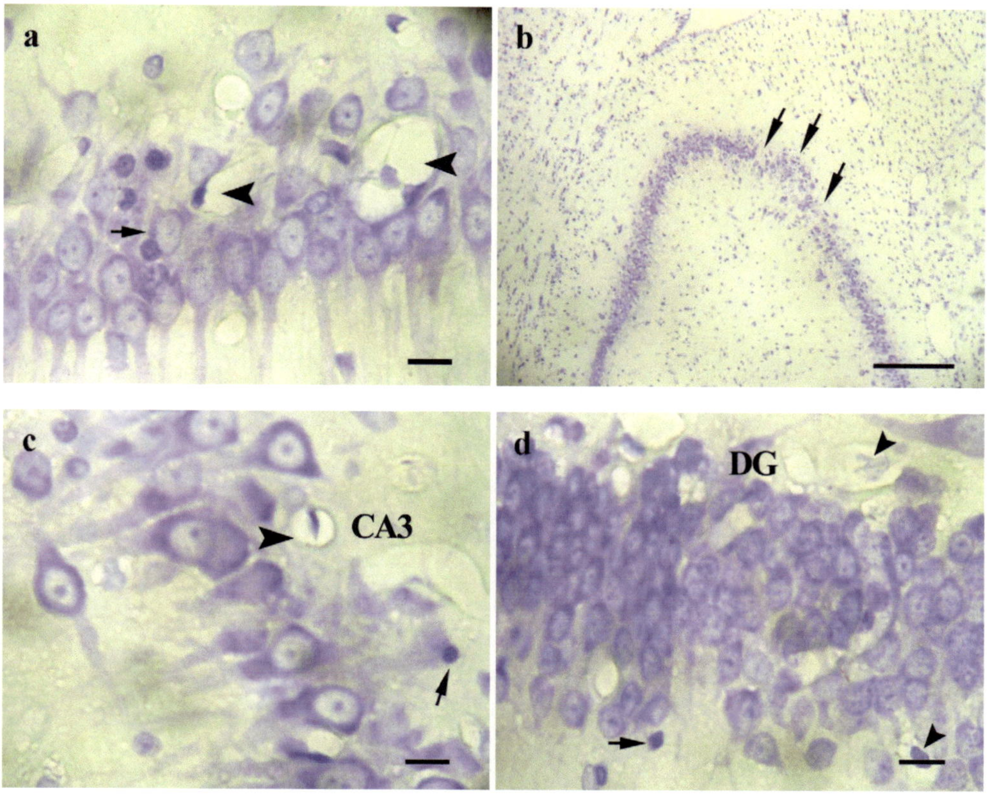

Fig. 2.1-3. Photomicrographs of Nissl-stained hippocampal regions, CA1(a), CA3 (b) and DG (c) in the control group (Fig.2. 1), acute-PTZ group (Fig. 2. 2) and the chronic-PTZ group (Fig.2. 3). Neuronal loss were seen Fig. 2. 2 and 2. 3. In the sections through hippocampus of the acute and chronic-PTZ groups showed a thinned, sparsely staining and a breach of staining in the CA3 pyramidal cell layer (arrows in).

with preservation of nuclear membrane continuity, and cell body shrinkage (Fig. 2. 4c). This feature could be distinguished from the signs of necrosis in CA1, including over swelling, cytolysis, and pyknotic nucleus with irregular contour of the chromatin (Fig. 2. 4b). These types of necrotic cells were observed in hippocampus of the chronic-PTZ group.

Groups/Hippocampal regions	CA1	CA3	DG
Control group	126.43±19.321	70.571±48.938	208.14±14.276
Acute-PTZ group	120±18.556	66.714±46.804	192.43±19.025*
Chronic-PTZ group	84.162±7.7766*	58.286±40.211	121.43±12.843*

*p< 0.001

Table 2. Number of neurons in the hippocampal regions of the brain from the control, acute- and chronic-PTZ groups. Neurons were counted in the 765x10^2 μm^2 fields of coronal sections.

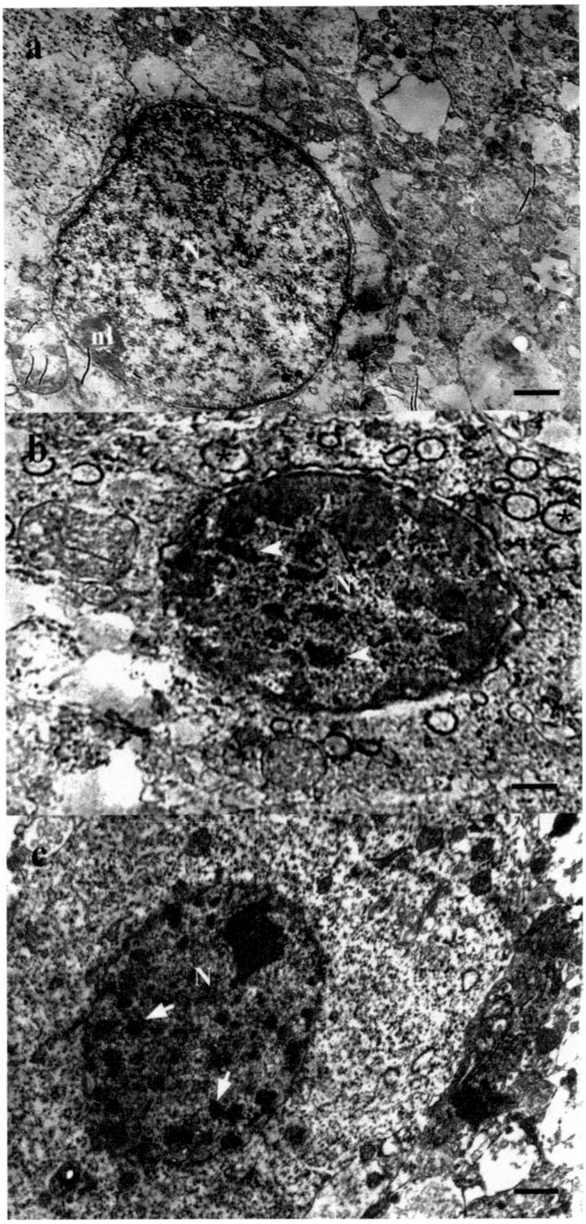

Fig. 2.4. Electron micrographs of hippocampal neurons in the control (a), acute- (b) and chronic-PTZ (c) groups. In a normal CA1 pyramidal neuron of control group, Nucleus (N) is euchromatic and exhibiting normal cytoplasmic features (a). Necrotic features (b), i.e., numerous small vacuoles throughout the cytoplasm as well as disruption of plasma membrane and a pyknotic nucleus with irregular contour of the chromatin clumps (arrowheads in) were seen the acute-PTZ group (b). In the chronic-PTZ group, CA1 neuron displayed apoptotic-like features such as chromatin condensation into a few round clumps (arrows in) and condensation of a relatively intact cytoplasm with preservation of plasma membrane (c)

5. Discussion

In this study, systemic injections of 55 mg/kg PTZ produced a high incidence of convulsions, and wild running. Yonekawa et al. (1980) have studied relationship between PTZ-induced seizures and brain PTZ levels in mice. PTZ is often used in experimental models of epilepsy. In their study examined this relationship anddetermined how different routes of PTZ administration affected brain PTZ uptake and seizure development. The critical brain PTZ level for onset of clonus ranged from 20 to 50 microg/g. This dose of PTZ was same with our experimental porcesses. Seizures activity is associated with neuronal damage.

The results of the present study demonstrated the presence of NSE immunoreactivity and Nissl staining in neurons of the different brain regions after PTZ-induced seizures. Although NSE (+) neurons significantly increased in the hypothalamic regions of the single dose PTZ-treated group, NSE (+) neurons were found to be increased only in the thalamic region of the repeated doses PTZ-treated group compared to the control group. In addition, NSE (+) neurons were found to be slightly increased in the cortex, thalamus and hippocampus in the single dose PTZ-treated group compared to the control group.

Several studies have demonstrated that seizure-associated brain damage was initiated by the release of excitatory amino acid neurotransmitters from excessively firing presynaptic terminals in ultimately neurotoxic concentrations. Additionally, pannecrosis through excessive focal tissue acidosis may also contribute to the neurotoxicity processes (Sloviter & Dempster, 1985; Auer & Siej¨o, 1988). Interictal energetic deficiency in the epileptogenic hippocampus could contribute to impaired glutamate reuptake and glutamate-glutamine cycling that resulting in persistently increased extracellular glutamate. Increased lactate production together with poor lactate and glucose utilization were the cause of worsening of energy metabolism, which then produced a glial and neuronal toxicity. It has been reported that in the epileptogenic hippocampus of patients with temporal lobe epilepsy the level of glutamate were increased (Cavus et al., 2005) and glial glutamine synthetase is down regulated (vander Hel et al., 2005). In this study, the number of NSE (+) neurons that was observed to be increased in investigated brain regions after single dose PTZ-induced seizure indicates that these regions become more active metabolically. These regions seems to be compatible with those that shown to have increased metabolic activity during epileptic seizures in previous studies (Siesjo et al., 1986; Rasmussen et al., 1994). NSE immunostaining in the brain is useful for evaluating brain damage from various cases. This also supports the notion that NSE immunoreactivity is spared in less necrotic neurons (Nogami et al., 1998a). The present findings, which showed an increase in NSE immunoreactivity and nissl staining in the single dose PTZ-treated group, were correlated with the results of Nogami et al. (1998a). In another study that quantified NSE (+) neurons in the frontal cortex and hippocampus of the rat brain after systemic administration of kainic acid, while the concentrations of NSE remained unaffected in the frontal cortex, NSE levels were found to be significantly decreased in the hippocampus (Ding et al., 2000). These results are in accordance with the present findings. In the present study, although the number of NSE (+) neurons in the cortex and hippocampus increased in the single dose PTZ-treated group, it was found to be decreased in the repeated doses PTZ-treated group. Neuronal damage from epileptic insults occurs predominantly in cortical lamina III and IV where thalamocortical afferents terminate, suggesting a transneuronal effect in producing cortical neuronal necrosis in SE (Auer & Siej¨o, 1988). In accordance with this study, in repeated doses PTZ-treated group, the present authors observed also a slight decrease in NSE (+) immunoreactivity and nissl amount in the cortical and hypothalamic neurons. It is well known that in comparison with the other brain regions, the hippocampal neurons appear to be more vulnerable to the excitatory damage caused by seizures (van Bogaert, 2001). In the present study, no statistically significant difference obtained in the hippocampus after repeated PTZ-induced seizures and number of NSE (+) neurons in these regions were close to the control values. Pavlova and collagues (2005) stated that the rats developed tolerance to PTZ kindled seizures, showing oxidative stress and neurodegeneration in hippocampal region. On the other hand, in the present study, the intensity of NSE staining in the hippocampal neurons was found to be decreased after repeated PTZ-induced seizures. Considering to the results of both studis, it can be suggested that oxidative damage of neurons resulting in neurodegeneration in the hippocampus was not directly related to the recurrence of a convulsive activity.

It has been stated that the focus of interest is the hippocampus, the brain area of great vulnerability in epilepsy (Holopainen, 2008).

According to our results, anti-NSE immunostaining might reflect the cellular damage of neurons during the antemortem period and could add further information about the integrity of neurons, which could be helpful determining the injured brain areas besides the morphological change of neurons assessed by CFV. Nissl substance can usually be seen in neuronal bodies stained with basophilic dyes and consist of rER and associated ribosomes. The ribosomes contain RNA and are the sites of protein synthesis. It imparts a light violet color to the rER. This stain gives a diffused coloration when the rER is less and spread out and imparts a granular appearance when the rER is abundant (Young & Health, 2000). Although neurons metabolically are highly active, the Nissl substance is often very prominent (Crossman & Neary, 2000). In this study the Nissl bodies were especially abundant in the perikarya of the cortical, thalamic, hypothalamic, and hippocampal neurons in the control and single PTZ-treated group, but the severity of Nissl staining was less in the repeated PTZ-treated group. The results also indicated that a significant increase in the number of NSE (+) neurons in the thalamus after repeated PTZ seizures comparing to the control group. This may be related with increase metabolically activity of the thalamic neurons after epileptic seizures. The increase in the number of NSE (+) cells in the brain were in parallel to the other studies that reported an increase in the serum levels of NSE following epileptic seizures (Sankar et al., 1997). NSE is a good indicator of neuronal damage. The level of sNSE and cNSE are increased in brain diseases such as anoxic brain injury and stroke (Hasegawa et al., 2002). The sNSE levels are reported to reach maximum levels between 1 and 6 h postictally (Tumani et al., 1999). It appears that this enzyme located in the neurons passes to the serum with the destruction of the BBB and reaches peak levels for a certain amount of time. The reversible opening of the BBB has already been shown during the PTZ-induced epileptic seizure (Sahin et al., 2003). As a result, while the number of NSE (+) neuron were increased in all investigated brain region of the single dose PTZ-treated group, the same increase was noted only in the thalamic region of repeated dose PTZ. Additionally a low NSE immunoreactivity was seen in the cortex, hypothalamus, and hippocampus of the repeated doses PTZ-treated group. These findings suggest that some adaptive changes may develop in the CNS after repeated seizures. On the other hand, a low immunoreactivity in the brain regions could reflect the lower metabolic state of damaged neurons (Nogami et al., 1998). Cellular and molecular mechanisms related with the metabolic changes that are observed following epileptic seizures may be responsible from the brain damage.

Both necrotic and apoptotic forms of cell death contribute to brain damage in the PTZ-induced epilepsy model. One of the most common stains used for nervous system tissues is CFV method, which binds very strongly to the RNA in the neuron's GER (rER), since it's a basic stain (Chan and Lowe, 2002; Damjanov, 1996). Therefore, CFV is a specific stain to show the GER in the neurons. It imparts a light violet color to the GER. This stain gives a diffused coloration when the GER is less and spread out and imparts a granular appearance when the GER is abundant (Young & Health, 2000). Perikaryal injury in cytoplasmic swelling and degranulation of ribosomes from the GER. This loss of RNA is seen as disappearance of cytoplasmic basophilia which is called chromatolysis. Necrotic and apoptotic neurons were observed in the chronic-PTZ treated group. Necrosis was observed extensively in the brains of the chronic-PTZ animals. Nadler and collagues (1980a) have used intraventricular injections of KA to destroy the hippocampal CA3-CA4 cells, thus denervating the inner third of the molecular layer of the fascia dentata and stratum radiatum and stratum oriens of area CA1. Their results showed a preferential ordering in the reinnervation of dentate granule cells (DGCs) that was not readily explained by proximity to the degenerating fibers and also that removal of CA3-CA4-derived innervation more readily elicits translaminar growth in the fascia dentata than in area CA1. These results might be relevant to clinical situations in which neurons of the hippocampal end-blade were lost. Nadler and collagues (1980b) have studied the degeneration of hippocampal CA3 pyramidal cells investigating by a light- and electron-microscopy after intraventricular injection of the potent convulsant, KA. EM revealed evidence of pyramidal cell degeneration within one hour. The earliest degenerative changes were confined to the cell body and proximal dendritic shafts. These included an increased incidence of lysosomal structures, deformation of the perikaryal and nuclear outlines, some increase in back ground electron density, and dilation of the cisternae of the endoplasmic reticulum accompanied by detachment of polyribosomes. Within the next few hours the pyramidal cells atrophied and became electron dense. Then these cells became electron lucent once more as ribosomes disappeared and their membranes and organelles broke up and disintegrated. The dendritic spines and the initial portion of the dendritic shaft became electron dense within four hours and degenerated rapidly, whereas the intermediate segment of the dendrites swelled moderately and became more electron lucent. We also a few necrotic neurons from the acute-PTZ group were seen in the CA3 regions. Our findings were similar to Nadler and collagues (1980b). We also observed necrotic degenerative changes including the deformation of nuclear and perikaryal outlines, dilatation of the cistarnae of endoplasmic reticulum in the acute-PTZ group. In the chronic-PTZ group, both a few necrotic and apoptotic neurons were observed in the CA3 regions. But we did not determined significant difference in the number of neurons in the CA3 region in the acute- and chronic-PTZ groups. Schwob and collagues (1980) have studied the effect of systemic and intracerebral injections of KA, a potent neuroexcitatory and neurotoxic analogue of glutamate, in the rat using a variety of light- and electron-microscopic techniques. The initial neuropathological reactions include dendritic and glial dilatations in discrete areas of the neuropil; affected neuronal somata either appear swollen and pale, or are shrunken with dark cytoplasm. In the most severely affected areas, the lesion progresses to severe disruption of the neuropil. The commonly affected areas include the olfactory cortex, amygdaloid complex, hippocampus, and related parts of the thalamus and neocortex. Intracerebral injections of 2–6 nmol produce extensive neuronal damage in distant structures, as well as at the injection site. The pattern of distant damage varies with the site of the injection and appears to reflect axonal connections between the affected areas near the injection and the distant areas of damage. Injections into the posterior part of the olfactory cortex which involve the entorhinal cortex (EC) tend to produce severe degeneration in field CA1 of the hippocampus, although field CA3 is more severely damaged following intraventricular, intrahippocampal or intrastriatal injections. Du and collagues (1993) have obtained specimens EC during a surgical treatment of intractable partial seizures and were studied by light microscopy in Nissl-stained sections. A distinct loss of neurons was observed in the anterior portion of the medial EC in the absence of apparent damage to temporal neocortical gyri. These observations provided neuropathological evidence for an involvement of the EC in temporal lobe epilepsy (TLE). Since the EC occupies a pivotal position in gating hippocampal inputs and outputs, their results further support previous suggestions that dysfunction of this region may contribute, either independently or in concert with Ammon's horn sclerosis, to epileptogenesis in humans. Du and collagues (1995) have examined the EC in three established rat models of epilepsy using Nissl staining. Adult male rats were either electrically stimulated in the ventral hippocampus for 90 minute or injected with KA or LiPC. At 24 hour, all animals that had exhibited a bout of acute SE showed a consistent pattern of neuronal loss in the EC in Nissl-stained sections. We also determined neuronal loss in hippocampal GD at 24 hours in Nissl stained sections of the acute-PTZ group. Du and collagues (1995) have also seen an identical pattern of nerve cell loss in the EC of rats killed 4 weeks following the treatments. This lesion was completely prevented by an injection of diazepam and pentobarbital, given one hour after KA administration. Taken together, these experiments indicated that prolonged seizures caused a preferential neuronal loss in layer III of the medial EC and that this lesion might be related to a pathological elevation of intracellular calcium ion concentrations. Isokawa (1998) has determined that dendritic degeneration was a common pathology in TLE animal models. In a study of a rat pilocarpine model, visualization of dendrites of the hippocampal DGCs by biocytin revealed a generalized spine loss immediately after an acute seizure induced by pilocarpine. The present finding suggests that initial acute seizures do not cause permanent damages in dendrites and spines of DGCs; instead, dendritic spines were dynamically maintained in the course of the establishment and maintenance of spontaneous seizures. Local dendritic spine degeneration, detected later in the chronic phase of epilepsy, was likely to have a separate cause from initial acute insults. We also detected both apoptotic and necrotic neurons ultrastructurally. Eid and collagues (1999) have been studied in the animals also develop hyperexcitability of the EC and the hippocampal region CA1. Pathologically swollen dendrites and electrondense neuronal profiles were present in the lesioned sector as well. The majority of the electron-dense profiles was identified as degenerating dendritic spines that were closely apposed to strongly glutamate-immunopositive axon terminals. These findings might be of relevance for the genesis and spread of temporal lobe seizures. Clinical, radiologic, and experimental evidence indicated that the EC region might be linked to the pathology of hippocampal sclerosis (HS) in patients with TLE (Du etal., 1993; Du &

Schwartz, 1992; Dawodu & Thom, 2005). Morphological analysis of hippocampal formation after pilocarpine-induced SE showed increased glucose utilization in most brain regions including the hippocampus during a period of continuous seizure activity (Clifford et al., 1987),an extensive loss of neurons within the hilar area of the GD (Mello et al., 1993; Cavalheiro,1995), and loss of interneurons in the CA1, CA3, and hilus regions (Cavalheiro,1995). In our study we also determined a significant neuronal loss in the CA1 region of the chronic-PTZ group compared with the control group. We determined significant difference between the numbers of CA1 neurons in the chronic-PTZ group and control group ($p<0.001$; Table 2). Chandler and collagues (2003) have observed that loss of interneurons could undoubtedly contribute to a decrease in GABA release. Several neurotransmitters, also including GABA, modulate glutamate release at synapses between hippocampal mossy fibers (MFs) and CA3 pyramidal neurons. Hilar mossy cells loss directly resulted in granule cell hyperexcitability (Toth et al., 1997; Santhakumar et al., 2001).

Impaired GABAergic inhibition might contribute to the development of hyperexcitability in epilepsy. Thus, decrease GABA activities, which an inhibitory neurotransmitter, leaded the removal of sinaptic inhibition on epileptic neurons and to epilepsy seizuresmaking neurons prone to more exitability. Heterotopic granule cells exhibit features often have been observed in epileptic tissue. Granule cell dispersion in the GD, similar to that seen in human epileptic hippocampi has also been observed in animal models of epilepsy, e.g., after KA injection into the dorsal hippocampus (Cavalheiro et al., 1982; Ben-Ari,1985) and in the pilocarpine model. The distribution of granule cells in the GD of the hippocampal formation has been studied in control autopsy and TLE specimens. Results contributed to the altered circuitry of the hippocampal formation in TLE (Houser,1990) and Houser (Houser,1999) stated that the neuronal loss and synaptic reorganization in TLE. It has remained unclear whether the appearance of heterotopic granule cells is related to granule cell loss in the epileptic hippocampus. Granule cell dispersion has not been observed when a cell loss is minimal (Lurton et al., 1997). We also determined decreased number of neurons in the GD region in the acute- and chronic-PTZ groups.There was a significant difference between the number of GD neurons between the gropus . Brevard and collagues (2006) have found that GD was twice as active as other hippocampal areas, but peaked just before seizure onset in the PTZ-induced seizure in rats. Neurons in this area might contribute to the neural network controlling the initiation of generalized tonic-clonic seizures. Some studies have also been shown that MFs were decreased in epilepsy. It is suggested that sprouting of MFs or their axon collaterals occurres in hippocampal epilepsy and the reorganized fibers contain at least one of the neuropeptides that are normally present in this system. Such fibers could form recurrent excitatory circuits and contribute to synchronous firing and epileptiform activity, as suggested in studies of experimental models of epilepsy (Houser et al.,1990). Hippocampal MFs represented a major input from DGCs to the hippocampal CA3 field. They exhibit several forms of presynaptic modulation of transmitter release, including marked short-term (Salin et al., 1996) and long-term (Harris& Cotman,1986) use-dependent plasticity. They are sensitive to several neurotransmitters that depress transmitter release, including glutamate (Kamiya et al., 1996), GABA (Min et al., 1998; Vogt & Nicoll,1999), and peptides (Weisskopf et al.,1993) acting on metabotropic receptors. MF transmission might be under such profound modulation because hippocampal principal cells are highly vulnerable to excitotoxicity (Meldrum, 1993). Nevertheless, these modulatory mechanisms could break down: excessive activity in the DG can spread into the hippocampus and result in neuronal loss that resembles similar to that seen after KA administration (Sloviter,1987; Sloviter,1991). An anatomic and neurobiologic study revealed functional abnormalities in the GD of epileptic KA-treated rats; however, lateral inhibition persists, suggesting that vulnerable hilar neurons were not necessary for generating lateral inhibition in the GD (Buckmaster & Dudek, 1997a; Buckmaster & Dudek, 1997b). Histological and quantitative stereological techniques were used to estimate numbers of neurons per GD of various classes and to estimate the extent of granule cell axon reorganization along the septotemporal axis of the hippocampus in control rats and epileptic KA-treated rats. Findings from the GD of epileptic KA-treated rats were strikingly similar to those reported for human TLE, and it was suggested that neuron loss and axon reorganization in the temporal hippocampus might be important in epileptogenesis (Buckmaster & Jongen-Relo,1999). Failure of modulation of MF transmission might also contributes to the delayed development of spontaneous seizures (Chandler, 2003). In their study, Sloviter and collagues (2006) in chronically epileptic rats demonstrated that DGCs were maximally hyperexitabl immediately after SE, prior to MF sprouting, and that synaptic reorganization following KAinduced injury was temporally associated with GABA (A) receptor-dependent granule cell hyper-inhibition rather than hypothesized progressive hyperexitability. Mortazavi and collagues (2005) have revealed that neuronal loss in the CA1 area and increased MF sprouting in the GD were similar to what was observed in human epilepsy. These results indicated that PTZ kindling provides a useful model of postseizure dysfunction, which can serve as a screen for potential treatments for those cognitive, emotional, and neuropathological deficits that resemble those symptoms observed in human epilepsy. In our study, hippocampal CA1 sections were examined in transmission EM samples of controls and PTZ-induced animals. According to ultrastructural appearance of the neuroplasm and nucleus, we determined that repeated-PTZ injections caused both necrotic and apoptotic neuronal death in the CA1 region of the hippocampus in the chronic-PTZ group. A few dying neurons were seen in the apoptotic morphology as described by Portera-Cailliau and collagues (1997a; 1997b), in PTZ-induced groups. This feature could be distinguished from the signs of necrosis in the CA1, including overt swelling, cytolysis, and pyknotic nucleus with irregular contour of the chromatin. In another study (Jerman et al., 2005), it was showen that cell loss was relatively uniform after ibotenic acid injections into areas CA1 and CA3, but variable after colchicine injections into the GD. CA1 and CA3 lesions appeared mostly localized to those relative subregions, and DG lesions appeared highly localized to the GD. Pavlova and collagues (2006) have suggested that, in a PTZ kindling model, oxidative damage of neurons resulting in neurodegeneration in the hippocampus was not directly related to the convulsive activity. PTZ-kindling in rats has been induced moderate neuronal cell loss in hippocampal fields CA1-4, and DG. They have suggested that PTZ-kindling might be a suitable model to study the mechanisms of seizure-induced neuronal death. Neuron death in the hippocampus is also accompanied by increases in oxidative stress, this is also being independent of the external manifestations of the brain seizure activity.

6. Conclusion

In conclusion, NSE immunoreactivity may be a valuable marker for determining the number of metabolically active neurons and the regions where these changes take place after single and repeated seizures. New studies investigating the neuronal activity changes during the epileptic seizure, including modulation of gene expression, will give a new insight for future research. While necrotic neurons of CA1 and CA3 regions were rarely seen, these cells were observed extensively in the GD region of a group treted with PTZ in our study. There was a significant difference between the number of GD neurons considering controls. CFV showed a decreased Nissl of hippocampal neurons in the chronic-PTZ group compared to controls. In the chronic-PTZ group, both necrotic and apoptotic neurons were observed in the CA1 region. There was significant difference between the numbers of CA1 neurons in the chronic-PTZ group and that of controlsIn the chronic-PTZ group, both necrotic and apoptotic neurons were observed in the GD region extensively. There was significant difference between the number of GD neurons in the control and chronic-PTZ group ($p<0.001$). According to the ultrastructural appearance of the cytoplasmic organelles and nuclear components of CA1 regions a few necrotic neurons was seen in the acute-PTZ group. Both necrotic and apoptotic neurons of the CA1 region were observed in the chronic-PTZ group. EM revealed that dying neurons at the CA1 region showed an apoptotic cells. These types of necrotic cells were observed in hippocampus of the chronic-PTZ group. The outcome of continuation of epilepsia seizures in the chronic-PTZ group was loss of hippocampal neurons with the decrease of GD and CA1 neurons. These finding might be result of excitocytotoxic sensitivity of these neurons especially. A decreasing in the numbers of CA1 neurons was determined only in the chronic-PTZ group. However, PTZ injections cause a decreasing of GD neurons in both acute- and chronic-PTZ groups. Our results showed that chronic-PTZ seizures cause neuronal degeneration and neuronal loss of hippocampus.

7. References

Ahmed MM, Arif M, Chikuma T, Kato T. 2005. Pentylenetetrazol-induced seizures affect the levels of prolyl oligopeptidase, thimet oligopeptidase and glial proteins in rat brain regions, and attenuation by MK-801 pretreatment. *Neurochem Int.* 47(4):248-59.

Auer, R. N., & Siej¨o, B. K. (1988). Biological differences between ischemia, hypoglycemia, and epilepsy. *Annals of Neurology, 24,* 699–707.

Babb, TL, Kupfer, WR, Pretorius, JK, Crandall, PH & Levesque, MF.1991. Synaptic reorganization by mossy fibers in human epileptic fascia dentata. *Neuroscience* 42: 351-363.

Barone, F. C., Clark, R. K., Price,W. J., White, R. F., Feuerstein, G. Z., Storer, B. L., &Ohlstein, E. H. (1993). Neuron-specific enolase increases in cerebral and systemic circulation following focal ischemia. *Brain Research, 623,* 77–82.

Ben-Ari Y. 1985. Limbic seizures and brain damage produced by kainic acid: Mechanisms and relevance to human temporal lobe epilepsy. *Neuroscience,* 14, 375–403.

Berg, A.T., Berkovic, S.F., Brodie, M.J., Buchhalter, J., Cross, J.H., Boas, W. van E., Engel, J., French, J., Glauser, T.A., Mathern, G.W., Moshe, S.L., Nordli, D., Plouin, P. and Scheffer, I.E. 2010. Revised terminology and concepts for organization of seizures and epilepsies: Report of the ILAE Commission on Classification and Terminology, 2005–2009. Epilepsia, 51(4):676–685.

Berg, A.T. and Scheffer, I.E. 2011. New concepts in classification of the epilepsies: Entering the 21st century Epilepsia, 52(6):1058–1062.

Bianchi M, Fone KF, Azmi N, Heidbreder CA, Hagan JJ, Marsden CA. 2006. Isolation rearing induces recognition memory deficits accompanied by cytoskeletal alterations in rat hippocampus. *Eur J Neurosci.* 20; [Epub ahead of print].

Bittermann, H. J., Felgenhauer, K., Paulus, W., & Markakis, E. (1999). Cisternal S100 protein and neuron-specific enolase are elevated and site-specific markers in intractable temporal lobe epilepsy. *Epilepsy Research, 36*, 75–82.

Bothwell, S., Meredith, G.E., Phillips, J., Staunton, H., Doherty, C., Grigorenko, E., Glazier, S., Deadwyler, S.A., O'Donovan, C.A. & Farrell, M. 2001. Neuronal hypertrophy in the neocortex of patients with temporal lobe epilepsy. *The Journal of Neuroscience*, 21, 4789–4800.

Brandel, J. P., Beaudry, P., Delasnerie-Laupretre, N., & Laplanche, J. L. (1999).

Brevard ME, Kulkarni P, King JA, Ferris CF. 2006. Imaging the neural substrates involved in the genesis of pentylenetetrazol-induced seizures. *Epilepsia.* 2006 Apr;47(4):745-54.

Buckmaster PS & Dudek FE. 1997a. Network properties of the dentate gyrus in epileptic rats with hilar neuron loss and granule cell axon reorganization. *Journal ofNeurophysiology*, 77, 2685–2696.

Buckmaster PS & Dudek FE. 1997b. Neuron loss, granule cell reorganization, and functional changes in the dentate gyrus of epileptic kainatetreated rats. *Journal of Comparative Neurology*, 385– 404.

Buckmaster PS, Jongen-Relo AL. 1999. Highly specific neuron loss preserves lateral inhibitory circuits in the dentate gyrus of kainate-induced epileptic rats. *TheJournal of Neuroscience*, 1999; 19: 9519-9529.

Buttner, T., Lack, B., J¨ager, M., Wunsche, W., Kuhn, W., Muller, T., Przuntek, H., &Postert, T. (1999). Serum levels of neuron-specific enolase and s-100 protein after single tonic-clonic seizures. *Journal of Neurology, 246*, 459–461.

Cavalheiro EA, Riche D & Le Gal La Salle G. 1982. Long-term effcts of intrahippocampal kainic acid injection in rats: a method for inducing spontaneous recurrent seizures. *Electroencephalography and ClinicalNeurophysiology*, 53,581-589.

Cavalheiro AE, Leite JP, Bortolotto ZA, Turski WA, Ikonomidou C & Turski L. 1991. Long-term effects of pilocarpine in rats: structural damage of the brain triggers kindling and spontaneous recurrent seizures. *Epilepsia*, 32: 778-782.

Cavalheiro AE. 1995. The pilocarpine model of epilepsy. *Italy Journal of Neurological Sciences*, 16, 33-37.

Cavus, I., Kasoff, W. S., Cassaday, M. P., Jacob, R., Gueorguieva, R., Sherwin, R. S.,Krystal, J. H., Spencer,D.D.,&Abi-Saab,W.M. (2005). Extracellularmetabolites in the cortex and hippocampus of epileptic patients. *Annals of Neurology, 57*, 226–235.

Commission on Classification and Terminology of the International eague against Epilepsy (ILE). 1989. Proposal for revised classification of epilepsies and epileptic syndromes. *Epilepsia*, 30(4):389-399.

Corda MG, Orlandi M, Lecca D, Giorgi O. 1992. Decrease in GABAergic function induced by pentylenetetrazol kindling in rats:antagonism by MK-801. *J PharmacolExp Ther* 262:792-800.

Chan, K., & Lowe, J. (2002). Techniques in neuropathology. In J. D. Bancroft & M.Gamble (Eds.), *Theory and practice of histological techniques.* (5th ed., chapter 18). (pp. 374–

375). New York, Edinburgh, London, Madrid, Melbourne, San Francisco, Tokyo: Churchill Livingstone.

Clifford DB, Olney JW, Maniotis A, Collins RC & Zorumski CF. 1987. The functional anatomy and pathology of lithium-pilocarpine and high-dose pilocarpine seizures. *Neuroscience,* 23, 953-968.

Chandler KE, Alessandra P. Princivalle, Ruth Fabian-Fine, Norman G. Bowery, Dimitri M. Kullmann & Matthew C. Walker. 2003. Plasticity of GABAB Receptor-Mediated Heterosynaptic Interactions at Mossy Fibers After Status Epilepticus. *The Journal of Neuroscience*, 23(36), 11382-11391.

Creutzfeldt-Jakob disease:Diagnostic value of protein 14-3-3 and neuronal specific enolase assay in cerebrospinal fluid. *Revue Neurologique (Paris), 155*(2), 148–151.

Correale, J., Rabinowicz, A. L., Heck, M. D., Smith, T. D., Loskota, W. J.,&DeGiorgio, C. M. (1998). Status epilepticus increases CSF levels of neuron-specific enolase and alters the blood-brain barrier. *Neurology, 50,* 1388–1391.

Crespo-Biel N, Canudas AM, Camins A, Pallas M. 2006. Kainate induces AKT, ERK and cdk5/GSK3beta pathway deregulation, phosphorylates tau protein in Mouse hippocampus. *Neurochem Int.* Nov 18; [Epub ahead of print].

Crossman, A. R., & Neary, D. (2000). Cells of the nervous system. *Neuroanatomy:An illustrated colour text* (2nd Ed). (pp. 33–36). Edinburgh, London, New York, Philadelphia, St. Louis, Sydney, Toronto: Churchill Livingstone.

Damjanov I. Histopathology. A color atlas and textbook. Chapter 19. Nervous system. International edition. Williams and Wilkins, 1996; pp. 459-490.

Dawodu S. & Thom Maria. 2005. Quantitative neuropathology of the entorhinal cortex region in patients with hippocampal sclerosis and temporal lobe epilepsy. *Epilepsy,* 46(1), 23-30.

DeGiorgio, C. M., Gott, P. S., Rabinowicz, A. L., Heck, C. N., Smith, T. D., & Correale, J. D. (1996). Neuron-specific enolase, amarker of acute neuronal injury, is increased in complex partial status epilepticus. *Epilepsia, 37*(7), 606–609.

DeGiorgio, C. M., Heck, M. D., Rabinowicz, A. L., Gott, P. S., Smith, T., & Correale, J. (1999). Serum neuron-specific enolase in the major subtype of status epilepticus.*Neurology, 52,* 746–749.

Ding, M., Haglid, K. G., & Hamberger, A. (2000). Quantitative immunochemistry on neuronal loss, reactive gliosis and BBB damage in cortex/striatum and hippocampus /amygdala after systemic kainic acid administration. *NeurochemistryInternational, 36*(4–5), 313–318.

Du F, Whetsell Jr WO, Abou-Khalil B, Blumenkopf B, Lothman EW & Schwarcz R. 1993. Preferential neuronal loss in layer III of the entorhinal cortex in patients with temporal lobe epilepsy. *Epilepsy Research,*16, 223-233.

Du F, Eid T, Lothman EW, Köhler C & Schwarcz R. 1995. Preferential neuronal loss in layer III of the medial entorhinal cortex in rat models of temporal lobe epilepsy. *The Journal of Neuroscience,* 15, 6301-6313.

Du F & Schwarcz R. 1992. Aminooxyacetic acid causes selective neuronal loss in layer III of the rat medial entorhinal cortex. *Neuroscience Letter,* 147, 185-188.

Eid T, Schwarcz R & Ottersen OP. 1999. Ultrastructure and immunocytochemical distribution of GABA in layer III of the rat medial entorhinal cortex following aminooxyacetic acid-induced seizures. *ExperimentalBrain Research,*125, 463-475.

Harris EW & Cotman CW. 1986. Long-term potentiation of guinea pig mossy fiber responses is not blocked by Nmethyl-D-aspartate antagonists. *Neuroscience Letter,* 70, 132-137. http://www.epilepsyfoundation.org

Hasegawa, D., Orima, H., Fujita, M., Hashizume, K., & Tanaka, T. (2002). Complex partial status epilepticus induced by a microinjection of kainic acid into unilateral amygdala in dogs and its brain damage. *Brain Research, 955,* 174–182. http://epilepsy.med.nyu.edu/epilepsy/frequently-asked-questions: NYU Langone Medical Center, 2011.

Hatfield, R. H., & McKernan, R. M. (1992). CSF neuron-specific enolase as a quantitative marker of neuronal damage in a rat stroke model. *Brain Research, 577,* 249–252.

Hay, E., Royds, J. A., Davies-Jones, G. A., Lewtas, N. A., Timperley, W. R., & Taylor, C. B. (1984). Cerebrospinal fluid enolase in stroke. *Journal of Neurology,Neurosurgery, and Psychiatry, 47,* 724–729.

Holopainen, I.E. (2008). Seizures in the developing brain: Cellular and molecular mechanisms of neuronal damage, neurogenesis and cellular reorganization. Neurochemistry International 52 (2008) 935–947.

Houser CR. 1990. Granule cell dispersion in the dentate gyrus of humans with temporal lobe epilepsy. *Brain Res* 10, 535(2):195-204.

Houser, CR, Miyashiro, JE, Swartz, BE, Walsh, GO, Rich, JR & Delgado-Escueta, AV. 1990. Altered patterns of dynorphin immunoreactivity suggest mossy fiber reorganization in human hippocampal epilepsy. *The Journal of Neuroscience,*1:267-282.

Houser CR. 1999. Neuronal loss and synaptic reorganization in temporal lobe epilepsy. *Advances in neurology,* 79, 743-761.

Ingvar, M., & Siej¨o, B. K. (1983). Local blood flow and glucose consumption in the rat brain during sustained bicuculline-induced seizures. *Acta Neurologica Scandinavica, 68,* 129–144.

Isokowa M. 1998. Remodeling dendritic spines in the rat pilocarpine model of temporal lobe epilepsy. *NeuroscienceLetter,* 18; 258(2), 73-6.

Isokawa M. & Mello LEAM. 1991. NMDA receptormediated excitability in dendritically deformed dentate granule cells in pilocarpine-treated rats. *Neuroscience Letter,* 129, 69-73.

Jacobi, C., & Reiber, H. (1988). Clinical relevance of increased neuron-specific enolase concentration in cerebrospinal fluid. *Clinica Chimica Acta, 177*(1), 49–54.

Kamiya H, Shinozaki H & Yamamoto C. 1996. Activation of metabotropic glutamate receptor type 2/3 suppresses transmission at rat hippocampal mossy fibre synapses. *The Journal of Physiology (London),* 493: 447-455.

Khanna, Bhalla S, Verma V, Sharma KK. 2000. Modulatory Effectes of Nifedipine and Nimodipine in Experimental Convulsions. *Indian Journal of Pharmacology* 2000; 32: 347-352.

Leite JP, Babb TL, Pretorius JK, Kulhman PA, Yeoman KM & Mathern GW. 1996. Neuronal loss, mossy fiber sprouting, and interictal spikes after intrahippocampal kainate in developing rats. *Epilepsy Research*, 26, 219-231.

Lima, J. E., Takayanagui, O. M., Garcia, L. V., & Leite, J. P. (2004). Use of neuronspecific enolase for assessing the severity and outcome of neurological disorders in patients. *Brazilian Journal of Medical and Biological Research, 37*(1), 19–26.

Lurton D, Sundstrom L, Brana C, Bloch B & Rougier A. 1997. Possible mechanisms inducing granule cell dispersion in humans with temporal lobe epilepsy. *Epilepsy Research*, 26, 351-361.

Meldrum BS. 1993. Excitotoxicity and selective neuronal loss in epilepsy. *Brain Pathology*, 3:405-412. [52]. Sloviter RS. 1987. Decreased hippocampal inhibition and a selective loss of interneurons in experimental epilepsy. *Science*, 235:73-76.

Mello LEAM, Cavalheiro EA, Tan AM, Kupfer WR, Pretorius JK, Babb TL & Finch DM. 1993. Circuit mechanisms of seizures in the pilocarpine model of chronic epilepsy: cell loss and mossy fiber sprouting. *Epilepsia*, 34:985-995.

Michel RP & Cruz-Orive LM. 1988. Application of the Cavalieri principle and vertical sections method to lung: estimation of volume and pleural surface area. *J Microsc*, 150(Pt 2):117-36.

Nadler JV, Perry BW & Cotman CW. 1980a. Selective reinnervation of hippocampal area CA1 and the fascia dentata after destruction of CA3-CA4 afferents with kainic acid. *Brain Research*, 182, 1–9.

Nadler JV, Perry BW, Gentry C, Cotman CW. 1980b. Degeneration of hippocampal CA3 pyramidal cells induced by intraventricular kainic acid. *J Comp Neurol*.15;192(2):333-59.

Nogami, M., Takatsu, A., Endo, N., & Ishiyama, I. (1998a). Immunohistochemistry of neuron-specific enolase in neurons of themedulla oblangata from human autopsies. *Acta Histochemica, 100*, 371–382.

Nogami, M., Takatsu, A., & Ishiyama, I. (1998b). Immunohistochemical study of neuron-specific enolase in human brains from forensic autopsies. *ForensicScienceInternatonal, 94*, 97–109.

Oztas, B., Kilic, S., Dural, E., & Ispir, T. (2001). Influence of antioxidants on the blood-brain barrier permeability during epileptic seizures. *Journal of NeuroscienceResearch, 66*, 674–678.

Paxinos, G., & Watson, C. (1994). *The rat brain in stereotaxic coordinates* (2nd ed.). San Diego, Boston, New York, London, Sydney, Tokyo, Toronto: Academic Pres Inc. Harcourt Brace Jovanovich, Publishers.

Pavlova, T. V., Iakovleva, A. A., Stepanichev, M. I. U., & Guliaeva, N. V. (2005). [Pentylenetetrazole kindling in rats: Whether neurodegeneration is associated with manifestations of seizure activity?]. *Rossiĭskii Fiziologicheskiĭ Zhurnal Imeni I.M.Sechenova, 91*(7), 764–775.

Pavlova TV, Yakovlev AA, Stepanichev MIu, Guliaeva NV. 2006. Pentylenetetrazol Kindling in Rats: Is Neurodegeneration Associated with Manifestations of Convulsive Activity? *Neuroscience and Behavioral Physiology*, 36(7):741-48.

Penfield, W. 1929. The mechanisms of cicatricial contraction in the brain. *Brain*, 50, 499–517.

Persson, L., Hardemark, H., Edner, G., Ronne, E., Mendel-Hartvig, I., & Pahlman, S. (1988). S-100 protein in cerebrospinal fluid of patients with subarachnoid haemorrhage: A potential marker of brain damage. *Acta Neurochirurgica, 93*, 116–122.

Portera-Cailliau C, Price DL, Martin LJ. 1997. Excitotoxic neuronal death in the immature brain is an apoptosisnecrosis morphological continuum. *J Comp Neurol.* 3; 378(1):70-87

Portera-Cailliau C, Price DL, Martin LJ. 1997. Non- NMDA and NMDA receptor-mediated excitotoxic neuronal deaths in adult brain are morphologically distinct: further evidence for an apoptosis-necrosis continuum. *J Comp Neurol.* 378 (1):88-104.

Rasmussen, C. V., Kragh, J., Bolwig, T. G., & Jorgensen, O. S. (1994). Repeated electroconvulsive shock selectively increases the expression of the neuron specific enolase in piriform cortex. *Neurochemical Research, 19* (12), 1527–1530.

Rodriquez-Nunez, A., Cid, E., Rodriguez-Garcia, J., Camina, F., Rodriguez-Segade, S., & Castro-Gago, M. (2000). Cerebrospinal fluid purine metabolite and neuronspecific enolase concentrations after febrile seizures. *Brain & Development, 22*(7), 427–431.

Sahin, D., Ilbay, G., & Ates, N. (2003). Changes in the blood-brain barrier permeability and in the brain tissue trace element concentrations after single and repeated pentylenetetrazole-induced seizures in rats. *Pharmacological Research, 1137*, 1–5.

Salin PA, Scanziani M, Malenka RC & Nicoll RA. 1996.Distinct short-term plasticity at two excitatory synapses in the hippocampus. *Proceedings of the National Academyof Sciences*, USA 93, 13304-13309.

Sankar, R., Shin, D. H., & Wasterlain, C. G. (1997). Serum neuron-specific enolase is a marker for neuronal damage following status epilepticus in the rat. *Epilepsy Research, 28*(2), 129–136.

Sankar R, Shin D, Liu H, Wasterlain C & Mazarati A. 2002. Epileptogenesis during development:injury, circuit recruitment and plasticity. *Epilepsia*, 43 (Suppl. 5):47-53.

Santhakumar V, Ratzliff AD, Jeng J, Toth K & Soltesz I. 2001. Long-term hyperexcitability in the hippocampus after experimental head trauma. *Annals of Neurology*, 50, 708-717.

Schmechel, D., Marangos, P. J., Zis, A. P., Brightman, M., & Goodwin, F. K. (1978). Brain enolases as specific markers of neuronal and glial cells. *Science, 199*(4326), 313–315.

Schreiber, S. S., Sun, N., Tocco, G., Baudry, M.,&DeGiorgio, C. M. (1999). Expression of neuron-specific enolase in adult rat brain following status epilepticus. *Experimental Neurology, 159*(1), 329–331.

Steven C. Schachter, M.D. ttp://www.epilepsy.com/EPILEPSY/Types_seizures Last Reviewed:12/15/06.

Schwob JE, Fuller T, Price JL & Olney JW. 1980. Widespread patterns of neuronal damage following systemic or intracerebral injections of kainic acid: a histological study. *Neuroscience*, 5: 991-1014.

Siesjo, B. K., Ingvar,M., &Wieloch, T. (1986). Cerebral blood flow and metabolic rate during seizures. Relationship to epileptic brain damage. *Annals of the New York Academy of Science, 462,* 194–206. http://www.searo.who.int.)

Sloviter, R. S., & Dempster, D. W. (1985). Epileptic brain damage is replicated qualitatively in the rat hippocampus by central injection of glutamate or aspartate but not by GABA or acetylcholine. *Brain Research Bulletin, 15*(1), 39–60.

Sloviter RS. 1991. Feedforward and feedback inhibition of hippocampal principal cell activity evoked by perforant path stimulation: GABA-mediated mechanisms that regulate excitability in vivo. *Hippocampus.* 1(1):31-40.

Sloviter RS, Zappone CA, Harvey BD, Frotscher M. 2006. Kainic acid-induced recurrent mossy fiber innervation of dentate gyrus inhibitory interneurons: possible anatomical substrate of granule cell hyper-inhibition in chronically epleptic rats. *Journal Comparative Neurology,* 20; 494 (6): 944-60.

Steinhoff, J. B., Tumani, H., Otto, M., Mursch, K., Wiltfang, J., Herrendorf, G., Tumani, H., Otto,M., Gefeller, O.,Wiltfang, J., Herrendorf, G.,Mogge, S., & Steinhoff, J. B. (1999). Kinetics of serum neuron-specific enolase and prolactin in patients after single epileptic seizures. *Epilepsia, 40*(6), 713–718.

Steward, O., Torre, E. R., Tomasulo, R., & Lothman, E. 1991. Neuronal activity up-regulates astroglial gene expression. *Proceedings of the National Academy of Sciences,* USA 88 (15):6819- 6823.

Toth Z, Hollrigel GS, Gorcs T & Soltesz I. 1997. Instantaneous perturbation of dentate interneuronal networks by a pressure wave-transient delivered to the neocortex. *The Journal of Neuroscience,* 17:8106-8117.

Xiang-ming Zha, Steven H. Green & Michael E. Dailey. 2005. Regulation of hippocampal synapse remodeling by epileptiform activity. *Molecular and Cellular Neuroscence,* 29 494–506.

van Bogaert, P., De Tiege, X., Vanderwinden, J. M., Damhaut, P., Schiffmann, S. N., & Goldman, S. (2001). Comparative study of hippocampal neuronal loss and in vivo binding of 5-HT1a receptors in the KA model of limbic epilepsy in the rat *Epilepsy Research, 47,* 127–139.

vander Hel,W. S., Notenboom, R. G., Bos, I.W., van Rijen, P. C., van Veelen, C.W., & de Graan, P. N. (2005). Reduced glutamine synthetase in hippocampal areas with neuron loss in temporal lobe epilepsy. *Neurology, 64,* 326–333.

Vogt KE & Nicoll RA. 1999. Glutamate and gammaaminobutyric acid mediate a heterosynaptic depression at mossy fiber synapses in the hippocampus. *Proceedings of the National Academy of Sciences,* USA 96, 1118-1122.

Wen HT, Rhoton AL Jr, de Oliveira E, Cardoso AC, Tedeschi H, Baccanelli M, Marino R Jr. 1999. Microsurgical anatomy and its vascular relationships as applied to amygdalohippocampectomy. *Neurosurgery,*(45):549-591.

Weisskopf MG, Zalutsky RA & Nicoll RA. 1993. The opioid peptide dynorphin mediates heterosynaptic depression of hippocampal mossy fibre synapses and modulates long-term potentiation. *Nature,* 365, 188.

http://www.medicinenet.com/seizures_symptoms_and_types/article.htm

1996-2011 Epilepsy: Type of Seizures and Their Symptoms, MedicineNet, Inc.

Yonekawa WD, Kupferberg HJ & Woodbury DM. 1980. Relationship between pentylenetetrazol-induced seizures and brain pentylenetetrazol levels in mice. *The Journal of Pharmacology and Experimental Therapeutics*, 214, (3):589-593.

Young, B., & Health, J. W. (2000). Nervous tissue. *Wheaters functional histology – A text and color atlas.* (4th Ed). (pp. 116–143). Edinburgh, London, New York, Philadelphia, St. Louis, Sydney, Toronto: Churchill Livingstone.

2

The Blood-Brain Barrier and Epilepsy

Gul Ilbay[1], Cannur Dalcik[2], Melda Yardimoglu[3],
Hakki Dalcik[3] and Elif Derya Ubeyli[4]
[1]*Kocaeli University, Faculty of Medicine, Department of Physiology, Umuttepe Campus, Kocaeli*
[2]*Kocaeli University, Faculty of Medicine, Department of Anatomy*
[3]*Kocaeli University, Faculty of Medicine, Department of Histology and Embryology*
[4]*Osmaniye Korkut Ata University, Faculty of Engineering, Department of Electrical and Electronics Engineering, Osmaniye, Turkey*

1. Introduction

The blood-brain barrier (BBB) is a dynamic interface between the blood and the central nervous system (CNS), that controls the exchanges between the blood and brain compartments. Therefore, the functional and structural integrity of the BBB is vital to maintain the homeostasis of the brain microenvironment and it provides protection against many toxic compounds and pathogens (Cardosa et al., 2010; Cucullo et al., 2011).

Anatomically, the BBB consists of microvascular endothelial cells (ECs) lining the brain microvessels together with the perivascular elements such as astrocytes, pericytes, neurons, and the extracellular matrix. The microcapillary endothelium is characterized by the presence of tight junctions (TJs), lack of fenestrations, and minimal pinocytotic vesicles. TJs between the cerebral ECs form a diffusion barrier, which selectively excludes most blood-borne substances from entering the brain, protecting it from systemic influences. Brain ECs also express plasma membrane transport protein and receptors, both of which provide selective routes of entry for polar nutrients such as glucose transport (GLUT-1), ions (Na, K-ATPase and Na, K) and some macromolecules (insulin and transferring receptors) and routes of exit for potentially toxic metabolic waste and macromolecules (ATP-binding cassette (ABC) transporters). Through restrictive barrier properties and polarized expression of selective transport proteins, the BBB effectively regulates solute and fluid exchange between the blood and brain parenchyma (Marchi et al., 2011a; Paolinelli et al., 2011).

The BBB is a dynamic system, capable of responding to local changes and requirements, and is regulated by a number of mechanisms and cell types, in both physiological and pathological conditions. Such regulation includes changes in tight junction function, and in expression and activity of many transporters and enzymes (Abbott et al., 2010). BBB breakdown has often been documented in the epileptic brains (Ilbay et al., 2003; Uzüm et al., 2006). Recently, as within many CNS diseases, also in epilepsy, BBB dysfunction have been described not only as a late event, but more interestingly as putatively involved in the early steps of disease progression. It is well acknowledged that epileptic seizures are the result of instant abnormal hypersynchronous electrical activity of neuronal network, originating from discharges of local brain tissue that caused by an imbalance between excitation and inhibition. However, in a substantial number of epilepsies, the etiology is almost unknown. The complex and multifactorial nature of epileptogenesis (occurrence of spontaneous epileptic seizures) has highlighted the fact that limited therapeutic solutions in clinical practice are left to a proportion of epilepsy patients that are refractory to all anti-epileptic drugs (AEDs). During the past several years, increasing interest has arisen in the role of the BBB in epilepsy. Recent advances in understanding nature of epileptogenesis have made it possible to realize the role of the BBB disruption in epileptogenesis (David et al., 2009; de Boer et al., 2008; Vezzani et al., 2008; Weis et al., 2009; You et al., 2011).

2. Blood- Brain Barrier: Structure and function

The BBB is localized at the interface between the blood and the cerebral tissue and, formed by the ECs of cerebral blood vessels which are characterized by the absence of fenestration correlating with the presence of intercellular TJs, the low level of non specific transcytosis and paracellular diffusion of hydrophilic compounds. ECs have a high number of mitochondria, associated with a strong metabolic activity and the polarized expression of membrane receptors and transporters, which are responsible for the active transport of blood-borne nutrients to the brain or the efflux of potentially toxic compounds from the neural tissues to the vascular compartment (Weis et al., 2009).

BBB properties are primarily determined by junctional complexes between the cerebral ECs. These complexes are comprised of TJs and adherens junctions (AJs). Such restrictive angioarchitecture at the BBB reduces paracellular diffusion, while minimal vesicle transport activity in brain ECs limits transcellular transport. Under normal conditions, this largely prevents the extravasation of macromolecules and most polar solutes (unless specific transporters are present) and prevents migration of any type of blood-borne cell. In AJs, cadherin proteins span the intercellular cleft and are linked into the cell cytoplasm by the scaffolding proteins alpha, beta and gamma catenin. The AJs hold the cells together giving the tissue structural support. They are essential for formation of TJs, and disruption of AJs leads to barrier disruption. The TJs consist of a complex of proteins spanning the intercellular cleft (occludin, nectin, claudins), and junctional adhesion molecules A, B and C (JAMs). TJs adhesion proteins are linked to a specific network of cytoskeletal and signaling proteins which include ZOs (zonula occludens proteins 1, 2, 3 and zonula occludens-1 associated nucleic-acid-binding protein (ZONAB), cingulin and others). The TJs are responsible for the severe restriction of the paracellular diffusional pathway between the ECs to ions and other polar solutes, and effectively block penetration of macromolecules by this route. The impediment to ion movement results in the high in vivo electrical resistance of the BBB, of $\sim 1800\ \Omega\ cm^2$. This high electrical resistance or low conductance of the potential paracellular pathway emphasises the extreme effectiveness of the TJs in occluding this pathway by effectively reducing the movement of ions (Abbott et al., 2010; Paolinelli et al., 2011; Stamatovic et al., 2008). However, the BBB is a dynamic structure capable of rapid modulation in response to physiological or pathological signals. The permeability can be controlled by a variety of signaling pathways that regulate TJs organization. Numerous substances and signaling tracks, such as Ca2+, protein kinase C, G protein, calmodulin, cAMP, phospholipase C, and tyrosine kinases, are believed to play a role in this regulation. Several cytoplasmic signaling molecules are concentrated at TJs complexes and are involved in signaling cascades that control assembly and disassembly of TJs. ZOs proteins link claudins and occludin via cingulin to intracellular actin cytoskeleton. Importantly, it has been shown that ZO-1 and ZONAB are a substrate of a serine kinase and protein kinase C, respectively, which are crucial for the formation and regulation of TJs. Failure to maintain the integrity of brain EC junctions can have profound effects on permeability control. (Paolinelli et al., 2011). In addition, the BBB is structurally associated with brain parenchymal cells. Pericytes, glial cells (especially astrocytes), neurons, together with the basal lamina (also called lamina basalis) ensheathing cerebral blood vessels, are indirectly involved in the establishment and maintenance of the BBB (Figure 1). These various cell types and basal lamina collectively constitute the neurovascular unit (NVU), a concept recently proposed to highlight the functional interactions that control the BBB integrity. The basal lamina of the cerebral endothelium is constituted by 3 apposed layers, one produced by ECs and containing laminin -4 and -5, one being astrocyte-derived, containing laminin -1 and -2 and the collagen IV-containing the middle one, contributed by both cell types. All three layers are also made of various types of collagen, glycoproteins and proteoglycans. The basal lamina of the cerebral endothelium shows strong labeling with cationic colloidal gold indicating that it forms a negatively charged screen or filter controlling the movement of charged solutes between blood and the brain interstitial fluid. Therefore large, charged molecules such as ferritin do not cross the basal lamina. Multiple basal lamina proteins, matrix metalloproteases and their inhibitors are involved in the dynamic regulation of the BBB in physiological as well as pathological conditions. (Cardosa et al., 2010; Marchi et al., 2011a; Nag, 2003a; Weis et al., 2009).

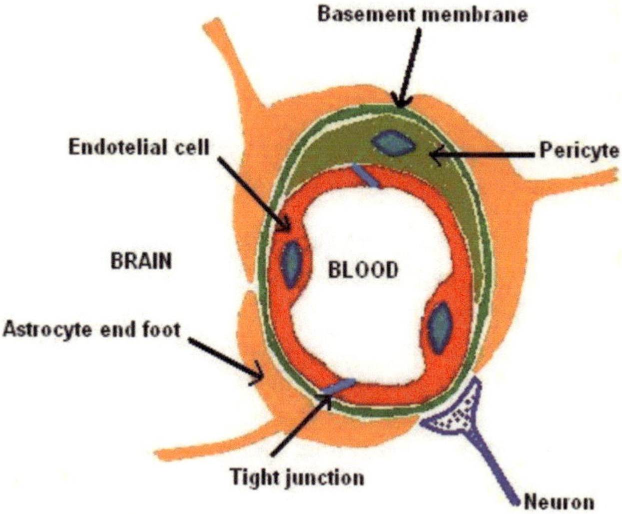

Fig. 1. Schematic representation of the BBB structure.

Induction and maintenance of many BBB properties depends on a close association with astrocytes. Astrocytes are one of the important components of NVU. Astrocytes are glial cells whose end feet form a lacework of fine lamellae closely apposed to the outer surface of the BBB endothelium and respective basement membrane (Figure 2). They play a major role in promoting proteoglycan synthesis with a resultant increase in brain microvascular ECs charge selectivity. On the other hand, a number of astrocyte-released and more generally glial-released factors have been suggested to contribute to BBB integrity, including glial-derived neurotrophic factor, angiopoietin-1 and more recently angiotensin II. Another important cell that contributes to the NVU is the pericytes. Pericytes are cells that are present along brain and non-brain microvessels, within the basal lamina surrounding ECs; interestingly, brain microvessels are notably rich in pericytes and the pericytes/ECs ratio has been correlated with the barrier capacity of the endothelium. Pericytes are actively involved in maintenance of the integrity of the vessel, vasoregulation and restricted BBB permeability. ECs, astrocytes and pericytes are in close contact with neuronal projections, allowing neuronal mediators to affect cerebral blood flow and vessel dynamics. However, the precise physiological or pathophysiological consequences of neuronal input onto the BBB remain largely unknown (Cardosa et al., 2010; Marchi et al., 2011a; Weis et al., 2009).

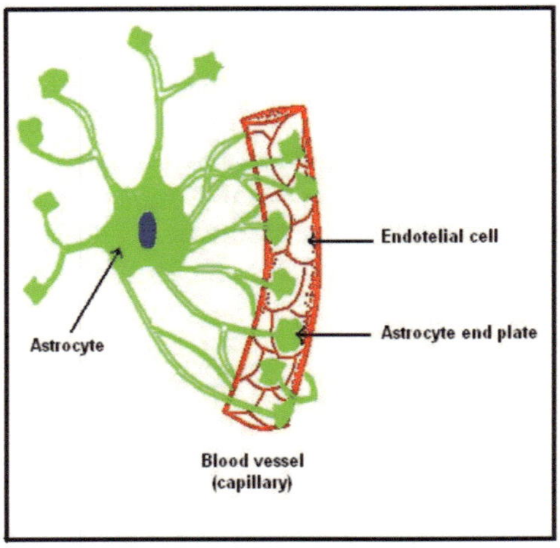

Fig. 2. Note that astrocytes's endfeet are in contact with the endothelial surface.

As mentioned above, BBB properties are primarily determined by junctional complexes between the cerebral ECs and NVU. However, it is necessary to put those junctions into the context of other BBB properties. These include the general paucity of vesicular transport at the BBB, the presence of enzymes that by degradation prevent the entry of a variety of compunds, and the presence of a wide range of transport systems. Because non-lipid soluble compounds only diffuse slowly across the BBB, the latter are necessary for both the entrance of nutrients into brain and the clearance of waste products from brain (Stamatovic et al., 2008).

In general, the brain ECs has a very low number of vesicles under normal conditions compared to other types of ECs. However, during some disease states the number of vesicles can increase. Fusion between vesicles may eventually lead to the formation of transendothelial channels and/or vesicle/vacuolar organelles (VVO). Transendothelial channels correspond to chains of two or more fused vesicles that are open simultaneously on the luminal and abluminal side of ECs. Besides that, channels made by one of the vesicle open on the both sides of ECs can occur. VVO on the other hand are large collection of interconnected vesicles and vacuoles. The fusion of vesicles, vacuole with luminal and abluminal plasma membranes creates transcellular pathways confirmed in the several ultrastructural studies (Stamatovic et al., 2008).

The BBB can also act as an enzymatic barrier, capable of metabolizing drug and nutrients. The ECs particularly in the BBB contain elevated concentrations of enzymes that are involved in metabolizing neuroactive blood-borne solutes. Such enzymes are γ-glutamyl transpeptidase, alkaline phosphatase, aromatic acid decarboxylase, and monoamine oxidases. These enzymes are often polarized between the luminal and abluminal membrane surface of brain ECs. We know that, the barrier to paracellular diffusion potentially isolates the brain from many essential polar nutrients such as glucose and amino acids. However, the BBB endothelium must contain a number of specific solute transporters to supply the CNS with these substances. The brain ECs forming the BBB expresses transport proteins for a wide variety of solutes and nutrients, mediating flux into and out of the brain. For example, the BBB has very high levels of the glucose transporter 1, GLUT1, and the large neutral amino acid transporter, LAT1, that facilitate movement of those nutrients from blood to brain. Other transporters are involved in ion homeostasis and the transport of signaling molecules between blood and brain. Ions or small molecules are mostly transferred through the brain ECs by transporters/carriers/channels present on the plasma membrane. Most macromolecules move across brain capillary endothelium by bulk phase non-receptor mediated endocytosis. For example, cationic macromolecules prefer uptake by clathrin coated membranes and pits and are subsequently delivered to and degraded in lysosomes. There are also efflux transporters that move compounds from brain to blood. Efflux transporters of the ATP Binding Cassette (ABC) family transport a large panel of lipophilic molecules, in particular xenobiotics, against a concentration gradient by ATP hydrolysis. The most active ABC-transporters at the BBB are P-glycoprotein (P-gp) encoded by the multidrug resistance gene (MDR1 or ABCB1), multidrug resistance proteins (MRPs, or ABCC proteins) and the breast cancer resistance protein (BCRP or ABCG2). The major role of the ABC transporters in the BBB is to function as active efflux pumps consuming ATP and transporting a diverse range of lipid-soluble compounds out of the brain capillary endothelium and the CNS. In this role, they are removing from the brain potentially neurotoxic endogenous or xenobiotic molecules and they are carrying out a vital neuroprotective and detoxifying function. In addition, many drugs are substrates for these ABC efflux transporters and their brain penetration is significantly reduced (Abbott et al., 2010; Stamatovic et al., 2008; Weiss et al., 2009).

3. Blood- Brain Barrier dysfunction in epilepsy: Experimental and clinical studies

The BBB breakdown has been recognized long ago in experimental and human epilepsy. There have been numerous reports indicating that seizures may produce an increase in cerebral capillary permeability (Ruth et al., 1984; Ilbay et al., 2003; Sheen et al., 2011). On the other hand, epileptogenic process may also be triggered by impaired function of BBB (Oby and Janigro, 2006). Generally, the duration of seizure activity is correlated with reduced BBB functions. However, with increased arterial blood pressure, the BBB becomes permeable to macromolecules under induced epileptiform seizures (Öztas et al., 1991; Sheen et al., 2011). Studies reveal that an acute increase in blood pressure or epileptic activity causes an increase in pinocytosis at the level of the cerebral endothelium (Ilbay et al., 2003;

Oby and Janigro, 2006). There is also information from earlier experiments indicating that this BBB alteration is reversible and confined to anatomically limited brain areas. BBB openings have been mapped for a variety of convulsive agents with different mechanisms of action (i.e., impairment of GABA transmission, increased glutamate neurotransmission, direct excitatory action) (Oby and Janigro, 2006; Vezzani et al., 2008). A direct link between the mechanism of action and the region where BBB breakdown was observed is not obvious. However, a few brain regions are easily affected by seizure activity irrespective of the means of induction. Other regions of the brain show BBB breakdown only under the influence of specific convulsants. It is shown that seizures induced by pentylenetetrazole primarily affect thalamus, hypothalamus, midbrain and cerebellum (Sahin et., 2003; Uzüm et al., 2006) (Figure 3). Also, the GABA-receptor blocker bicuculline induces loss of barrier in hippocampus whereas kainic acid induces alterations in capillaries of the neocortical brain areas (Nitsch et al., 1983). The results of our previous study demonstrated that generalised tonic-clonic convulsions, together with elevated blood pressure resulted in BBB opening in multiple areas (Ilbay et al., 2003; Sahin et al., 2003), (Figures 4). Unfortunately, while several models are available to study the BBB in animal models, a suitable (i.e., microscopic, quantitative, and minimally invasive) technique for evaluating BBB integrity in humans does not exist. However, studies on human epileptic tissue show clear BBB abnormalities, including increased micropinocytosis and fewer mitochondria in ECs, a thickening of the basal membrane, and the presence of abnormal Tjs. Finally, it can be stated that opening of BBB is present before, during or after epileptic seizures, is a hallmark of vascular injury in the brain and there are multiple factors involved in etiology of BBB breakdown in epilepsy (Oby and Janigro., 2006; Stamatovic et al., 2008; Vezzani et al., 2008;).

It has been demonstrated that, seizures and epilepsy often develop after traumatic, ischemic, or infectious brain injury. After injury, local compromise of BBB integrity is common (Abbott et al., 2010; Cacheaux et al 2009; Oby and Janigro, 2006), as revealed by ultrastructural studies of animal and human epileptic tissue in multiple forms of epilepsy. (Cacheaux et al 2009; Marchi et al., 2007a; van Vliet et al., 2007), and BBB opening may specifically serve as a trigger event leading to epilepsy. However, the research field has recently considered the fact that damage to the cerebral vasculature could be involved in the pathogenesis of seizures (Oby & Janigro, 2006).

While large amount of published data support a correlation between occurrence of spontaneous epileptic seizures, seizures and abnormal BBB, direct evidence for the involvement of BBB breakdown in epileptogenesis has been only recently confirmed by the studies, in which opening of the BBB was sufficient to induce delayed epileptiform activity (Cacheaux et al., 2009; Seiffert et al., 2004).

Epileptogenesis in the BBB-disrupted brain seems to be mediated by exposure of the brain cortex to serum albumin, mediated via its action on brain astrocytes. The possible involvement of albumin in astrocytic activation and proliferation is supported by previous studies. Based on the studies, it was concluded that there is a specific glial receptor (TGF-β) and signaling pathway for the action of albumin. Thus, it seems plausible that damage to the microvasculature during brain insults leads to the extravasation of serum proteins, leading to the transformation of the neighboring astrocytes as the primary step in the epileptogenic process. In addition, it is notable that in many cases of epilepsy, as well as in most animal models, neuronal damage is observed. In this respect it is interesting to point out that BBB breakdown has been associated with early or delayed neuronal damage to what extent this damage contributes to epileptogenesis, a result of abnormal activity or not related, is not as yet clear (David et al., 2009; Friedman et al., 2011; vanVliet et al., 2007).

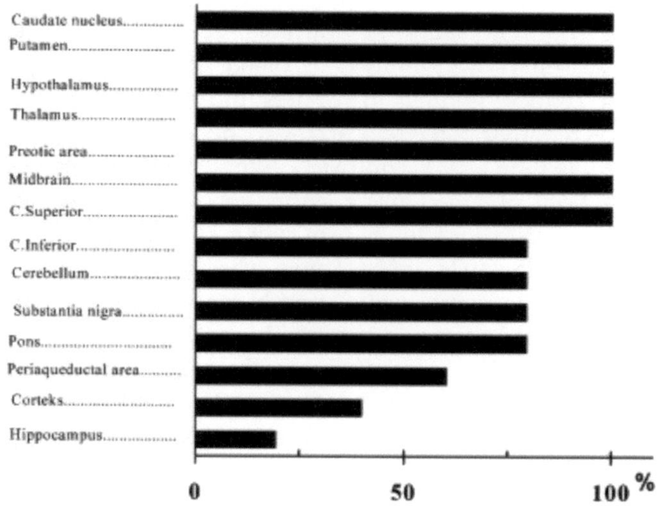

Fig. 3. Patern of BBB opening after PTZ-induced seizures is demonstrated. Regional distribution indicated as percent of total number of animal exhibiting EB leakage. BBB opening is most pronounced and frequently observed in the deep brain areas (Sahin et al., 2003).

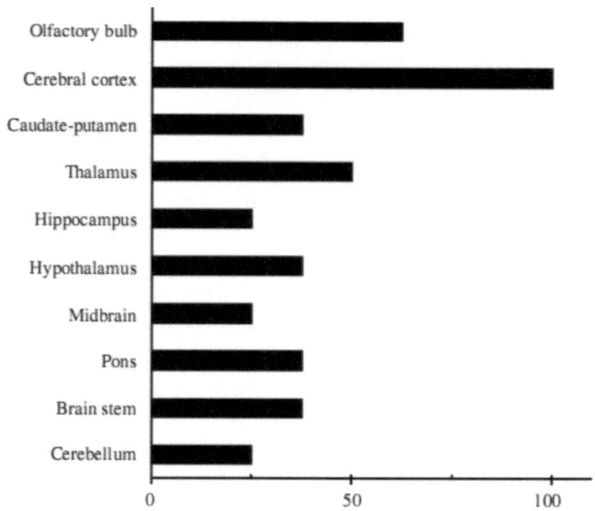

Fig. 4. Pattern of BBB opening after hot water-induced epilepsy is demonstrated. Regional distribution indicated as percent of total number of animal exhibiting EB leakage. BBB opening is most pronounced in cortical areas (Ilbay et al., 2003).

Experimental evidence also supports a role of intravascular inflammation, often associated with BBB damage, in seizure disorders (Marchi et al., 2007a; Marchi et al., 2009; Seiffert et al., 2004; van Vliet et al., 2007). Recently, the association between circulating immune cells, their interaction with the BBB, and seizure propensity was proposed; leukocyte adhesion was shown to directly support ictogenesis in the pilocarpine model of seizures (Fabene et al., 2008). A recent study conducted on patients affected by temporal lobe epilepsy demonstrated an increase in circulating natural killer cells and cytotoxic T lymphocytes (Bauer et al., 2008). The results were limited by blood analysis at one single time point, making the relationship of these findings to seizures uncertain. These results are, however, comparable to what was reported in the pilocarpine model of epilepsy (Marchi et al., 2007b), where an increase in circulating CD8+ cytotoxic T lymphocytes was measured. In general, concordant data have been obtained using either models of peripheral inflammation, regardless of the initiating trigger; activation of white blood cells was observed prior to development of seizures. Moreover, when kainic acid was used to induce seizures, a contribution of the immune response was demonstrated, suggesting the influence of the adaptive immune response on kainic acid induced hippocampal neurodegeneration (Chen et al., 2004; Silverberg et al., 2010). Taken together, these findings support a link between seizures and immunity in both animal models and in a clinical scenario. The molecular players linking white blood cells activation to BBB damage and seizures are, however, still unknown (Marchi et al., 2011b; Marchi et al., 2009).

BBB dysfunction is a major etiological event of seizure disorder in GLUT-1 deficiency syndrome. Glucose is the principal energy source for the brain and a continuous supply of this substrate is essential to maintain normal cerebral function. GLUT-1 ensures nutrient delivery, supplying glucose for the brain, which is its main energy source. GLUT-1 is highly restricted to the capillary ECs in the brain. The GLUT-1 deficiency syndrome is attributed to a defect of GLUT1 causing impaired transport of glucose across the BBB, interfering with cerebral energy metabolism and brain function, ultimately leading to seizures. GLUT1 deficiency syndrome has led to the investigation of the properties of GLUT1 in epilepsy. Many clinical studies showed that by using fluoro-2-deoxy-D-glucose positron emission tomography (FDG-PET) of patients with complex partial seizures yielded interictal scans that exhibit hypometabolic regions including the epileptic focus as identified by electroencephalography (EEG). In contrast, ictal scans showed focal, multifocal and generalized increases in the metabolic rate. However, ictal hypermetabolism did not always correspond to regions of interictal hypometabolism. Recent studies concluded that in epileptogenic temporal cortex both the physiology of blood flow and glucose metabolism are altered (Oby & Janigro, 2006). It is hypothesized that the interictal zone of hypometabolism coincides with altered BBB transporter activity and demonstrated that the zone of reduced metabolism according to a FDG-PET corresponds to a region of decreased BBB glucose transporter activity (Cardoso et al., 2010; Cornford et., 1999; Oby & Janigro, 2006). Moreover, human and animal epilepsy studies suggest that, a ketogenik diet leads to anticonvulsant consequences (Masino et al., 2011). The findings of these studies underline that BBB-related hypometabolism may be a common feature of focal epilepsy and not only limited to molecular changes in GLUT1 transporter expression.

Another important aspect of BBB function is its control of drug transport to brain. Permeability of the BBB determines the bioavailability of therapeutic drugs and resistance to chemically different AEDs. ABC efflux transporters at the BBB influence the brain uptake of a variety of therapeutic agents, including many AEDs. The permeability of the BBB becomes particularly relevant in drug resistant patients. Also effective concentrations of the AEDs are not attained in the brain, because of aberrant functioning of multidrug transporters and changes in drug efflux transporters include the overexpression of P-gp (MDR1), MRP1 and MRP2 (Oby & Janigro, 2006). The ABC transporter P-gp or MDR1 recognizes a wide range of substrates, including a number of AEDs, and is thought to play a major role in drug extrusion. MDR1 overexpression has been demonstrated in a variety of cells in both the BBB and the parenchyma in patients with intractable epilepsy. Many researchers have shown that, there is an overexpression of P-gp in ECs, astrocytes and neurons in a variety of animal models of epilepsy. Seizures have been shown to induce overexpression of other transporters as well. In many cases the overexpression of efflux drug transporters contributes to reduced efficacy of AEDs (Bartmann et al., 2010; Oby & Janigro, 2006; Seegers et al., 2002).

4. Detection of Blood- Brain Barrier dysfunction in epilepsy

As addressed above, epileptic activity influences the BBB and can result in disturbances of its integrity and functionality. From the visible and fluorescence to electron microscope, microscopy constitutes as a widely used tool to study the permeability properties of cerebral vessels in epilepsy. Conventional visible microscopy is used in detecting focal changes in BBB permeability such as in the detection of exogenous tracers or plasma proteins immunoreactivity around blood vessels in brain sections. BBB related proteins can therefore, be detected by immunocytochemical analysis using a flourescent microscope.

Transcellular permeability to exogenous tracers can yield valuable information regarding barrier integrity. Tracers such as horseradish peroxidase (HRP) and Evans blue were used extensively in epilepsy studies (Nag, 2003b). Horseradish peroxidase (HRP) is a plant enzyme having a molecular weight of 40,000 daltons and a diameter of 5 nm. When HRP is reacted with the oxidizable substrate 3 -3' diaminobenzidine, an insoluble brown reaction product is produced that is easily visualized by light microscopy, and after exposure to osmium ions, and is easily detected by electron microscopy. Multifocal areas of HRP extravasation from cerebral vessel has been observed in rat brains after epileptic seizures (Nag, 2003c).

Evans blue, a dye with high affinity to serum albumin that is not expected to permeate through the BBB, has been used as an indicator of BBB breakdown in epilepsy. Following EB injection in an animal, the staining of the brain is evaluated macroscopically against an arbitrary staining scale, providing a qualitative evaluation of the BBB permeability (Figures 5 and 6). Although this approach has several pitfalls as evaluation is inherently subjective, it is still used nowadays, at least as a first approach to assess BBB permeability in vivo. Such assessment can be improved by semiquantitative analysis of brain slices by fluorescence microscopy or by quantitative determination of the dye in brain homogenates (Ahishali et al., 2010; Cardosa et al., 2010; Uzüm et al., 2006; Ilbay et al., 2003; You et al 2011). Most tracers are labeled by a fluorescent dye or radioactive label that helps the quantification of the molecule (Marchi & Teng, 2010). To determine the limiting size for permeability, different molecular weight tracers can be used, such as fluorescent conjugated dextran (20-,10- and 4-kDa FITC-dextran), propidiumiodide (668 Da), and sodium fluorescein (376 Da). (Cardosa et al., 2010). The main features of the BBB are the presence of tight intercellular junctions, which strictly limit the diffusion of blood-borne solutes and cells into the brain. Several studies have been performed regarding these junctions, based on commonly used methodologies such as Western blot and PCR, as well as several types of microscopy, as addressed above. Using such approaches, increased BBB permeability associated with a loss of TJ integrity was also demonstrated in models of epilepsy (Ahishali et al., 2010; Rigau et al., 2007).

Membrane transporters and vesicular mechanisms protect the healthy brain, shielding it from toxic substances and allowing the entrance of others that are necessary. Some of the most recent studies regarding this barrier include the regulation of major efflux transporters in epileptic brains (Bartmann et al., 2010). Immunohistochemical methods are used in evaluating P-gp expression induced by seizure in post mortem. Positron emission tomography and Single-photon emission computed tomography (SPECT) studies are currently describing to asses changes in P-gp expression and functionality in vivo (Bartmann et al., 2010; Seegers et al., 2002). The impermeable properties of the BBB constitute an obstacle to AEDs delivery to the CNS, essential to treat brain epilepsy. Therefore, transport of therapeutic agents across the BBB is an area of intensive research. In addition, qualitative evaluation of BBB disruption in humans is available using imaging modalities (magnetic resonance imaging, computerized tomography, and SPECT) following the peripheral administration of non-permeable contrast agents. Although MRI imaging, due to its' high spatial resolution, is considered the best available method for studying anatomical lesions, it is regarded as relatively insensitive for detecting small changes in contrast agent accumulation as compared with SPECT. A quantitative evaluation of BBB permeability functioning in patients may be obtained using analysis of the cerebro-spinal fluid for serum proteins or brain constitutes (e.g. S100β) in the peripheral blood. Opening of the BBB provides molecules normally present in blood with open passage into the CNS. Proteins normally present in the blood freely diffuse into the CNS, and in turn, molecules and protein normally present in high concentrations in the CNS freely diffuse into the blood. Such markers of BBB opening can be detected in the blood in order to evaluate the permeability characteristics of the BBB. However, these methods do not offer spatial information, are invasive, may give false positive results in the presence of intracerebral hemorrhage and S100 levels may depend on the extent of injury or activation of brain astrocytes (Friedman et al., 2009; Marchi et al., 2010; Marchi et al., 2011a).

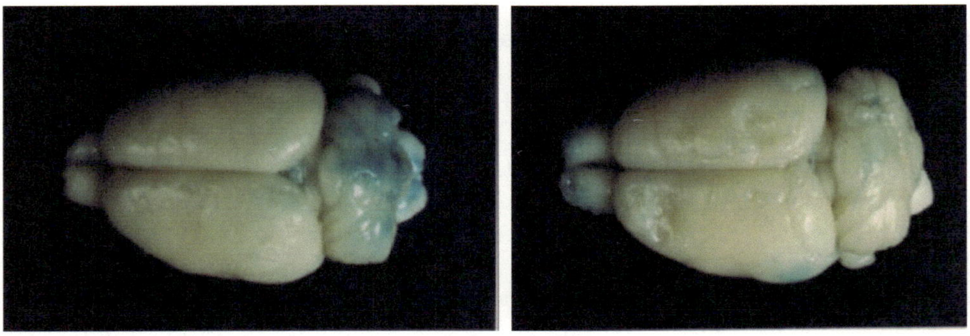

Fig. 5. Gross evaluation of BBB opening in rat brains. Evan's blue staining is noticable by visual inspection in cerebellum (left), temporal cortex and olfactory bulbs (right).

On the other hand, there are studies suggest that a correlation between disrupted BBB and abnormal neural activity. Quantitative EEG analysis may offer spatial information of cortical dysfunction due to BBB disruption. Spectral EEG analysis reveals slow (delta band) activity in regions showing BBB disruption. The usage of quantitative EEG analysis and the detection of abnormal BBB permeability may give information not only about the identification of the epilectic region, but also for targeting population at risk to develop epilepsy (Friedman et al., 2009; Mairinger et al., 2011; Marchi et al., 2010; Tomkins et al., 2007)

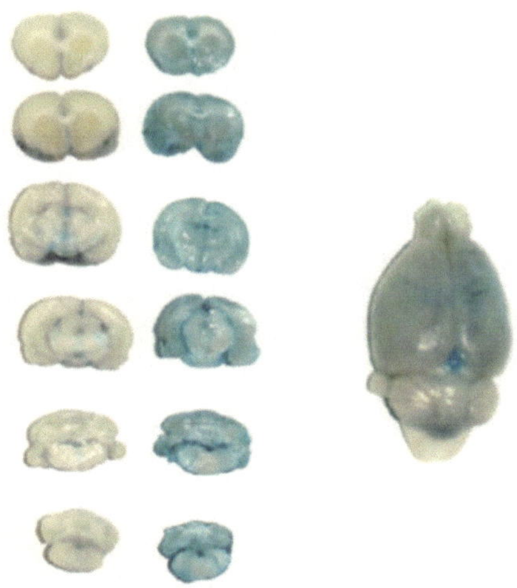

A B

Fig. 6. Photographs of rat brains are depicted. Evan's blue staining is seen in the whole brain and corresponding slices (B) whereas no stain is present in the other brain slices (A).

5. Spectral nalysis of EEG signals

Spectral analysis methods are used for determining the spectral content (distribution of power over frequency) of a time series from a finite set of measurements (Kay & Marple, 1981; Kay, 1988; Proakis & Manolakis, 1996; Stoica & Moses, 1997; Akay, 1998; Akay et al., 1990). The power spectrum is showing abnormal neural activity caused by BBB disruption in the frequency bands.

The power spectral density (PSD) of the signal is estimated by applying spectral analysis methods. The PSD estimates represent the changes in frequency with respect to time. The classical methods (nonparametric or fast Fourier transform-based methods), model-based methods (autoregressive, moving average, and autoregressive moving average methods), time-frequency methods (short-time Fourier transform, wavelet transform), eigenvector methods (Pisarenko, multiple signal classification, Minimum-Norm) can be used to obtain PSD estimates of the signals (Kay & Marple, 1981; Kay, 1988; Proakis & Manolakis, 1996; Stoica & Moses, 1997; Akay, 1998; Akay et al., 1990; Übeyli, 2009a; Übeyli, 2009b; Übeyli, 2010). The obtained PSD estimates provide the features which are well defining the signals. Therefore, spectral analyses of the signals are important in representing, interpreting and discriminating the signals (normal and abnormal signals).

The frequency resolutions of the fast Fourier transform based (FFT-based) methods are limited by the data record duration, independent of the characteristics of the data. The spectral leakage occurs due to windowing that are seen in finite-length data records. The effect of windowing processing with the FFT-based methods is obtaining smear or smooth estimated spectrum (Kay & Marple, 1981; Kay, 1988; Proakis & Manolakis, 1996; Stoica & Moses, 1997). The modelling approach for the model-based methods eliminates the need for window functions. The statistical stability and spectral resolution properties of the model-based methods are better and the resolution is less dependent on the length of the record. The disadvantages of the model-based methods compared to the FFT-based methods are: the FFT-based methods are more widely available and are more widely used approach in spectrum analysis; the model-based spectra are slower to compute; the model-based methods are not reversible; the model-based methods are slightly more complicated to code; the model-based methods are more sensitive to round-off errors, and the orders of the model-based methods depend on the characteristics of the signal and the methods for model order determination are not sufficient (Kay & Marple, 1981; Kay, 1988; Proakis & Manolakis, 1996; Stoica & Moses, 1997). The results of the studies existing in the literature are indicating high performance of the model-based methods for the time-varying biomedical signals such as EEG signals (Übeyli, 2010).

The time-frequency analysis methods show better performance in spectral analysis of nonstationary signals by comparing with the classical and model-based methods. Time and frequency resolutions of the short-time Fourier transform (STFT) are fixed over the entire time-frequency plane. The STFT assumes that the data are quasistationary for the duration of each analyzed segment. The computation of the FFT of a short segment of the signals is forming a spectral estimate distortion and leakage of signal energy into spurious side lobes due to the sharp truncation of the signal. In order to reduce this distortion, a window function is used which reduces the amplitude of the analyzed signal toward the beginning and end of the data segment. Using short data windows lead to the distortion and poor spectral resolution and using longer data windows lead to the spectral broadening that arises from nonstationary characteristics of the signal. In order to solve this problem, a scalable window, which is more flexible approach, can be used (Akay, 1998). The WT solves the problem of fixed resolution by using base functions that can be scaled. The wavelets are better suited to analyzing nonstationary signals, since they are well localized in time and frequency. The property of time and frequency localization is one of the most important features of the WT. The WT has a varying window size, being broad at low frequencies and narrow at high frequencies, and therefore leading to an optimal time-frequency resolution in all frequency ranges. The windows are adapted to the transients of each scale and then the wavelets do not require stationarity. The EEG signals are time-varying and random and therefore, the WT has become a powerful alternative to the STFT in analysis of the time-varying biomedical signals (Akay, 1998; Übeyli 2009a).

The estimation of frequencies and powers of signals from noise-corrupted measurements can be performed by the eigenvector methods. Eigen-decomposition of the correlation matrix of the noise-corrupted signal is the basic process of the eigenvector methods. The resolution of the PSD estimates obtained by these methods is high for the signals having low signal-to-noise ratio (SNR). These methods have a good performance for obtaining PSD estimates of the signals that can be assumed to be composed of several specific sinusoids buried in noise (Akay et al., 1990; Übeyli, 2009b). Therefore, the eigenvector methods have good performance in spectral analysis of the EEG signals.

6. Conclusion

BBB is present at the interface between the blood and the CNS. It controls the exchanges between the blood and brain compartments, by actively transporting nutrients to the brain, and protecting CNS from many toxic compounds and pathogens. Its extremely low permeability is due to the endothelial tight junctions and the activity of multiple efflux transport systems. BBB dysfunction or BBB opening has long been known in seizure disorders and epilepsy. Data from experimental and clinical studies indicates that epileptic seizures produce significant alterations in BBB permeability. These BBB alterations are reversible and confined to anatomically limited brain areas. Epileptic seizure induced disturbances in cerebrovascular permeability involve an interaction between global systemic factors and more localised molecular phenomena occuring in the microenvironment of recruited neurones. However, cerebrovascular dysfunction has been recently proposed to have an etiological role in epilepsy. It is suggested that astrocytic activation, inflammation and neuronal damage contribute to epileptogenesis in the BBB- disrupted brain. There are several applications used to understand BBB dysfunction in the epileptic brains. The exogenous tracers have been used to detect BBB permeability. The endogeneous protein extravasations caused by BBB disruption have been detected by immunohistochemistry. Magnetic resonance imaging, positron emission tomography, computerized tomography, and SPECT studies are currently described to assess changes in BBB permeability. In addition, evaluation of BBB disruption is available using EEG analysis. Spectral analyses of the signals are important in representing, interpreting and discriminating the signals (normal and abnormal signals). Finally, spectral analysis methods can be used for obtaining the power spectrum those showing abnormal neural activity caused by BBB disruption in the frequency bands.

7. References

Abbott, N. J.; Patabendige, A. A.; Dolman, D. E.; Yusof, S. R. & Begley, D. J. (2010). Structure and function of the blood-brain barrier. *Neurobiology of Disease*, Vol.37, No.1, 13-25, ISSN: 0969-9961

Ahishali, B.; Kaya, M.; Orhan, N.; Arican, N.; Ekizoglu, O.; Elmas, I.; Kucuk, M.; Kemikler, G.; Kalayci, R. & Gurses C. (2010). Effects of levetiracetam on blood-brain barrier disturbances following hyperthermia induced seizures in rats with cortical dysplasia. *Life Sciences*, Vol. 87, No. 19-22, 609-19, ISSN:0024-3205

Akay, M.; Semmlow, J.L. ; Welkowitz, W.; Bauer, M.D. & Kostis, J.B. (1990). Noninvasive detection of coronary stenoses before and after angioplasty using eigenvector methods, *IEEE Transactions on Biomedical Engineering*, Vol. 37, No. 11, 1095-1104.

Akay, M. (1998). Time Frequency and Wavelets in Biomedical Signal Processing, Institute of Electrical and Electronics Engineers, Inc., New York.

Bartmann, H.; Fuest, C.; la Fougere, C.; Xiong, G.; Just, T.; Schlichtiger, J.; Winter, P.; Böning, G.; Wängler, B.; Pekcec, A.; Soerensen, J.; Bartenstein, P.; Cumming, P.& Potschka, H. (2010). Imaging of P-glycoprotein-mediated pharmacoresistance in the hippocampus: proof-of-concept in a chronic rat model of temporal lobe epilepsy. *Epilepsia*, Vol.51, No.9, 1780-90, ISSN: 0013-9580

Bauer, S.; Koller, M.; Cepok, S.; Todorova-Rudolph, A.; Nowak, M.; Nockher, W. A.; Lorenz, R.; Tackenberg, B.; Oertel, W.H.; Rosenow, F.; Hemmer, B. Hamer & HM. (2008). NK and CD4+ T cell changes in blood after seizures in temporal lobe epilepsy. *Experimental Neurology*, Vol.211, No.2, 370-377, ISSN: 0014-4886

Cacheaux, L. P.; Ivens, S.; David, Y.; Lakhter, A. J.; Bar-Klein, G.; Shapira, M.; Heinemann, U.; Friedman, A. & Kaufer, D.J. (2009). Transcriptome profiling reveals TGF-beta signaling involvement in epileptogenesis. *The Journal of Neuroscience*, Vol.29, No.28, 8927-35, ISSN 0270-6474

Cardoso, F. L.; Brites, D. & Brito, M. A. (2010). Looking at the blood-brain barrier: molecular anatomy and possible investigation approaches. *Brain Research Review,* Vol.64, No.2, 328-63, ISSN:0165-0173

Chen, Z.; Yu, S.; Concha, H. Q.; Zhu, Y.; Mix, E.; Winblad, B.; Ljunggren, H. G. & Zhu J. (2004). Kainic acid-induced excitotoxic hippocampal neurodegeneration in C57BL/6 mice: B cell and T cell subsets may contribute differently to the pathogenesis. *Brain, Behavior, and Immunity*, Vol.18, No.2, 175–185, ISSN:0889-1591

Cornford, E. M. (1999). Epilepsy and the blood brain barrier: endothelial cell responses to seizures. *Advances in Neurology*, Vol. 79, 845–862, ISSN:0091-3952

Cucullo, L.; Hossain, M.; Puvenna, V.; Marchi, N. & Janigro, D. (2011). The role of shear stress in Blood-Brain Barrier endothelial physiology. *BMC Neuroscience,* Vol.11, No.12, 40. ISSN: 1471-2202

David, Y.; Cacheaux, L. P.; Ivens, S.; Lapilover, E.; Heinemann, U.; Kaufer, D. & Friedman A. (2009). Astrocytic dysfunction in epileptogenesis: consequence of altered potassium and glutamate homeostasis? *The Journal of Neuroscience,* Vol.29, No.34,10588-99, ISSN:0270-6474

de Boer, H. M.; Mula, M. & Sander, J. W. (2008).The global burden and stigma of epilepsy. *Epilepsy & Behavior*, Vol.12, No. 4, 540-6, ISSN:1525-5050

Fabene, P. F.; Navarro, M. G.; Martinello, M.; Rossi, B.; Merigo, F.; Ottoboni, L.; Bach, S.; Angiari, S.; Benati, D.; Chakir, A.; Zanetti, L.; Schio, F.; Osculati, A.; Marzola, P.; Nicolato, E.; Homeister, J. W.; Xia, L.; Lowe, J. B.; McEver, R. P.; Osculati, F.; Sbarbati, A.; Butcher, E. C. & Constantin G. (2008). A role for leukocyte-endothelial adhesion mechanisms in epilepsy. *Nature Medicine*, Vol.14, No.12, 1377–1383. ISSN:1078-8956

Friedman, A. .; Cacheaux, L. P.; Ivens, S. & Kaufer, D. (2011). Elucidating the Complex Interactions between Stress and Epileptogenic Pathways. *Cardiovascular Psychiatry Neurology*, Vol. 2011, 461263, ISSN:2090-0163

Friedman, A., Kaufer, D. & Heinemann, U. (2009). Blood-brain barrier breakdown-inducing astrocytic transformation: novel targets for the prevention of epilepsy. *Epilepsy Research*, Vol. 85, No. 2-3, 142-9, ISSN:0920-1211

Ilbay, G.; Sahin, D. & Ates N. (2003). Changes in blood-brain barrier permeability during hot water-induced seizures in rats. *Neurological Sciences*, Vol.24, No,4, 232-5, ISSN: 1590-1874

Mairinger, S.; Erker, T.; Müler, M. & Langer, O. (2011). PET and SPECT Radiotracers to Assess Function and Expression of ABC Transporters in Vivo. *Current Drug Metabolism*, Epub ahead of print, ISSN:1389-2002

Kay, S.M. & Marple, S.L. (1981). Spectrum analysis - A modern perspective, *Proceedings of the IEEE*, Vol. 69, 1380-1419.

Kay, S.M. (1988). Modern Spectral Estimation: Theory and Application, Prentice Hall, New Jersey.

Marchi, N.; Tierney, W.; Alexopoulos, A.V.; Puvenna, V., Granata, T. & Janigro D. (2011a). The etiological role of blood-brain barrier dysfunction in seizure disorders.*Cardiovascular Psychiatry Neurology*, Vol. 2011, 482415, ISSN:2090-0163

Marchi, N.; Johnson, A.J.; Puvenna, V.; Johnson, H.L.; Tierney, W.; Ghosh, C.; Cucullo, L.; Fabene, P.F. & Janigro D. (2011b). Modulation of peripheral cytotoxic cells and ictogenesis in a model of seizures. *Epilepsia.*,doi: 10.1111/j.1528-1167.2011.03080, ISSN: 0013-9580.

Marchi, N.; Teng, Q.; Ghosh, C.; Fan, Q.; Nguyen, M.T.; Desai, N.K.; Bawa, H.; Rasmussen, P.; Masaryk, T.K. & Janigro D. (2010) Blood-brain barrier damage but not parenchymal white blood cells, is a hallmark of seizure activity. *Brain Research,* Vol.1353,176-186, ISSN: 0006-8993.

Marchi, N.; Fan, Q.; Ghosh, C.; Fazio, V.; Bertolini, F.; Betto, G.; Batra, A.; Carlton, E.; Najm, I.; Granata, T. & Janigro, D. (2009). Antagonism of peripheral inflammation reduces the severity of status epilepticus. *Neurobiol of Disease,* Vol.33, No 2, :171-181, ISSN:0969-9961.

Marchi, N.; Angelov, L.;Masaryk, T.; Fazio, V.; Granata, T.; Hernandez, N.; Hallene, K.; Diglaw, T.; Franic, L.;, Najm, I. & Janigro, D. (2007a). Seizure-promoting effect of blood-brain barrier disruption. *Epilepsia,* Vol, 48, No 4,732-742, ISSN: 0013-9580

Marchi, N.; Oby, E.; Batra, A.; Uva, L.; de Curtis, M.; Hernandez, N.; Boxel-Dezaire, A.; Najm, I. & Janigro D. (2007b). In vivo and in vitro effects of pilocarpine: relevance to ictogenesis. *Epilepsia,* Vol, 48, No 10, 1934-1946, ISSN: 0013-9580.

Masino, S. A.; Li, T.; Theofilas, P., Sandau, U. S.; Ruskin, D. N.; Fredholm, B. B.; Geiger, J. D.; Aronica, E. & Boison, D. (2011). A ketogenic diet suppresses seizures in mice through adenosine A1 receptors. *The Journal of Clinical Investigation.* Vol.121, No.7, 2679-83, ISSN: 0021-9738

Nag, S. (2003a). Morphology and Molecular Properties of Cellular Components of Normal Cerebral Vessels. Nag, S. In: *The Blood-Brain Barrier : Biology and Research Protocols,* (3-36), Humana Press, ISBN: 1588290735

Nag, S. (2003b). Pathophysiology of Blood-Brain Barrier Breakdown. Nag, S. In: *The Blood-Brain Barrier : Biology and Research Protocols,* (97-119), Humana Press, ISBN: 1588290735

Nag, S. (2003c). Blood-Brain Barrier Permeability Using Tracers and Immunohistochemistry.

Nag, S. In: *The Blood-Brain Barrier : Biology and Research Protocols,* (133-144), Humana Press, ISBN: 1588290735

Nitsch, C. & Klatzo, I. (1983). Regional patterns of blood-brain barrier breakdown during epileptiform seizures induced by various convulsive agents. *Journal of the neurological Sciences,* Vol. 59, No.3, 305-22, ISSN: 0022-510X

Oby, E. & Janigro, D. (2006). The blood-brain barrier and epilepsy. *Epilepsia,* Vol.47, No.11, 1761-74, ISSN: 0013-9580

Öztas, B. & Kaya, M. (1991). The effect of acute hypertension on blood-brain barrier permeability to albumin during experimentally induced epileptic seizures. *Pharmacological Research,* Vol. 23, No.1, 41-46, ISSN: 1043-6618

Paolinelli, R.; Corada, M.; Orsenigo, F. & Dejana, E. (2011). The molecular basis of the blood brain barrier differentiation and maintenance. Is it still a mystery? *Pharmacological Research,* Vol.63, No.3, 165-71, ISSN: 1043-6618

Proakis, J.G. & Manolakis, D.G. (1996). Digital Signal Processing Principles, Algorithms, and Applications, Prentice Hall, New Jersey.

Rigau, V.; Morin, M.; Rousset, M. C.; de Bock, F.; Lebrun, A.; Coubes, P.; Picot, M. C.; Baldy-Moulinier, M.; Bockaert, J.; Crespel, A. & Lerner-Natoli, M. (2007). Angiogenesis is associated with blood-brain barrier permeability in temporal lobe epilepsy. *Brain,* Vol. 130, No.7,1942-56, ISSN:0006-8950.

Ruth, R.E. (1984). Increased cerebrovascular permeability to protein during systemic kainic acid seizures. *Epilepsia,* Vol. 25, No. 2, 259-268, ISSN: 0013-9580

Sahin, D.; Ilbay, G. & Ates, N. (2003). Changes in the blood-brain barrier permeability and in the brain tissue trace element concentrations after single and repeated pentylenetetrazole-induced seizures in rats. *Pharmacological Research,* Vol.48, No.1, 69-73, ISSN: 1043-6618

Seegers, U.; Potschka, H. & Löscher, W. (2002). Transient increase of P-glycoprotein expression in endothelium and parenchyma of limbic brain regions in the kainate model of temporal lobe epilepsy.*Epilepsy Research,* Vol. 51, No.3, 257-68, ISSN:0920-1211

Seiffert, E.; Dreier, J. P.; Ivens, S.; Bechmann, I.; Tomkins, O.; Heinemann, U. & Friedman, A. (2004). Lasting blood-brain barrier disruption induces epileptic focus in the rat somatosensory cortex. *The Journal of Neuroscience,* Vol.24, No. 36, 7829–7836, ISSN:0270-6474

Sheen, S. H.; Kim, J. E.; Ryu, H. J.; Yang, Y.; Choi, K. C. & Kang T C. (2011). Decrease in dystrophin expression prior to disruption of brain-blood barrier within the rat piriform cortex following status epilepticus. *Brain Research,* Vol.19, No.1369,173-83, ISSN: 0006-8993

Silverberg, J.; Ginsburg, D.; Orman, R.; Amassian, V.; Durkin, H. G. & Stewart, M. (2010). Lymphocyte infiltration of neocortex and hippocampus after a single brief seizure inmice. *Brain, Behavior, and Immunity,* Vol. 24, No. 2, 263–272, ISSN: 0889-1591.

Stamatovic, S.M.; Keep, R.F. & Andjelkovic, A.V. (2008). Brain endothelial cell-cell junctions: how to "open" the blood brain barrier. *Current Neuropharmacology,* Vol. 6, No.3, 179-92, ISSN: 1570-159X

Stoica, P. & Moses, R. (1997). Introduction to Spectral Analysis, Prentice Hall, New Jersey.

Tomkins, O.; Shelef, I.; Kaizerman, I.; Eliushin, A.; Afawi, Z.; Misk, A.; Gidon, M.; Cohen, A.; Zumsteg, D.& Friedman A. (2008). Blood-brain barrier disruption in post-traumatic epilepsy. *Journal of Neurology, Neurosurgery, and Psychiatry,* Vol.79, No.7, 774-7, ISSN: 0022-3050.

Uzüm, G.; Sarper Diler, A.; Bahçekapili, N. & Ziya Ziylan, Y. (2006). Erythropoietin prevents the increase in blood-brain barrier permeability during pentylentetrazol induced seizures. *Life Sciences,* Vol. 78, No.22, 2571-6, ISSN:0024-3205.

Übeyli, E.D. (2009a). Combined neural network model employing wavelet coefficients for EEG signals classification, Digital Signal Processing , Vol. 19, No. 2, 297-308.

Übeyli, E.D. (2009b). Modified mixture of experts employing eigenvector methods and Lyapunov exponents for analysis of EEG signals, Expert Systems, Vol. 26, No. 4, 339-354.

Übeyli, E.D. (2010). Least squares support vector machine employing model-based methods coefficients for analysis of EEG signals, Expert Systems with Applications, Vol. 37, No. 1, 233-239, 2010.

VanVliet, E. A.; daCostaArau ´ jo, S.; Redeker, S.; van Schaik, R.;Aronica, E. & Gorter, J. A. (2007). Blood-brain barrier leakage may lead to progression of temporal lobe epilepsy. *Brain,* Vol.130, No. 2, 521–534, ISSN: 0006-8950

Vezzani, A.; Peltola, J. & Janigro D. (2008). Inflammation. In: *Epilepsy: A Comprehensive Textbook,* Engel, J. Jr. & Pedley, T. A., (267-276), Philadelphia: Lippincott Williams & Wilkins, ISBN:978-0-7817-5777-5, USA

Weiss, N.; Miller, F.; Cazaubon, S. & Couraud, P.O. (2009).The blood-brain barrier in brain homeostasis and neurological diseases. *Biochimica et Biophysica Acta,* Vol.1788, No.4, 842-57, ISSN: 0006-3002

You, Y.; Bai, H.; Wang, C.; Chen, L.W.; Liu, B.; Zhang, H. & Gao, G.D. (2011). Myelin damage of hippocampus and cerebral cortex in rat pentylenetetrazol model. *Brain Research,* Vol.24, No.1381, 208-16, ISSN: 0006-8993

Part 2

Electroencephalographic Aspects

3

EEG Signal Processing for Epilepsy

Carlos Guerrero-Mosquera[1], Armando Malanda Trigueros[2] and Angel Navia-Vazquez[1]
[1]*University Carlos III of Madrid, Signal Theory and Communications Department Avda, Universidad, 30 28911 Leganes*
[2]*Public University of Navarre, Electrical and Electronic Engineering Department Campus Arrosadia, 31006 Pamplona Spain*

1. Introduction

Neural activity in the human brain starts from the early stages of prenatal development. This activity or signals generated by the brain are electrical in nature and represent not only the brain function but also the status of the whole body.

At the present moment, three methods can record functional and physiological changes within the brain with high temporal resolution of neuronal interactions at the network level: the electroencephalogram (EEG), the magnetoencephalogram (MEG), and functional magnetic resonance imaging (fMRI); each of these has advantages and shortcomings. MEG is not practical for experimental work when subjects may move freely, because of the large size of magnetic sensors. For image sequences, fMRI has a time resolution very low and many types of EEG activities, brain disorders and neurodegenerative diseases cannot be recorded. On the other hand the spatial resolution of the EEG is limited to the number of electrodes, as described in Ebersole & Pedley (2003); Sanei & Chambers (2007).

Much effort has been made to integrate information of multiple modalities during the same task in an attempt to establish an alternative high-resolution spatiotemporal imaging technique. The EEG provides an excellent tool for the exploration of network activity in the brain associated to synchronous changes of the membrane potential of neighboring neurons. Understanding of neuronal functions and neurophysiological properties of the brain together with the mechanisms underlying the generation of biosignals and their recordings is important in the detection, diagnosis, and treatment of brain disorders.

Cerebral sources of electroencephalography potentials are three-dimensional volumes of cortex. These sources produce three-dimensional potential fields within the brain. From the surface of the scalp, these can be recorded as two-dimensional fields of time-varying voltage. The physical and functional factors that determine the voltage fields that these sources produce could be appreciated in order to locate and characterize cortical generators of the EEG.

Electroencephalography enables clinician to study and analyze electrical fields of brain activity recorded with electrodes placed on the scalp, directly on the cortex (e.g., with subdural electrodes), or within the brain (with depth electrodes). For each type of recordering, the specialist attempts to determine the nature and location of EEG patterns and whether they correspond to normal or abnormal neural activity.

In this chapter will introduce several typical methods in which EEG signal pre-processing and processing in EEG signals with epilepsy. The chapter is organized as follows: Section 2 presents a brief outline of electroencephalography, Section 3 introduces to EEG waveform analysis, Section 4 is an overview of different alternatives in EEG signal modeling and feature extractions, Section 5 presents the state of art in EEG epilepsy detection and classification, Section 6 shows different methods to dimensionality reduction for EEG signals and Section 7 gives a summary and conclusions of this chapter.

2. Outline of electroencephalography

The nervous system is an organ system containing a network of specialized cells called neurons that gathers, communicates, and processes information from the body and send out both internal and external instructions that are handled rapidly and accurately. In most animals the nervous system is divided in two parts, the central nervous system (CNS) and the peripheral nervous system (PNS). CNS contain the brain and the spinal cord, and the PNS consists of sensory neurons, grouping of neurons called ganglia, and nerves cells that are interconnected and also connect to the CNS. The two systems are closely integrated because sensory input from the PNS is processed by the CNS, and responses are sent by the PNS to the organs of the body. Neurons transmit electrical potentials to other cells along thin fibers called axons, which cause chemicals called neurotransmitters that permit the neuronal function called synapses. These electrical potentials, called as "action potentials" is the information transmitted by a nerve that, in one cell, cause the production of action potentials in another cell at the synapse. A potential of 60-70 mV with some polarity may be recorded under the membrane of the cell body. This potential changes with variations in the synaptic process. In this sequence, the first cell to produce actions potentials is called the *presynaptic cell*, and the second cell, which responds to the first cell across the synapse, is called the *postsynaptic cell*. Presynaptic cells are typically neurons, and postsynaptic cells are typically other neurons, muscle cells, or gland cells. A cell that receives a synaptic signal may be inhibited, excited or otherwise modulated. The Fig.1 shows the synaptic activities schematically.

The CNS is a major site for processing information, initiating responses, and integrating mental processes. It is analogous to a highly sophisticated computer with the ability to receive inputs, process and store information, and generate responses. Additionally, it can produce ideas, emotions, and other mental processes that are not automatic consequences of the information input.

2.1 Neural activities

Cells of the nervous system include neurons and nonneural cells. *Neurons* or *nerve cell* communicate information to and from the brain. They are organized to form complex networks that perform the functions of the nervous systems. All nerve cells are collectively referred to as neurons although their size, shape, and functionality may differ widely. Neurons can be classified with reference to morphology or functionality. Using the latter classification scheme, three types of neurons can be defined: *sensory neurons*, connected to sensory receptors, *motor neurons*, connected to muscles, and *interneurons*, connected to other neurons.

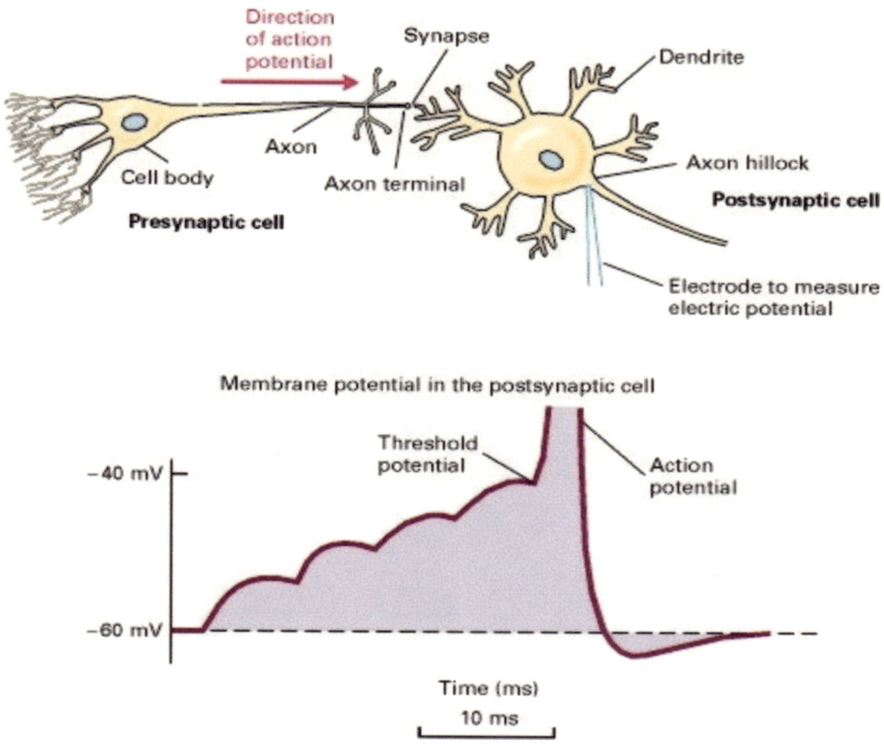

Fig. 1. Presynaptic and postsynaptic activities in the neurons. An action potential that travels along the fibre ends in an excitatory synapse. This process causes an excitatory postsynaptic potential in the following neuron.

The cell body is called the *soma*, from which two types of structures extend: the *dendrites* and the *axon*. Dendrites are short and consist of as many as several thousands of branches, with each branch receiving a signal from another neuron. The axon is usually a single branch which transmits the output signal of the neuron to various parts of the nervous system. Each axon has a constant diameter and can vary in size from a few millimeters to more than 1 m in length; the longer axons are those which run from the spinal cord to the feet. Dendrites are rarely longer than 2 mm. and are connected to either the axons or dendrites of other cells. These connexions receive impulses from other nerves or relay the signals to other nerves. The human brain has approximately 10,000 connexions between one nerve and other nerves, mostly through dendritic connections.

Neurons are, of course, not working in splendid isolation, but are interconnected into different circuits ("neural networks"), and each circuit is tailored to process a specific type of information.

2.2 Cerebral cortex

The cerebral cortex constitutes the outermost layer of the cerebrum and physically it is a structure within the brain that plays an important role in memory, perceptual awareness, attention, thought, consciousness and language. Normally, it is called "grey matter" for its grey color and it is formed by neurons and "gray fibers" covered by a dielectric called myelin.

Myelinated axons are white in appearance, this characteristic is the origin of the name "white matter," and it is localized below the grey matter of the cortex. Their composition is formed predominantly by myelinated axons interconnecting different regions of the nervous central system.

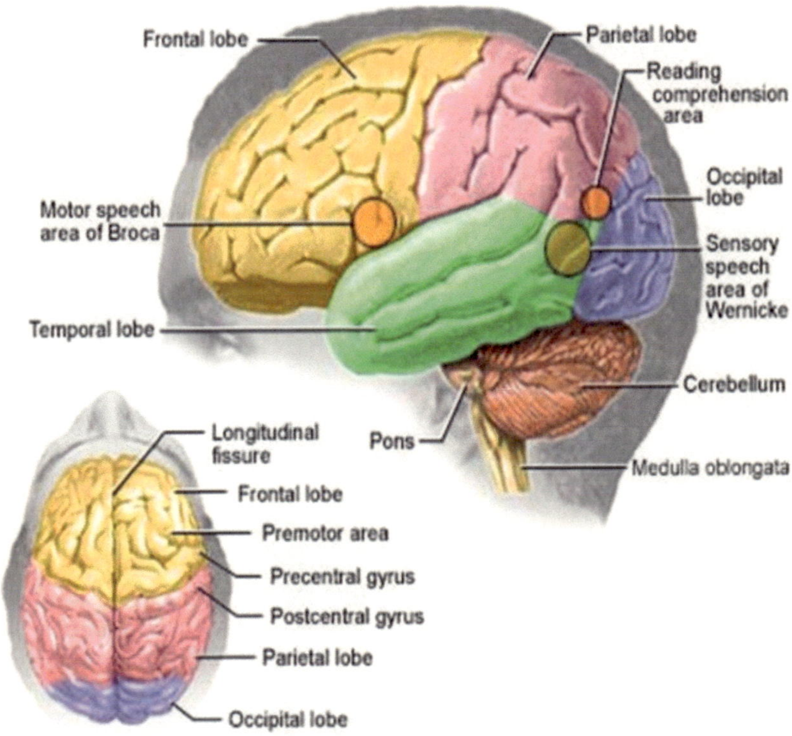

Fig. 2. Cerebral cortex and its four lobes.

The human cerebral cortex is 2-4 mm thick. The cortical surface is highly convoluted by ridges and valleys of varying sizes and thus increases the neuronal area; the total area is as large as 2.5 m^2 and includes more than 10^{10} neurons. The cortex consists of two symmetrical hemispheres–left and right–which are separated by the deep longitudinal fissure (the central sulcus). Each cerebral hemisphere is divided into lobes, which are named for the skull bones overlying each one: the frontal lobe, involved with decision-making, motor speech, problem solving, and planning; temporal lobe, involved with memory, sensory speech, emotion, hearing, and language; the parietal lobe, involved in the reception, reading comprehension and processing of sensory information from the body; and the occipital lobe, involved with vision, see Fig.2.

3. Introduction to the EEG waveform analysis

Most of the brain disorders are diagnosed by visual inspection of EEG signals and the analysis is a rational and systematic process requiring a series of orderly steps characterizing the recorded electrical activity in terms of specific descriptors or *features* and measurements as viewed in Table.1.

1. Frequency or wavelength
2. Voltage
3. Waveform.
4. Regulation
   a. Frequency
   b. Voltage
5. Manner of occurrence (random, serial, continuous)
6. Locus
7. Reactivity (eye opening, mental calculation, sensory stimulation, movement, affective state)
8. Interhemispheric coherence (homologous areas)
   a. Symmetry
      i. Frequency
      ii. Voltage
   b. Synchrony
      i. Wave
      ii. Burst

Table 1. Essential features of EEG analysis described in Ebersole & Pedley (2003).

For example, an EEG from an 8 year old child, some 2 Hz waves are identified in the awake EEG. This activity must then be characterized according to their location, voltage, waveform, manner of occurrence, frequency, amplitude modulation, synchrony and symmetry. A change in any of these features might entirely change the significance of the 2 Hz waves finding this difference as abnormal. Some clinical information is required before the EEG analysis is begun, by example the patient's *age* and *state*. Both age and birth date should be part of the EEG record. For example, there are clearly defined differences between the EEG of a premature infant with a conceptional age of 36 weeks, but there are no important or sharply delineated differences between the EEG of a 3 year old child and that 4 year old child described in Ebersole & Pedley (2003).

The clinical experts in the fields are familiar with manifestation of brain rhythms in the EEG signals and it is important to recognize that the identification of a particular activity or phenomenon may depend on its "reactivity" (see Table.1). An important element of the recording and its analysis is the testing of the reactions, or responses, of the various components of the EEG to certain physiological changes.

Specification of the reactivity of a given activity, rhythm or pattern is essential for the identification and subsequent analysis of the activity and may clearly differentiate it from another activity with similar characteristics. For example, in healthy adults, the amplitudes and frequencies of brain rhythms change from one state of the human to another, such as wakefulness and sleep. Similarly, a series of rhythmic, high voltage 3 to 4 Hz waves in the prefrontal leads (just over the eyes) occurring in association with arousal in a young child may be normal, but a similar burst occurring spontaneously and not associated with arousal may be abnormal.

3.1 Brain rhythms and waveforms

The electrical activity of the cerebral cortex is often called as *rhythm* because this recorded signals exhibit oscillatory, repetitive behavior. The diversity of EEG rhythms is enormous and depends, among many other things, on the mental state of the subject, such as the degree of attentiveness, waking, and sleeping. The rhythms usually are conventionally characterized by their frequency range and relative amplitude.

On the other hand, there are five brain waves characterized by their frequency bands. These frequency ranges are alpha (α), theta (θ), beta (β), delta (δ), and gamma (γ) and their frequencies range from low to high frequencies respectively. The alpha and beta waves were introduced in 1929 by Berger. In 1938, Jasper and Andrews found waves above 30 Hz that labeled as " gamma" waves. A couple years before, in 1936, Walter introduced the delta rhythm to designate all frequencies below the alpha range and he also introduced theta waves as those frequencies within the range of 4-7.5 Hz. In 1944 the definition of a theta wave was introduced by Wolter and Dovey in Sanei & Chambers (2007).

Alpha waves are over the occipital region of the brain and appear in the posterior half of the head. The normal range for the frequency of the occipital alpha rhythm in adults is usually given as 8 to 13 Hz, and commonly appears as a sinusoidal shaped signal. However, sometimes it may manifest itself as sharp waves. In such cases, the alpha wave consist of a negative and positive component that appears to be sharp and sinusoidal respectively. In fact, this wave is very similar to the morphology of the brain wave called rolandic mu (μ) rhythm.

Delta waves lie within the range of 0.5-4 Hz. These waves appear during deep sleep and have a large amplitude. It is usually not encountered in the awake, normal adult, but is indicative of, e.g., cerebral damage or brain disease.

Theta waves are the electrical activity of the brain varying the range of 4-7.5 Hz and its name might be chosen to origin assumption from thalamic region. The theta rhythm occurs during drowsiness and in certain stages of sleep or consciousness slips towards drowsiness. Theta waves are related to access to unconscious material, creative inspiration and associated to deep meditation.

Beta waves are within the range of 14-26 Hz and consists in a fast rhythm with low amplitude, associated with an activated cortex and observed during certain sleep stages. This rhythm is mainly present in the frontal and central regions of the scalp.

Gamma waves (sometimes called the fast beta waves) are those frequencies above 30 Hz (mainly up to 45 Hz) related to a state of active information processing of the cortex. The observation of gamma rhythm during finger movement is done simply by using an electrode located over the sensorimotor area and connected to a high-sensitivity recording system.

Other waves frequencies much higher than the normal activity range of EEG have been found in the range of 200-300 Hz. The localization of these frequencies take place in cerebellar structures of animals, but they do not play any role in clinical neurophysiology. Most of the above rhythms may persist up to several minutes, while others occur only for a few seconds, such as the gamma rhythm.

Fig.3 shows typical normal brain waves. There are also less common rhythms introduced by researchers such as Phi (φ), Kappa (κ), Sigma (σ), Tau (τ), Chi (χ), Lambda (λ) and transient waveforms associated to two sleep states, commonly referred to as non-REM (Rapid Eye Movement) and REM sleep: vertex waves, sleep spindles, and K complexes described in Sanei & Chambers (2007); Sörnmo & Laguna (2005).

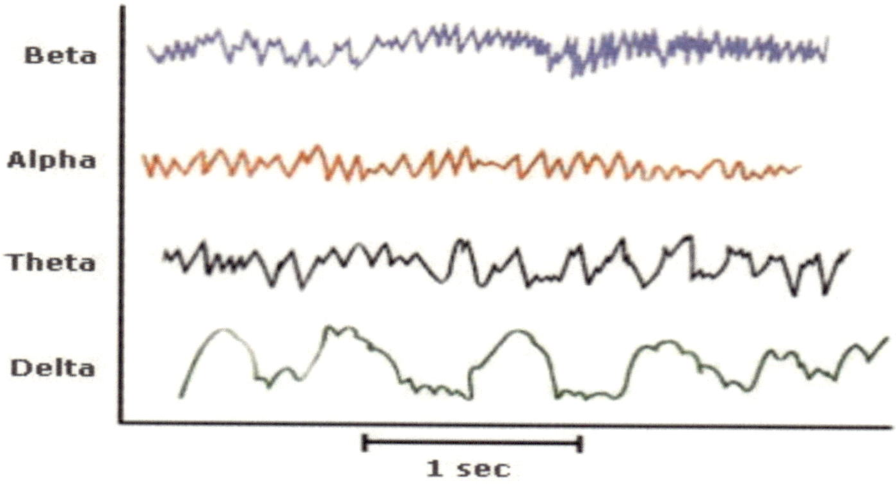

Fig. 3. Typical normal brain waves in the EEG

It is often difficult to understand and detect the brain rhythms and waves from the scalp EEGs, even with trained eyes. New applications in advanced signal processing tools, however, should enable analysis and separation of the desired waveforms from the EEGs. Definitions such as foreground and background EEG are very subjective and totally depends on the abnormalities and applications. Possibly it is more useful to divide the EEG signal into two general categories: the spontaneous brain activity (the "background EEG"); and brain potentials which are evoked by various sensory and cognitive stimuli (evoked potentials, EPs).

3.2 Artifacts

Analysis of EEG activity usually raises the problem of differentiating between genuine EEG activity and that which is introduced through a variety of external influence. These *artifacts* may affect the outcome of the EEG recording. Artifacts originate from a variety of sources such as eyes movement, the heart, muscles and line power. Their recognition, identification, and eventual elimination are a primary responsibility of the EEG expert. Even the most experienced neurophysiologist cannot always eliminate all artifacts in EEG records. However, it is always a major goal to identify the artifactual activity and be sure that it is not of cerebral origin and should not be misinterpreted as such.

Following Ebersole & Pedley (2003); Fisch (1999), artifacts are generally divided into two groups: physiological and non-physiological. Physiological artifacts usually arise from generator sources within the body but not necessarily the brain, for example, eye movements; electrocardiographic and electromyographic artifacts, galvanic skin response and so on. Biological generators present in the body may produce artifacts when an EEG recording is made directly from the surface of the brain. Nonphysiological artifacts come from a variety of sources such as instrumental and digital artifacts (electronic components, line power, inductance, etc.), electrode artifacts, environment, etc.

As technology expands and additional equipment is developed and put into clinical use, novel artifacts will apper. Then, a correct artifact filtering strategy should on the one hand eliminate unnecessary amount of information that has to be eliminated, and on the other hand maintain or ensure that the resulting information is not affected by undetected artifacts. Sometimes visual artifacts inspections could be a good alternative in cases when the artifacts are relatively easy detected by the EEG experts. However, there is the possibility that during the analysis of EEG databases these patterns from artifacts cause serious misinterpretation and then reduce the clinical usefulness of the EEG recordings.

3.3 Abnormal EEG patterns

Any variation in EEG patterns for certain states of the subject indicate abnormality. This may be due to many causes such as distortion and loss of normal patterns, increased occurrence of abnormal patterns, or disappearance of all patterns. In most abnormal EEGs, the abnormal EEG patterns do not entirely replace normal activity: they appear only intermittently, only in certain head regions, or only superimposed on a normal background.

An EEG is considered abnormal if it contains (a) generalized intermittent slow wave abnormalities, commonly associated in the delta wave range and brain dysfunctions, (b) bilateral persistent EEG, often associated with impaired conscious cerebral reactions, and (c) focal persistent EEG usually associated with focal cerebral disturbance.

The classification of the three categories presented before is not easy and needs to be extended to several neurological diseases and any other available information. A precise characterization of the abnormal patterns leads to a clearer insight into some specific neurodegenerative diseases such as epilepsy, Parkinson, Alzheimer, dementia and sleep disorders, or specific disease processes, for example Creutzfeldt-Jakob disease (CJD) described in Sanei & Chambers (2007). However, following Fisch (1999), recent studies have demonstrated that there is correlation between abnormal EEG patterns, general cerebral pathology and specific neurological diseases.

4. Modelling and segmentation

4.1 Modelling the EEG signals

Modelling the brain activities is not an easy task as compared with modelling any other organ. First literature related to EEG signal generation includes physical model such as the model proposed by Hodgkin and Huxley, linear models such as autoregressive (AR) modelling, AR moving average (ARMA), multivariate AR (MVAR), Prony methods and so on. There are also methods based on no-linear models such as autoregressive conditional heteroskedasticity (GARCH), Wiener modeling and local EEG model method (LEM). More details about the methods described above can be found in Celka & Colditz (2002); Sanei & Chambers (2007).

Following Senhadji & Wendling (2002), other model relates a sampled EEG signal $X(n)$ with relevant activities as elementary waves, background activity, noise and artifacts as:

$$X(n) = F(n) + \sum_{i=1}^{n_p} P_i(n - t_{pi}) + \sum_{j=1}^{n_a} R_j(n - t_{aj}) + B(n) \tag{1}$$

where $F(n)$ is the background activity; the P_i terms represent brief duration potentials corresponding to abnormal neural discharges; the R_j terms are related to artifacts (discussed later in section 3.2) and $B(n)$ is the measurement noise which is modeled as a stationary process. This model shows all the EEG information including the abnormal EEG signal. This is a mathematical model rather than an EEG generation signal model, but facilitates the manipulation of concepts that are introduced in the next sections.

4.2 Signal segmentation

Signal segmentation is a process that divides the EEG signal by segments of similar characteristics that are particularly meaningful to EEG analysis. Traditional techniques of signal analysis, for example, spectrum estimation techniques, assume time-invariant signals but in practice, this is not true because the signals are time-varying and parameters such as amplitude, frequency and phase change over time. Furthermore, the presence of short time events in the signal causes a nonstationarity effect.

Non-stationary phenomena are present in EEG usually in the form of transient events, such as sharp waves, spikes or spike-wave discharges which are characteristic for the epileptic EEG, or as alternation of relatively homogenous intervals (segments) with different statistical features (e.g., with different amplitude or variance). The transient phenomena have specific patterns which are relatively easy to identify by visual inspection in most cases, whereas the identification of the homogeneous segments of EEG, known as quasi-stationary, requires a certain theoretical basis. Usually each quasi-stationary segment is considered statistically stationary with similar time and frequency statistics. This eventually leads to a dissimilarity measurement denoted as $d(m)$ between the adjacent EEG frames, where m is a integer value indexing the frame and the difference is calculated between the m and $(m-1)$th (consecutive) signal frames.

There are different dissimilarity measures such as autocorrelation, high-order statistics, spectral error, autoregressive (AR) modelling and so on, presented in Sanei & Chambers (2007). These methods are effective in EEG analysis but can not be efficient for detection of certain abnormalities due to the impossibility of obtaining segments completely stationary. It is then necessary to take into account a different group of methods potentially useful for detecting and analyzing non-stationary EEG signals where the segmentation does not play a fundamental role such as the time-frequency distributions (TFDs).

4.3 Denoising and filtering

Biomedical signals in general, but more particularly EEG signals, are subject to noise and artifacts which are introduced through a variety of external influences. These undesired signals may affect the outcome of the recording procedure, being necessary a method that appropriately eliminates then without altering original brain waves. EEG denoising methods try to reject artifacts originated in the brain or body such as ocular movements, muscle artifacts, ECG etc.

Filtering is a signal processing operation whose objective is to process a signal in order to manipulate the information contained in the signal. In other words, a filter is a device that maps an input signal to an output signal facilitating the extraction (or elimination) of information (or noise) contained in the input signal. In our context, the filtering process is oriented to eliminate electrical noise generated by electrical power line or extracting certain frequency bands.

4.3.1 Lowpass filtering

Most frequently EEG signals contain neuronal information below 100 Hz, for example, epileptic waves lie below 30 Hz, it is possible to remove frequency components above this value simply using lowpass filters. In the cases where the EEG data acquisition system is unable to remove electrical noise as 50 or 60 Hz line frequency, it is necessary to use a notch filter to remove it. Although digital filters could introduce nonlinearities or distortions to the signal in both of amplitude and phase, there are digital EEG process that allow corrections of these distortions using commercial hardware devices. However, it should be better to know the characteristics of the internal and external noises that affect the EEG signals but these information usually is not available.

4.4 Independent component analysis (ICA)

ICA is of interest to scientists and engineers because it is a mathematical tool able to reveal the driving forces which underlie a set of observed phenomena. These phenomena may well be the firing of a set of neurons, mobile phone signals, brain images such as fMRI, stock prices, or voices, etc. In each case, a set of complex signals are measured, and it is known that each measured signal depends on several distinct underlying factors, which provide the driving forces behind the changes in the measured signals. These factors or *source signals* (that are primary interest) are buried within a large set of measured signals or *signal mixtures*. Following Stone (2004), ICA can be used to extract the source signals underlying a set of measured signal mixtures.

ICA belongs to a class of blind source separation (BSS) methods for estimating or separating data into underlying informational components. The term "blind" is intented to imply that such methods can separate data into source signals using only the information of their mixtures observed at the recording channels. BSS in acoustics is well explained in the "cocktail party problem," which aims to separate individual sounds from a number of recordings in an uncontrolled environment such as a cocktail party. So, simply knowing that each voice is statistically unrelated to the others suggests a strategy for separating individual voices from mixtures of voices. The property of being unrelated is of fundamental importance, because it can be generalized to separate not only mixtures of sounds, but mixtures of other kind of signals such as biomedical signals, images, radio signals and so on.

The informal notion of unrelated signals can be associated to the more precise concept of *statistical independence*. If two or more signals are statistically independent of each other then the value of one signal provides no information regarding the value of the other signals. ICA works under this assumption and this concept plays a crucial role in separating and denoising the signals.

4.4.1 ICA fundamentals

The basic BSS problem that ICA attempts to solve assumes a set of m measured data points at time instant t, $\mathbf{x}(t) = [x_1(t), x_2(t), ..., x_m(t)]^T$ to be a combination of n unknown underlying sources $\mathbf{s}(t) = [s_1(t), s_2(t), ..., s_n(t)]^T$. The combination of the sources is generally assumed to be linear and fixed, and the mixing matrix describing the linear combination of $\mathbf{s}(t)$ is given by the full rank $n \times m$ matrix $\mathbf{A}$ such that

$$\mathbf{x}(t) = \mathbf{A}\mathbf{s}(t) \tag{2}$$

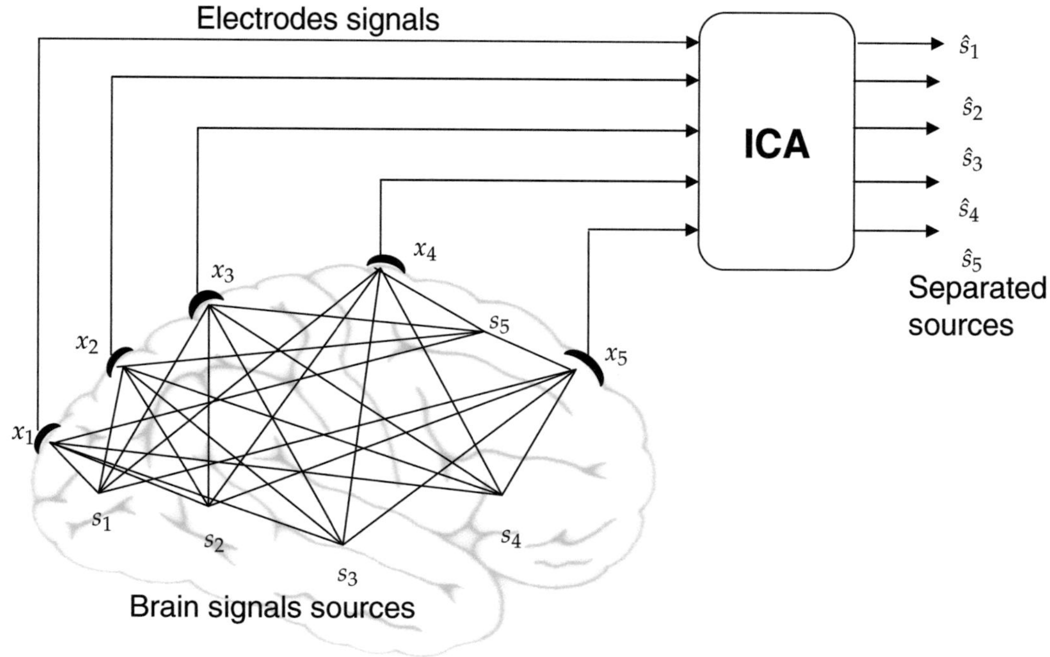

Fig. 4. General ICA process applied to EEG signals

It is also generally assumed that the number of underlying sources is less than or equal to the number of measurement channels ($n \leq m$).

The task of the ICA algorithms is to recover the original sources $\mathbf{s}(t)$ from the observations $\mathbf{x}(t)$ and this is generally equivalent to that of finding a separating (de-mixing matrix) $\mathbf{W}$ such that

$$\hat{\mathbf{s}}(t) = \mathbf{W}\mathbf{x}(t) \tag{3}$$

given the set of observed values in $\mathbf{x}(t)$ and where $\hat{\mathbf{s}}(t)$ are the resulting *estimates* of the underlying sources. This idealistic representation of the ICA problem is described in Fig.4.

In reality the basic mixing model assumed in Eq.2 is simplistic and assumed for the ease of implementation. In fact, a perfect separation of the signals requires taking into account some assumptions and the structure of the mixing process:

- *Linear mixing*: The first traditional assumption for ICA algorithms is that of linear mixing, a realistic model can be formulated as

  $$\mathbf{x}(t) = \mathbf{A}\mathbf{s}(t) + \mathbf{n}(t) \tag{4}$$

  where $\mathbf{A}$ is the linear mixing matrix described earlier and $\mathbf{n}(t)$ is additive sensor noise corrupting the measurements $\mathbf{x}(t)$ (generally assumed to be i.i.d. spatially and temporally white noise, or possibly temporally colored noise), as described in James & Hesse (2005).

  In a biomedical signal context, linear mixing assumes (generally instantaneous) mixing of the sources using simple linear superposition of the attenuated sources at the measurement channel.

- *Noiseless mixing*: If observations $\mathbf{x}(t)$ are noiseless (or at least the noise term $\mathbf{n}(t)$ is negligible) then Eq.4 reduces to Eq.2. Whilst this is probably less realistic in practical terms, it allows ICA algorithms to separate sources of interest even if the separate sources themselves remain contaminated by the measurement noise.
- *Square mixing matrix*: So far it has been assumed that the mixing matrix $\mathbf{A}$ may be non-square ($n \times m$); in fact most classical ICA algorithms assume a *square-mixing* matrix, i.e. $m = n$, this makes the BSS problem more tractable. From a biomedical signal analysis perspective the square-mixing assumption is sometimes less than desirable, particularly in situations where high-density measurements are made over relatively short periods of time such as in most MEG recordings or fMRI.
- *Stationary mixing*: Another common assumption is that the statistics of the mixing matrix $\mathbf{A}$ do not change with time. In terms of biomedical signals this means that the physics of the mixing of the sources as measured by the sensors is not changing.
- *Statistical independence of the sources*: The most important assumption in ICA is that the sources are mutually independent. Two random variables are statistically independent if there is a joint distribution of functions of these variables. This means, for example, that independent variables are uncorrelated and have no higher order correlations. In the case of time-series data, it is assumed that each source is generated by a random process which is independent of the random processes generating the other sources.

4.5 Feature extraction

Feature extraction consist in finding a set of measurements or a block of information with the objective of describing in a clear way the data or an event presents in a signal. These measurements or *features* are the fundamental basis for detection, classification or regression tasks in biomedical signal processing and is one of the key steps in the data analysis process.

These features constitute a new form of expressing the data, and can be binary, categoricals or continuous, and also represent attributes or direct measurements of the signal. For example, features may be age, health status of the patient, family history, electrode position or EEG signal descriptors (amplitude, voltage, phase, frequency, etc.).

More formally, feature extraction assumes we have for N samples and D features, a matrix $N \times D$, where D represents the dimension of the feature matrix. That means, at the sample n from the feature matrix, we could obtain an unidimensional vector $x = [x_1, x_2, \ldots, x_D]$ called as "pattern vector." Several methods in EEG feature extractions can be found in the literature, see Guyon et al. (2006).

More specifically in EEG detection and classification sceneries, features based on power spectral density are introduced in Lehmanna et al. (2007); Lyapunov exponents are introduced in Güler & Übeyli (2007); wavelet transform are described in Hasan (2008); Lima et al. (2009); Subasi (2007) and Xu et al. (2009); sampling techniques are used in Siuly & Wen (2009) and time frequency analysis are presented on Boashash (2003); Guerrero-Mosquera, Trigueros, Franco & Navia-Vazquez (2010); Tzallas et al. (2009) and Boashash & Mesbah (2001). Other approach in feature extraction based in the fractional Fourier transform is described in Guerrero-Mosquera, Verleysen & Navia-Vazquez (2010). It is important to add that features extracted are directly dependent on the application and also to consider that there are important properties of these features to have into account, such as noise, dimensionality, time information, nonstationarity, set size and so on (Lotte et al. (2007)).

This section emphasizes methods oriented to frequency analysis, without excluding the time domain that permits to justify the importance of the frequency analysis and their shortcomings in front of nonstationary signals like the EEG.

4.5.1 Classical signal analysis tools

A signal could be represented in different forms being for example in time and frequency. While time domain indicates how a signal changes over time, frequency domain indicates how often such changes take place. For example, let us consider a signal with a linear frequency modulation varying from 0 to 0.5 Hz and with constant amplitude (see Fig.5). Looking at the time domain representation (Fig.5 upper) it is not easy to say what kind of modulation is contained in the signal; and from the frequency domain representation (see Fig.5 bottom), nothing can be said about the evolution in time of the frequency domain characteristics of the signal.

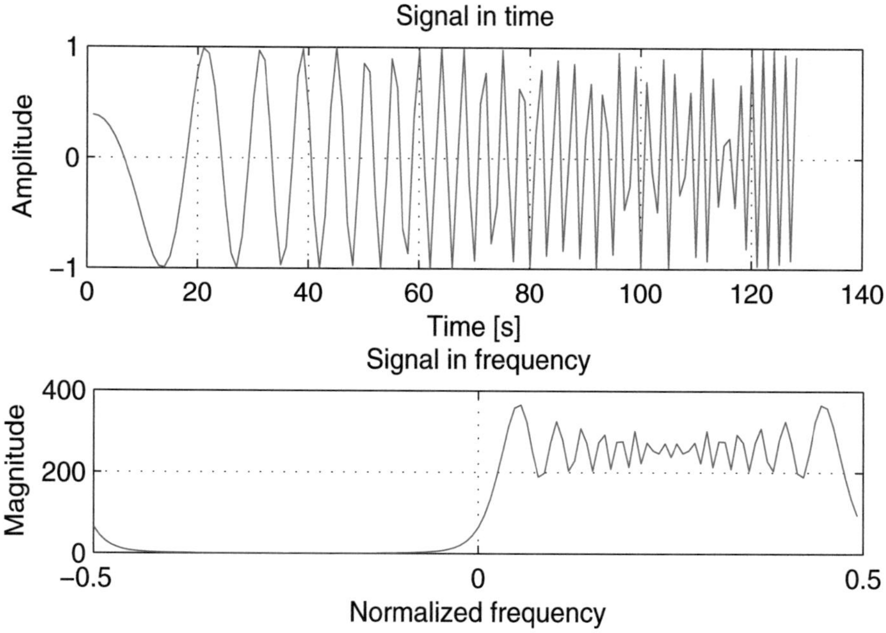

Fig. 5. Chirp signal using time domain (upper) and frequency domain (bottom).

The two representations are related by the *Fourier transform* (FT) as:

$$X(\omega) = \int_{-\infty}^{\infty} x(t)e^{-j\omega t}dt \tag{5}$$

or by the *inverse Fourier transform* (IFT) as:

$$x(t) = \int_{-\infty}^{\infty} X(\omega)e^{-j\omega t}d\omega \tag{6}$$

Eq.6 indicates that signal $x(t)$ can be expressed as the sum of complex exponentials of different frequencies, whose amplitudes are the complex quantities $X(\omega)$ defined by Eq.5.

The squared magnitude of the Fourier transform , $|X(\omega)|^2$, is often taken as the frequency representation of the signal $x(t)$, which allows in some sense easier interpretation of the signal nature than its time representation.

Better interpretation is obtained using a domain that directly represents frequency content while still keeping the time description parameter. This characteristic is the aim of time frequency analysis. To illustrate this, let us represent the chirp signal explained above using the spectrogram (more details about this in the following). Note how it is possible to see the linear progression with time of the frequency components, from 0 to 0.5 (Fig.6).

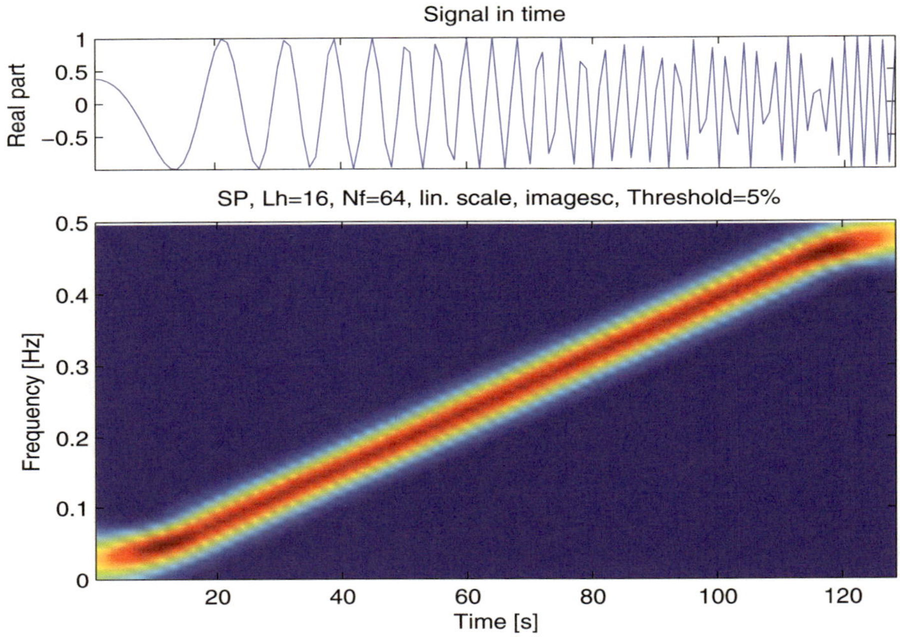

Fig. 6. Spectrogram representation of the chirp

4.5.2 Time-frequency distributions (TFD)

In a series of papers (Akay (1996); Cohen (1995)), Cohen generalized the definition of time-frequency distributions (TFDs) in such a way that a wide variety of distributions could be included in the same framework. Specifically the TFD of a real signal $x(n)$ is computed as:

$$P(t,\omega) = \frac{1}{2\pi} \int_{-\infty}^{\infty} \int_{-\infty}^{\infty} A(\theta,\tau)\mathbf{\Phi}(\theta,\tau)e^{-j\theta t - j\omega\tau} d\theta d\tau \quad (7)$$

where,

$$A(\theta,\tau) = \frac{1}{2\pi} \int_{-\infty}^{\infty} x(u+\frac{\tau}{2})x^*(u-\frac{\tau}{2})e^{j\theta u} du \quad (8)$$

is the so-called ambiguity function and the weighting function $\mathbf{\Phi}(\theta,\tau)$ is a function called the kernel of the distribution that, in general, may depend on time and frequency.

If $\Phi(\theta, \tau) = 1$ in Eq.(7), we have

$$P(t,w) = \frac{1}{2\pi}\int_{-\infty}^{\infty}\int_{-\infty}^{\infty} x(u+\frac{\tau}{2})x^*(u-\frac{\tau}{2})e^{-j\omega\tau} \frac{1}{2\pi}\int_{-\infty}^{\infty} e^{-j\theta\,(t-u)}d\theta du d\tau \quad (9)$$

where

$$\frac{1}{2\pi}\int_{-\infty}^{\infty} e^{-j\theta(t-u)}d\theta = \delta(t-u) \quad (10)$$

and we know that

$$\int_{-\infty}^{\infty} x(u+\frac{\tau}{2})x^*(u-\frac{\tau}{2})\delta(t-u)du = x(t+\frac{\tau}{2})x^*(t-\frac{\tau}{2}) \quad (11)$$

If we substitute the Eq.(10) and Eq.(11) in Eq.(9), then we have the Wigner-Ville distribution (WV) defined as:

$$WV(\omega,t) = \frac{1}{2\pi}\int_{-\infty}^{\infty} x(t+\frac{\tau}{2})x^*(t-\frac{\tau}{2})e^{-j\omega\tau}d\tau \quad (12)$$

Following Hammond & White (1996), the recurrent problem of the WV is the so-called crossterm interference, due to bilinear nature of its definition. These crossed terms tend to be located mid-way between the two auto terms and are oscillatory in nature.

When $\Phi(\theta, \tau) = 1$, we have the Wigner-Ville distribution $WV(t, \omega)$. The Smooth Pseudo Wigner-Ville (SPWV) distribution is obtained by convolving the $WV(t, \omega)$ with a two-dimensional filter in t and ω. This transform incorporates smoothing by independent windows in time and frequency, namely $W_w(\tau)$ and $W_t(t)$:

$$SPWV(t,\omega) = \int_{-\infty}^{\infty} W_w(\tau)\Big[\int_{-\infty}^{\infty} W_t(u-t)x(u+\frac{\tau}{2}) x^*(u-\frac{\tau}{2})du\Big]e^{-j\omega\tau}d\tau \quad (13)$$

Eq.(13) provides great flexibility in the choice of time and frequency smoothing, but the length of the windows should be determined empirically according to the type of signal analyzed and the required cross term suppression, discussed in Afonso & Tompkins (1995).

As proved in Hlawatsch & Boudreaux-Bartels (1992), the SPWV in Eq.(13) does not satisfy the marginal properties, that is, the frequency and time integrals of the distribution do not correspond to the instantaneous signal power and the spectral energy density, respectively. However, it is still possible for a distribution to give the correct value for the total energy without satisfying the marginals, described in Cohen (1989; 1995). Therefore the total energy can be a good feature to detect signal events in the SPWV representation because the energy in EEG seizure is usually larger than the one during normal activity.

The TFDs offer the possibility of analyzing relatively long continuous segments of EEG data even when the dynamics of the signal are rapidly changing. Taking the most of these, it can extract features from the time frequency plane such as ridges energy, frequency band values and so on. However, three considerations have to be taken, presented in Cohen (1989; 1995) and Durka (1996):

- A TFD will need signals as clean as possible for good results.
- A good resolution both in time and frequency is necessary and as the "uncertainty principle" states, it is not possible to have a good resolution in both variables simultaneously.
- It is also required to eliminate the spurious information (i.e. cross-term artifacts) inherent in the TFDs.

The first consideration implies a good pre-processing stage to eliminate artifacts and noise. Second and third considerations have motivated the TFD selection or design, then it is important and necessary to choose a suitable TFD for seizure detection in EEG signals as well as for a correct estimation of frequencies on the time-frequency plane. Indeed, it is desirable that the TFD has both low cross-terms and high resolution. Choosing a distribution depends on the information to be extracted and demands a good balance between good performance, low execution time, good resolution and few and low-amplitude cross terms.

One consideration before using the TFD is to convert each EEG segment into its analytic signal for a better time-frequency analysis. The analytic signal is defined to give an identical spectrum to positive frequencies and zero for the negative frequencies, and shows an improved resolution in the time-frequency plane, discussed in Cohen (1989). It associates a given signal $x(n)$ to a complex valued signal $y(n)$ defined as: $y(n) = x(n) + jHT\{x(n)\}$, where $y(n)$ is the analytic signal and $HT\{.\}$ is the Hilbert transform.

4.5.3 Wavelet coefficients

The EEG signals can be considered as a superposition of different structures occurring on different time scales at different times. As presented in Latka & Was (2003), the Wavelet Transform (WT) provides a more flexible way of time-frequency representation of a signal by allowing the use of variable size windows and can constitute the foundation of a relatively simple yet effective detection algorithm. Selection of appropriate wavelets and the number of decomposition levels is very important in the analysis of signals using the WT. The number of decomposition levels is chosen based on the dominant frequency components of the signals. Large windows are used to get a finer low-frequency information and short windows are used to get high-frequency resolution. Thus, WT gives precise frequency information at low frequencies and precise time information at high frequencies. This makes the WT suitable for EEG analysis of spikes patterns or epileptic seizures.

Wavelets overcome the drawback of a fixed time-frequency resolution of short time Fourier transforms. The WT performs a multiresolution analysis, $W_\Psi f(a,b)$ of a signal, $x(n)$ by convolution of the mother function $\Psi(n)$ with the signal, as given in Latka & Was (2003), and Mallat (2009) as:

$$W_\Psi x(b,a) = \sum_{n'=0}^{N-1} x(n')\Psi^* \left(\frac{n'-b}{a}\right) \tag{14}$$

$\Psi(t)^*$ denote the complex conjugate of $\Psi(n)$ (basis function), a the scale coefficient, b the shift coefficient and $a, b \in \Re$, $a \neq 0$.

Wavelets overcome the drawback of a fixed time-frequency resolution of short time Fourier transforms. The WT performs a multiresolution analysis, $W_\Psi f(a,b)$ of a signal, $x(n)$ by convolution of the mother function $\Psi(n)$ with the signal, as given in Latka & Was (2003), and Mallat (2009) as:

$$W_{\Psi}x(b,a) = \sum_{n'=0}^{N-1} x(n')\Psi^{*}\left(\frac{n'-b}{a}\right) \tag{15}$$

$\Psi(t)^{*}$ denote the complex conjugate of $\Psi(n)$ (basis function), a the scale coefficient, b the shift coefficient and $a, b \in \Re$, $a \neq 0$.

In the procedure of multiresolution decomposition of a signal $x(n)$, each stage consists of two digital filters and two downsamplers by 2. The bandwidth of the filter outputs are half the bandwidth of the original signal, which allows for the downsampling of the output signals by two without loosing any information according to the Nyquist theorem. The downsampled signals provide detail D1 and approximation A1 of the signal, this procedure is described in Hasan (2008).

Once the mother wavelet is fixed, it is possible to analyze the signal at every possible scale a and translation b. If the basis function $\Psi(n)$ is orthogonal, then the original signal can be reconstructed from the resulting wavelet coefficients accurately and efficiently without any loss of information. The Daubechies' family of wavelets is one of the most commonly used orthogonal wavelets to non-stationary EEG signals presenting good properties and allowing reconstruction of the original signal from the wavelet coefficients, as described Mallat (2009).

4.5.4 Fractional Fourier transform

Fourier analysis is undoubtedly one of the most used tools in signal processing and other scientific disciplines and this technique uses harmonics for the decomposition of signals with time-varying periodicity. Similarly, TFDs are very frequently used in signal analysis especially when it is necessary to eliminate the windowing dependence on non-stationary signals.

In 1930, Namias employed the fractional Fourier transform (FrFT) to solve partial differential equations in quantum mechanics from classical quadra-tic Hamiltonians[1]. The results were later improved by McBride and Kerr in Tao et al. (2008). They developed operational calculus to define the FRFT. The FrFT is a new change in the representation of the signal which is an extension of the classical Fourier transform. When fractional order gradually increases, the FrFT of a signal can offer much more information represented in an united representation than the classical Fourier transform and it provides a higher concentration than TFDs, avoiding the cross terms components produced by quadratics TFDs.

FrFT has established itself as a potential tool for analyzing dynamic or time-varying signals with changes in very short time and it can be interpreted as the representation of a signal in neutral domain by means of the rotation of the signal by the origin in counter-clockwise direction with rotational angle α in time-frequency domain as shown in Fig.7. The FrFT with

[1] A development based on a concept called *fractional operations*. For example, the n-th derivative of $f(x)$ can be expressed as $d^{n}f(x)/dx^{n}$ for any positive integer n. If another value derived is required, i.e. the 0.5-th derivative, it is necessary to define the operator $d^{a}f(x)/dx^{a}$, where the value a could be an any real value. The function $[f(x)]^{0.5}$ is the square root of the function $f(x)$. But $d^{0.5}f(x)/dx^{0.5}$ is the 0.5-th derivative of $f(x)$ ($a = 0.5$), $(df(x)/dx)^{0.5}$ being the square root of the derivative operator d/dx. As it can be seen, fractional operations is a concept that goes from the whole of an entity to its fractions.

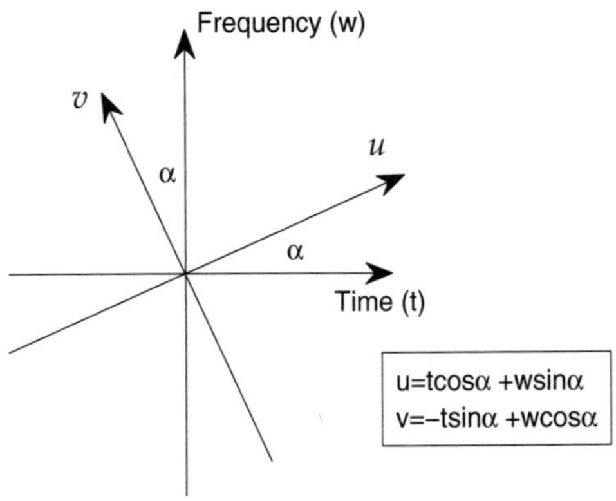

Fig. 7. The relation of fractional domain (u, v) with traditional time-frequency plane (t, w) rotated by an angle α.

angle α of a signal $x(t)$, denoted as $X_\alpha(u)$ is defined in Almeida (1994) as:

$$X_\alpha(u) = \int_{-\infty}^{\infty} x(t) K_\alpha(t, u) dt \tag{16}$$

where $K_\alpha(u, t)$ is a linear kernel function continuous in the angle α, which satisfies the basic conditions for being interpretable as a rotation in the time-frequency plane. The kernel has the following properties

$$K_\alpha(t, u) = K_\alpha(u, t) \tag{17}$$

$$K_{-\alpha}(t, u) = K^*_{-\alpha}(t, u) \tag{18}$$

$$K_\alpha(-t, u) = K_\alpha(t, -u) \tag{19}$$

$$\int_{-\infty}^{\infty} K_\alpha(t, u) K_\beta(u, z) du = K_{\alpha+\beta}(t, z) \tag{20}$$

$$\int_{-\infty}^{\infty} K_\alpha(t, u) K^*_\alpha(t, u') dt = \delta(u - u') \tag{21}$$

The FrFT is given by

$$X_\alpha(u) = \begin{cases} \sqrt{(\frac{1 - j\cot\alpha}{2\pi}} e^{j\frac{u^2}{2}\cot\alpha} \int_{-\infty}^{\infty} x(t) e^{j\frac{t^2}{2}\cot\alpha} e^{jut\csc\alpha} dt, \\ \text{if } \alpha \text{ is not a multiple of } \pi \\ \\ x(t), \quad \text{if } \alpha \text{ is multiple of } 2\pi \\ \\ x(-t), \quad \text{if } \alpha + \pi \text{ is multiple of } 2\pi \end{cases}$$

More detailed definitions, proof and further properties of the kernel can be found in Almeida (1994).

In summary, the FrFT is a linear transform, continuous in the angle α, which satisfies the basic conditions for being interpretable as a rotation in the time-frequency plane.

5. The detection problem in EEG signals

Epilepsy is considered the disease with major prevalence within disorders with neurological origin. The recurrent and sudden incidence of seizures can lead to dangerous and possibly life-threatening situations. Since disturbance of consciousness and sudden loss of motor control often occur without any warning, the ability to predict epileptic seizures would reduce patients' anxiety, thus improving quality of life and safety considerably.

Intractable epilepsy is one of the most physically and emotionally destructive neurological disorders affecting population of all ages. It is generally accepted that surgical rejection of epileptic foci is the best solution. However, before conducting neurosurgery, it is necessary to study the presence of epileptiform activity, which is distinct from background EEG activity. The analysis of EEG data and the extraction of information is not an easy task. EEG recording may be contaminated by extraneous biologically generated (human body) and externally generated signals (power line, electrode movement etc.). The presence of this kind of noise or "artifacts" makes it difficult to discriminate between original brain waves and noise. This problem motivates a preprocessing step to obtain clean signals before the detection task.

Another important problem in EEG processing is to figure out which kind of information or "patterns" we want to extract from the signal. This procedure is known as feature extraction. Extracted features depend considerably on the method used, which are usually transformations to other domains that permit the extraction of hidden information in the signal. Care has to be taken not to extract similar or irrelevant features that could reduced the detector performance or increase the computational load. Therefore, a feature selection procedure is also necessary to complement the features extraction procedure.

Other important task in the medical environment to diagnose, classify or detect abnormalities, is to obtain ictal and interictal patterns. This usually involves monitoring of the patient during several weeks. Continuous observation or patient monitoring is a care activity that requires time and expensive work, being necessary specialized personnel for alerting of possible changes that a patient may have. When information is stored, there is another activity equally important: the analysis of the EEG registers. The specialists have to analyze waveforms, spectrum and peaks, and based on this analysis try to determine the pathology that the patient suffers. Usually they use a video unit. In many instances, there are disagreements among specialists about the same record due to the subjective nature of the analysis.

The introduction of new techniques and mathematical algorithms in the EEG analysis can be helpful to design new supporting methods in medical decision and diagnosis, thus avoiding tedious analysis of long-term records and doubts about the brain pathology that a patient suffers.

Nowadays there are many published studies about neurological diseases detection but these results are very focused on private institutional databases or rely on impractical numerical methods which are difficult to implement in a hospital environment. Therefore, the implementation and design of practical and reliable detection systems are very important in hospitals. This doctoral thesis, tries to narrow the gap that exists between EEG signal theory and practical implementation for the medical practice.

5.1 Summary of previous work in epileptic detection on EEG signals

Some methods of seizure detection were based on detecting strong rhythmic movements of the patient, but these methods had a limitation: seizures do not always present strong movements. This limitation led the detection problem to methods based on EEG signal analysis, for example, detection of large seizures discharges in several EEG channels by amplitude discrimination was described by J.R. Ives & Woods (1974); T.L. Babb & Crandall (1974) designed an electronic circuit for seizures detection from intracraneal electrodes. However, some seizures do not present EEG changes, therefore seizure detection only based on EEG analysis was not at all reliable and it was necessary to combine it with other methods. For example, P.F. Prior & Maynard (1973) identified on the EEG signal a large increase followed by a clear decrease in the amplitude and at the same time by large electromyogram (EMG) activity; *A.M. Murro& Meador* (1991) described a method based on spectral parameters and discriminant analysis.

New alternatives for this detection problem are addressed from the point of view of pattern recognition. Gotman (1982) presented an automatic detection system based on seizure patterns. The drawback of this method is the necessity of traditional visual inspection of the patterns, being necessary a careful examination of them by a specialist.

Presently, EEG epileptic detectors have evolved including new techniques such as neural networks, non-linear models, independent component analysis (ICA), Bayesian methods, support vector machines and variance-based methods, as described in Guerrero-Mosquera, Trigueros, Franco & Navia-Vazquez (2010). Other group of methods potentially useful for detecting and analyzing non-stationary signals are time-frequency distributions (TFDs) Cohen (1995). These methods allow us to visualize the evolution of the frequency behavior during some non-stationary event by mapping a one dimensional (1-D) time signal into a two-dimensional (2-D) function of time and frequency. Therefore, from the time-frequency (TF) plane it is possible to extract relevant information using methods such as peak matching, filter banks, energy estimation, etc.

On the other hand, most of the detection methods proposed in the literature assume a clean EEG signal free of artifacts or noise, leaving the preprocessing problem open to any denoising algorithm such as digital filters, independent component analysis (ICA) or adaptive schemes using the electrooculogram (EOG) as reference signal, as in Guerrero-Mosquera & Navia-Vazquez (2009).

5.2 Classification algorithms for EEG signals

Unlike many theoretical approaches that solve certain problems using some model or formula, many classifiers are based on statistical learning. In such cases the system should be trained to obtain a good classifier taking into account that, under the following considerations described in Sanei & Chambers (2007), classification algorithms do not perform efficiently when:

- the number of features is high,
- there is limited execution time for a classification task,
- the classes or labels from feature matrix are unbalanced,
- there are nonlinearities between inputs and outputs,
- data distribution is unknow,

- there is no convergence guarantee to best solution (problem not convex or monotonic).

Up-today, several algorithms in EEG signal classification and detection have been propose in the literature. For example, Multiple signal classification (MUSIC) combining EEG and MEG for EEG source localization described in Mosher & Leahy (1998); classification of patients with Alzheimer using Support Vector Machine (SVM) and neural networks (NNs) described in Lehmanna et al. (2007); Güler & Übeyli (2007) introduced the multiclass SVM for EEG. Lotte et al. (2007) describes several applications for BCI using methods such as Hidden Markov Modelling (HMM), Linear Discriminant Analysis (LDA) and fuzzy logic; Chiappa & Barber (2006) used the Bayes's rule to discriminate mental tasks; detection of ERPs using SVM described in Thusalidas et al. (2006); Fuzzy SVM (FSVM) is utilized in Xu et al. (2009); Fisher's discriminant is introduced in Müller et al. (2003). Applications in epilepsy classification such as Artificial Neural Networks (ANN) described in Subasi (2007); *k*-NN classifier and logistic regression with TFDs are used in Tzallas et al. (2009); Least Square SVM (LS-SVM) in Siuly & Wen (2009); Learning Vector Quantization with NN (LVQ-NN) described in Hasan (2008); Mixture of Experts (ME) and Multilayered Perceptron (MLP) in Subasi (2007); an automatic EEG signal classification using Relevance Vector Machine (RVM) is proposed by Lima et al. (2009).

As showed Guyon et al. (2006), SVM and its variants have more applications in different classification scenarios and are powerful approach for pattern recognition, showing to be a good alternative for EEG signal classification due to their high performance, good generalization and adaptability in stationary and nonstationary environments compared to other methods such as NN.

6. Dimensionality reduction for EEG signals

After feature extraction, it is necessary to select the subset of features that present better performance or are most useful for a problem at hand, such as regression, classification or detection. The data acquisition in environments such as biomedical signals leads to define each problem by hundreds or thousands of measurements leading to obtain high dimensional data with high computational cost.

As discussed in Guyon & Elisseeff (2003), feature selection is based on the principle that choosing a smaller number of variables among the original ones, leads to an easier interpretation. In fact, under the assumption that reducing the *training data*[2] might improve the performance task, the feature selection methods also allows a better data understanding and visualization together with reduction in data measurement and storage.

Feature selection could be summarized in two main tasks: choosing the relevant features and searching the best feature subset. The first one tries to solve the question: is a feature (or subset of features) relevant for the problem? And the second one tries to search the best feature subset among all the possible subsets extracted from the initial task[3]. The application of these two

[2] Concept related to the fact of using a data set (also called data points, samples, patterns or observations) in order to gain knowledge, learn a task associated with desired outcomes.

[3] Although feature extraction and feature selection are different aspects of the pattern recognition process, it is important to distinguish the difference between them. The first one aims at building a good feature representation based on several measurements, and the second one tries to reduce the feature matrix by selecting subsets of features more useful in determined tasks.

tasks to high dimensional data causes a reduction in the data dimension, process known as dimensionality reduction.

Besides feature selection, there is another set of methods known as *projection methods* that perform the same task but in practice could retain the problems suffered by high dimensional data, presented in Rossi et al. (2007; 2006). Typical projection algorithms are Principal Component Analysis (PCA), Sammon's Mapping, Kohonen maps, Linear Discriminant Analysis (LDA), Partial Least Squares (PLS) or Projection pursuit, amongst others (see Duda et al. (2009)).

6.1 Subset relevant assessment

This step is mainly based on a *relevance criterion* that has to be chosen by some measurement. The best choice for the criterion is certainly to estimate the performances of the model itself, i.e., an individual feature ranking could be appropriate at scenarios where the features provide a good performance by itself and there is the possibility of choosing features associated to high ranks.

The idea of the "individual relevance ranking" can be clarified by the following example: Fig.8 shows a situation where the feature X_2 is more relevant individually to predict the output Y than the feature Y_1. Notice the importance of choosing the right features to improve the performance of a task, which in this example is related to prediction of Y. There are different alternatives in relevance criteria, such as the Pearson correlation

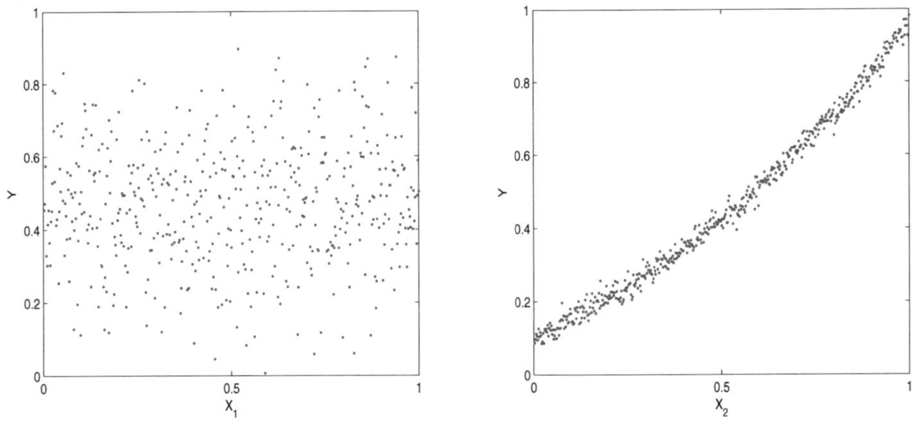

Fig. 8. Simple prediction problem. The horizontal axis represents the feature and the vertical axis the output. It can see that feature X_2 (left) is more relevant individually than feature X_1 (right) in this simple prediction problem.

coefficient, mutual information (MI) and wrapper methodology. Although each method has its advantages and disadvantages, mutual information has proven to be an appropriate measure in several applications such as selection of spectral variables, spectrometric nonlinear modelling and functional data classification, see Gomez-Verdejo et al. (2009); Rossi et al. (2007; 2006). Moreover, as discussed in Cover & Thomas (1991), correlation does not measure nonlinear relations among features and wrapper approach presents a high computational load. Furthermore, MI could be seen as a correlation measure applied to determine the nonlinearity among features.

Next section focuses on the well-known concept of MI and shows why this relevance criterion is applicable for feature selection.

6.2 Mutual information (MI)

Mutual information (MI) measures the relevance between a group of features X and the variable or output Y. This relationship is not necessarily linear. As described in Cover & Thomas (1991), the mutual information between two variables is the amount of uncertainty (or entropy) that is lost on one variable when the other is known, and vice-versa. The variables X and Y could be multidimensional, solving the drawback in correlation measurements that are based on individual variables.

Let $p_X(x)$ and $p_Y(y)$ be the marginal of probability density function (pdf) of X and Y respectively, and the joint probability density function of X and Y is $p_{X,Y}(x,y)$. If X has $\mathcal{X}$ alphabets, the entropy of X is defined as

$$H(X) = -\sum_{x\in\mathcal{X}} p_X(x)\log p_X(x) \tag{22}$$

The base of the logarithm determines the units in which information is measured. Particularly, if the logarithm is base 2 the entropy is expressed in bits.

The joint entropy $H(X,Y)$ of a pair of discrete random variables (X,Y) with a joint distribution $p_{X,Y}(x,y)$ is defined as

$$H(X,Y) = -\sum_{x\in\mathcal{X}}\sum_{y\in\mathcal{Y}} p_{X,Y}(x,y)\log p_{Y|X}(y|x) \tag{23}$$

And the MI between two variables is calculated as

$$I(X,Y) = \sum_{x\in\mathcal{X}}\sum_{y\in\mathcal{Y}} p_{X,Y}(x,y)\log\frac{p_{X,Y}(x,y)}{p_X(x)p_Y(y)} \tag{24}$$

Eq.24 gives the relation between X and Y, meaning that $I(X,Y)$ is large (small) the variables are closely (not closely) related. The MI and entropy have the following relation, see Cover & Thomas (1991):

$$I(X,Y) = H(Y) - H(Y|X) \tag{25}$$

For continuous variables, the entropy and MI are defined as

$$H(X) = -\int_{-\infty}^{\infty} p_X(x)\log p_X(x)dx \tag{26}$$

$$I(X,Y) = \int_{-\infty}^{\infty}\int_{-\infty}^{\infty} p_{X,Y}(x,y)\log\frac{p_{X,Y}(x,y)}{p_X(x)p_Y(y)}dxdy \tag{27}$$

Note in Eq.24 and Eq.27 that it is necessary to know the exact pdf's for estimating the MI and this is the most sensitive part in the MI estimation. Several methods have been proposed in the literature to estimate such joint densities, see Duda et al. (2009); Lotte et al. (2007).

7. Summary and conclusions

In this chapter the fundamental concepts in the nervous system and different tools for EEG signal processing have been briefly explained. Several concepts in visual analysis of the EEG, brain rhythms, artifacts and abnormal EEG patterns, including EEG applications such as epilepsy detection, EEG modelling, EEG feature extraction, epilepsy detection and classification with methods oriented to dimensionality reduction have been reviewed. The chapter also provides key references for further reading in the field of EEG signal processing.

Although all methods have been described in a brief way, they are introduced to give a good theoretical grounding in EEG processing and to better understand the methods proposed and their performance. Signal processing algorithms for EEG applications have specific requirements in filtering, feature extractions and selections.

EEG is widely used as a diagnostic tool in clinical routine with an increasing develop of both analytical and practical methods. Its simplicity, low cost and higher temporal resolution of EEG maintains this tool to be considered in applications such as epilepsy seizures detection, sleep disorders and BCI.

Future work implies the design of new EEG artifacts elimination methods, feature extraction to obtain possible hidden information and dimensionality data reduction.

In medical environment, the steps that we follow in a classification problem are: (i) denoising and artifacts removal, (ii) LFE features extractions, (iii) detection/classification using these features and their combinations, (iv) if we do not have conclusive results, we add features from wavelets and fractional Fourier transform, (v) detection/classification using these features and their combinations, (vi) we apply dimensionality reduction if necessary.

An additional potential field of research is the **EEG Integration with other techniques such as fMRI.** The principal drawback of the EEG is its low spatial resolution because it depends on the number of electrodes. MEG has a better temporal resolution than EEG but suffers the same disadvantage. fMRI solves this problem and its spatial resolution is on the order of milimeters. The integration of these techniques is of vital importance in neuroscience studies because this could improve the detection of other neurodegenerative diseases like Alzheimer, Parkinson, depression or dementia.

8. References

Afonso, V. & Tompkins, W. (1995). Detecting ventricular fibrillation, *IEEE Engineering in Medicine and Biology* 14: 152–159.

Akay, M. (1996). *Detection and estimation methods for biomedical signals*, Vol. 1, 1 ed Academic Press.

Almeida, L. (1994). The fractional fourier transform and time-frequency representation, *IEEE. Trans. on Signal Proc.* 42: 3084–3091.

A.M. Murro, D.W. King, J. S. B. G. H. F. & Meador, K. (1991). Computerized seizure detection of complex partial seizures, *Electroencephalography and Clinical Neurophysiol.* 79: 330–333.

Boashash, B. (2003). *Time Frequency Signal Analysis and processing. A comprehensive reference*, Vol. 1, Elsevier.

Boashash, B. & Mesbah, M. (2001). A time-frequency approach for newborn seizure detection, *IEEE Eng. in Med. and Biol. Magazine* 20: 54–64.

Celka, P. & Colditz, P. (2002). Nonlinear nonstationary wiener model of infant eeg seizures, *IEEE Trans. On Biomed. Eng.* 49: 556–564.

Chiappa, S. & Barber, D. (2006). Eeg classification using generative independent component analysis, *Neurocomputing* 69: 769–777.

Cohen, L. (1989). Time-frequency distributions-a review, *Proceedings of the IEEE* 77: 941–981.

Cohen, L. (1995). *Time-Frequency Analysis*, Vol. 1, Prentice Hall.

Cover, T. & Thomas, J. (1991). *Elements of Information Theory*, Vol. 2, Wiley.

Duda, R., Hart, P. & Stork, D. (2009). *Pattern classification*, Vol. 2, Elsevier.

Durka, P. (1996). *Time-Frequency analysis of EEG*, Thesis Institute of Experimental Physics, Warsaw University.

Ebersole, J. & Pedley, T. (2003). *Current Practice of Clinical Electroencephalography*, Vol. 3, Lippincott Williams & Wilkins.

Fisch, B. J. (1999). *Fisch and Spehlmann's EEG primer: Basic principles of digital and analog EEG*, Vol. 3, Springer.

Güler, I. & Übeyli, E. D. (2007). Multiclass support vector machines for eeg-signals classification, *IEEE Trans. on Inf. Tech. in Biomed.* 11: 117–126.

Gomez-Verdejo, V., Verleysen, M. & Fleury, J. (2009). Information-theoretic feature selection for functional data classification, *Neurocomputing* 72: 3580–3589.

Gotman, J. (1982). Automatic recognition of epileptic seizures in the EEG, *Electroencephalography and Clinical Neurophysiol.* 54: 530–540.

Guerrero-Mosquera, C. & Navia-Vazquez, A. (2009). Automatic removal of ocular artifacts from eeg data using adaptive filtering and independent component analysis, *Proceedings of the 17th European Signal Processing Conference (EUSIPCO)* pp. 2317–2321.

Guerrero-Mosquera, C., Trigueros, A. M., Franco, J. I. & Navia-Vazquez, A. (2010). New feature extraction approach for epileptic eeg signal detection using time-frequency distributions, *Med. Biol. Eng. Comput.* 48: 321–330.

Guerrero-Mosquera, C., Verleysen, M. & Navia-Vazquez, A. (2010). Eeg feature selection using mutual information and support vector machine: A comparative analysis, *Proceedings of the 32nd Annual EMBS International Conference* pp. 4946–4949.

Guyon, I. & Elisseeff, A. (2003). An introduction to variable and feature selection, *Journal of Machine Learning Research* 3: 1157–1182.

Guyon, I., Gunn, S., Nikravesh, M. & Zadeh, L. (2006). *Feature Extraction, Foundations and Applications*, Vol. 1, Springer.

Hammond, J. & White, P. (1996). The analysis of non-stationary signals using time-frequency methods, *Journal of Sound and Vibration* 3: 419–447.

Hasan, O. (2008). Optimal classification of epileptic seizures in eeg using wavelet analysis and genetic algorithm, *Signal processing* 88: 1858–1867.

Hlawatsch, F. & Boudreaux-Bartels, G. (1992). Linear and quadratic time-frequency signal representation, *IEEE SP Magazine* 9: 21–67.

James, C. & Hesse, C. (2005). Independent component analysis for biomedical signals, *Physiol. Meas.* 26: R15–R39.

J.R. Ives, C.J. Thompson, P. G. A. O. & Woods, J. (1974). The on-line computer detection and recording of spontaneous temporal lobe epileptic seizures from patients with implanted depth electrodes via radio telemetry link, *Electroencephalography and Clinical Neurophysiol.* 37: 205.

Latka, M. & Was, Z. (2003). Wavelet analysis of epileptic spikes, *Physical Rev. E* 67: 052902.

Lehmanna, C., Koenig, T., Jelic, V., Prichep, L., John, R., Wahlund, L., Dodgee, Y. & Dierks, T. (2007). Application and comparison of classification algorithms for recognition of alzheimerŠs disease in electrical brain activity (eeg), *Journal of Neuroscience Methods* 161: 342–350.

Lima, C., Coelho, A. & Chagas, S. (2009). Automatic eeg signal classification for epilepsy diagnosis with relevance vector machines, *Expert Systems with Applications* 36: 10054–10059.

Lotte, F., Congedo, M., Lécuyer, A., Lamarche, F. & Arnaldi, B. (2007). A review of classification algorithms for eeg based brain-computer-interface, *Journal of Neural Eng.* 4: R1–R13.

Mallat, S. (2009). *A wavelet tour of signal processing, Third edition: The sparse way*, Vol. 3, Elsevier.

Müller, K.-R., Anderson, C. & Birch, G. (2003). Linear and nonlinear methods for brain-computer interfaces, *IEEE Trans. on Neural Systems and Rehabilitation Eng.* 11: 165–168.

Mosher, J. & Leahy, R. (1998). Recursive music: A framework for eeg and meg source localization, *IEEE Trans. On Biomed. Eng.* 45: 1342–1354.

P.F. Prior, R. V. & Maynard, D. (1973). An EEG device for monitoring seizure discharges, *Epilepsia* 14: 367–372.

Rossi, F., Françoise, D., Wertz, W., Meurens, M. & Verleysen, M. (2007). Fast selection of spectral variables with b-spline compression, *Chemometrics and Intelligent Laboratory Systems* 86: 208–218.

Rossi, F., Lendasse, A., Françoise, D., Wertz, W. & Verleysen, M. (2006). Mutual information for the selection of relevant variables spectrometric nonlinear modelling, *Chemometrics and Intelligent Laboratory Systems* 80: 215Ű226.

Sanei, S. & Chambers, J. (2007). *EEG signal processing*, Vol. 1, Wiley.

Senhadji, L. & Wendling, F. (2002). Epileptic transient detection: wavelets and time-frequency approaches, *Neurophysiol Clin.* 32: 175–192.

Siuly, Y. & Wen, P. (2009). Classification of eeg signals using sampling techniques and least square support vector machine, *Lectures notes in Computer Science* 5589: 375–382.

Sörnmo, L. & Laguna, P. (2005). *Bioelectrical signal processing in cardiac and neurological applications*, Vol. 1, Elsevier Academic Press.

Stone, J. (2004). *Independent Component Analysis: A tutorial introduction*, Vol. 1, The MIT Press.

Subasi, A. (2007). Eeg signal classification using wavelet feature extraction and a mixture of expert model, *Expert Systems with Applications* 32: 1084–1093.

Tao, R., Deng, B., Zhang, W. & Wuang, Y. (2008). Sampling and sampling rate conversion of band limited signals in the fractional fourier transform domain, *IEEE. Trans. on Signal Proc.* 56: 158–171.

Thusalidas, M., Guan, C. & Wu, J. (2006). Robust classification of eeg signal for brain-computer interface, *IEEE Trans. on Neural Systems and Rehabilitation Eng.* 14: 24–29.

T.L. Babb, E. M. & Crandall, P. (1974). An electronic circuit for detection of EEG seizures recorded with implanted electrodes, *Electroencephalography and Clinical Neurophysiol.* 37: 305–308.

Tzallas, A., Tsipouras, M. & Fotiadis, D. (2009). Epileptic seizure detection in eegs using time-frequency analysis, *IEEE Trans. on Inf. Tech. in Biomed.* 13: 703–710.

Xu, Q., Zhou, H., Wang, Y. & Huang, J. (2009). Fuzzy support vector machine for classification of eeg signals using wavelet-based features, *Medical Eng. & Physics* 31: 858–865.

4

Automated Epileptic Seizure Detection Methods: A Review Study

Alexandros T. Tzallas, Markos G. Tsipouras,
Dimitrios G. Tsalikakis, Evaggelos C. Karvounis,
Loukas Astrakas, Spiros Konitsiotis and Margaret Tzaphlidou
Department of Medical Physics, Medical School, University of Ioannina, Ioannina, Greece

1. Introduction

Epilepsy is a neurological disorder with prevalence of about 1-2% of the world's population (Mormann, Andrzejak, Elger & Lehnertz, 2007). It is characterized by sudden recurrent and transient disturbances of perception or behaviour resulting from excessive synchronization of cortical neuronal networks; it is a neurological condition in which an individual experiences chronic abnormal bursts of electrical discharges in the brain. The hallmark of epilepsy is recurrent seizures termed "**epileptic seizures**". Epileptic seizures are divided by their clinical manifestation into partial or focal, generalized, unilateral and unclassified seizures (James, 1997; Tzallas, Tsipouras & Fotiadis, 2007a, 2009). Focal epileptic seizures involve only part of cerebral hemisphere and produce symptoms in corresponding parts of the body or in some related mental functions. Generalized epileptic seizures involve the entire brain and produce bilateral motor symptoms usually with loss of consciousness. Both types of epileptic seizures can occur at all ages. Generalized epileptic seizures can be subdivided into absence (petit mal) and tonic-clonic (grand mal) seizures (James, 1997).

Monitoring brain activity through the electroencephalogram (EEG) has become an important tool in the diagnosis of epilepsy. The EEG recordings of patients suffering from epilepsy show two categories of abnormal activity: **inter-ictal**, abnormal signals recorded between epileptic seizures; and **ictal**, the activity recorded during an epileptic seizure (Fig. 1). The EEG signature of an inter-ictal activity is occasional transient waveforms, as either isolated spikes, spike trains, sharp waves or spike-wave complexes. EEG signature of an epileptic seizure (ictal period) is composed of a continuous discharge of polymorphic waveforms of variable amplitude and frequency, spike and sharp wave complexes, rhythmic hypersynchrony, or electrocerebral inactivity observed over a duration longer than the average duration of these abnormalities during inter-ictal periods (McGrogan, 2001).

Given that ictal recordings (recording during an epileptic seizure) are rarely obtained, EEG analysis of patients suffering from epilepsy usually relies on inter-ictal findings. In those inter-ictal EEG recordings, epileptic seizures are usually activated with photo stimulation, hyperventilation and other methods (McGrogan, 2001). However, one weakness of these stimulation techniques is that provoked epileptic seizures do not necessarily have the same behaviour as the spontaneous ones. The introduction of long-term video-EEG recordings has been an important milestone providing not only the possibility to capture and analyze ictal events, but also contributing to valuable clinical information, especially in those candidates evaluated for epilepsy surgery. Prior to the advent of portable recording devices all EEG recording took place in special hospital units. The introduction of portable recording systems (ambulatory EEG), however, has allowed outpatient EEG recording to become more common. This method has advantages that patients are recorded in their normal environment without the reduction in seizure frequency usually seen during a long (and expensive) in-patient sessions. Many studies have shown that ambulatory EEG recordings generally increase the yield of useful diagnostic information and improve the overall medical management of patients (Casson, Yates, Smith, Duncan, & Rodriguez-Villegas, 2010; Waterhouse, 2003).

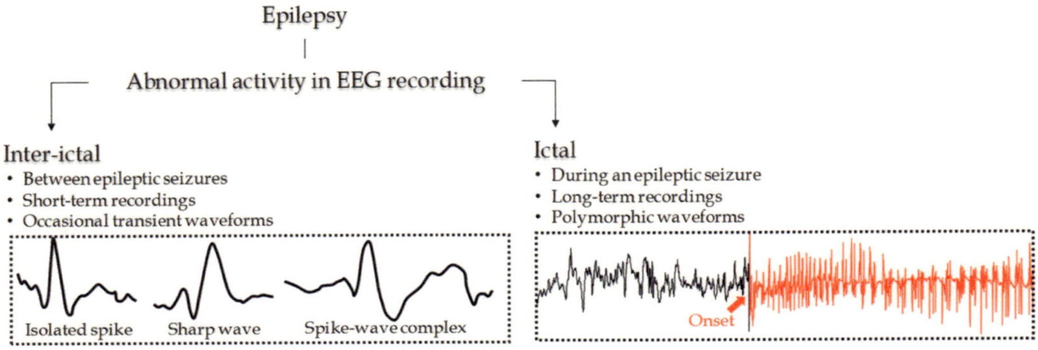

Fig. 1. During inter-ictal periods, or between epileptic seizures, EEG recordings of patients affected by epilepsy will exhibit abnormalities like isolated spike, sharp waves and spike-wave complexes (usually all termed as inter-ictal spikes or spikes). In ictal periods, or during epileptic seizures, the EEG recording is composed of a continuous discharge of one of these abnormalities, but extended over a longer duration and typically accompanied by a clinical correlate (Exarchos, Tzallas, Fotiadis, Konitsiotis & Giannopoulos, 2006; Oikonomou, Tzallas & Fotiadis, 2007; Tzallas et al., 2006; Tzallas, Oikonomou, & Fotiadis, 2006; Tzallas, et al., 2007a; Tzallas, Tsipouras & Fotiadis, 2007b; Tzallas, et al., 2009).

Generally, the detection of epilepsy can be achieved by visual scanning of EEG recordings for inter-ictal and ictal activities by an experienced neurophysiologist. However, visual review of the vast amount of EEG data has serious drawbacks. Visual inspection is very time consuming and inefficient, especially in the case of long-term recordings. In addition, disagreement among the neurophysiologists on the same recording is possible due to the subjective nature of the analysis and due to the variety of inter-ictal spikes morphology. Moreover, the EEG patterns that characterize an epileptic seizure are similar to waves that are part of the background noise and to artefacts (especially in extracranial recordings) such as eye blinks and other eye movements, muscle activity, electrocardiogram, electrode "pop" and electrical interference. For these reasons, methods for the automated detection of inter-ictal spikes and epileptic seizures can serve as valuable clinical tools for the scrutiny of EEG data in a more objective and computationally efficient manner.

2. Automated analysis of epileptic EEG recordings

Automated analysis of EEG recordings for assisting in the diagnosis of epilepsy started in the early 1970s (Gotman, 1999; Tzallas, et al., 2007a, 2007b, 2009; Wilson & Emerson, 2002). From the beginning, the automated analysis of epileptic EEG recordings has progressed in two main directions:

- inter-ictal spike detection or spike detection analysis, and
- epileptic seizure analysis.

2.1 Automated spike detection analysis

The automatic spike detection problem can be simply transferred to the detection of the presence of inter-ictal spikes in the multichannel EEG recording with high sensitivity and selectivity (James, 1997; Oikonomou, et al., 2007). That means that high proportion of true events must be detected with a minimum number of false detections. Although desirable, it is not realistic to expect high sensitivity and selectivity due to the imprecise definition among neurophysiologists of what constitutes a spike varies. Several studies evaluated this issue by extracting features from the raw EEG recordings that best describe the spike morphology. On the other hand, other studies have chosen to use machine learning techniques (usually artificial neural networks) as a means of using the raw EEG without having to make any decision concerning what parameters are more important than others in detecting spikes (James, 1997). Whatever the method used, the spike detection problem seems to be divided into two main stages: feature extraction and classification (Fig. 2).

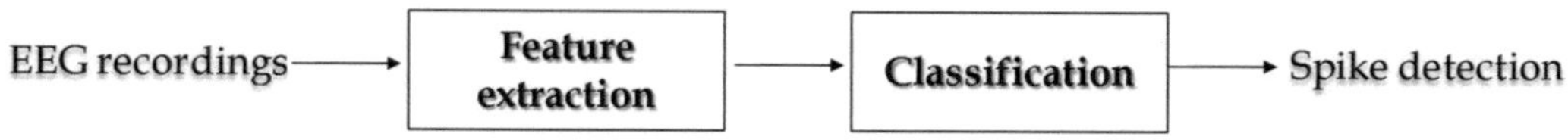

Fig. 2. The spike detection problem seems to be broken down into two main stages: feature extraction and classification. This can be viewed as mapping the *N*-dimensional EEG pattern space to a *F*-dimensional feature space (where $N \geq F$) and then performing classification in the feature space. In the case of use of raw EEG recordings without feature extraction, this can be seen as the case where the *N*-dimensional EEG space is mapped onto an identical *N*-dimensional feature space where classification then takes place (James, 1997).

It is well established that, apart from the spike detection on a single channel itself, other contextual information (spatial and temporal) is also vital to neurophysiologists when identifying candidate transient waveforms as spikes (James, Jones, Bones, & Carroll, 1999; Tzallas, Karvelis, et al., 2006). This information is related to other channels waveforms that take place at the same time. Based on the above, the spike detection problem depicted in Fig. 2, can now be modified, as shown in Fig. 3, to incorporate the use of spatio-temporal information in helping detect spikes in the multichannel EEG recordings.

The following provides a short summary of the most common methods to the spike detection problem in the literature (Gotman, 1999; Wilson & Emerson, 2002). These methods have been grouped according to their spike detection criterion into nine (9) categories:

a. methods based on traditional recognition techniques, known as **mimetic** techniques,
b. methods based on **morphological** analysis,

c. methods using **template matching** algorithms,
d. methods based on **parametric** approaches
e. methods based on **independent component** analysis
f. methods based on **artificial neural networks,**
g. methods based on **clustering** techniques,
h. methods employed **data mining** and **other classification techniques**, and
i. methods utilizing **knowledge-based rules**.

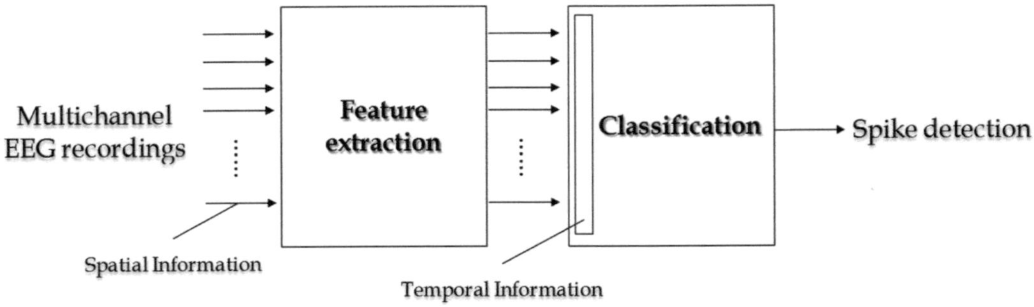

Fig. 3. The spatial and temporal information (contextual) is important in the spike detection problem. The *N*-dimensional EEG pattern space is mapped onto a *F*-dimensional feature space for each channel in the EEG recording. The multichannel features introduce spatial information into the method. The classification of candidate spikes then takes place using features extracted from the pattern space. Temporal information can then be introduced to the classification process by considering the presence of previous spikes in the EEG throughout the multichannel recording and allowing this to strengthen or weaken the outcome due to spatial information alone (James, 1997).

a. Methods based on traditional recognition techniques, known as mimetic techniques

Mimetic methods are based on the hypothesis that the process of identifying a transient waveform in EEG recordings as spike could be divided into well-defined steps representing the reasons and expertise of a neurophysiologist (Gotman, 1982; Gotman & Gloor, 1976; Guedes de Oliveira, 1983; Ktonas, 1983; Ktonas, Luoh, Kejariwal, Reilly, & Seward, 1981). Distinctive attributes of the spikes such as slope, height, duration and sharpness are compared with values provided by the neurophysiologists. Gotman and Gloor (1976) decomposed the waveform into two half-waves with opposite directions. Similar methods for decomposing the EEG waveform into half-waves have been used by many authors (Davey, Fright, Carroll, & Jones, 1989; Faure, 1985; Webber, Litt, Wilson, & Lesser, 1994). Faure (1985) introduced a concept where the duration, amplitude, and slope attributes of half-waves were used to classify them into states.

b. Methods based on morphological analysis

Methods based on morphological analysis characterize the waveforms, frequency bands, or time-frequency representations of spikes (Gotman, 1990, 2003; Michel, Seeck, & Landis, 1999). Morphological analysis has proven an efficient tool in EEG signal processing since it can decompose raw EEG signal into several physical parts. Background activity and spike component are separated and the main morphological characteristic of spikes is retained.

Pon and coworkers (2002) selected a circle structure element and utilized mathematical morphology and wavelet transform to detect bi-directional spikes in epileptic EEG recordings. Nishida and coauthors (1999) presented a detection method based on morphological filter, in which open–closing operation was selected as the basic algorithm and the general structure elements are designed by second-order polynomial functions. Using a morphological filter with proper morphological operation and structure elements, it was possible to restrain the background activity completely. Xu and coworkers (2007) presented a method for automatic spike detection by using an improved morphological filter. The basic idea of the improved morphological filter was to separate spikes and its background activity by the differences of their geometric characteristics.

c. Methods using template matching algorithms,

In the template matching algorithms, the user manually selects spikes from a set of test EEG recordings that are averaged to create a template (El-Gohary, McNames, & Elsas, 2008; Lopes da Silva, A., & H., 1975; Sankar & Natour, 1992). Many researchers (Goelz, Jones, & Bones, 2000; Schiff, Aldroubi, Unser, & Sato, 1994; Senhadji, Dillenseger, Wendling, Rocha, & Kinie, 1995; Senhadji & Wendling, 2002) used wavelets to obtain features of the signal for template building and spike detection.

d. Methods based on parametric approaches

In the parametric approaches, researchers (Birkemeier, Fontaine, Celesia, & Ma, 1978; Diambra & Malta, 1999; Lopes da Silva, et al., 1975) assume local stationarity of the noise and spikes are detected as deviation from that stationarity. Tzallas and coauthors (2006) presented a new technique based on a time-varying autoregressive model that made use of the nonstationarities of EEG. The autoregressive parameters were estimated via Kalman filter. The signal was first processed to accentuate the spikes and attenuate background activity and then passed through a thresholding function to determine spikes locations.

e. Methods based on independent component analysis

Various spike detection approaches based on independent component analysis (ICA) have been proposed in applications to EEG recordings (Hesse & James, 2007; Ossadtchi et al., 2004). Kobayashi and coauthors (1999) performed both model based and real data demonstrations of the use of ICA to isolate spikes from multichannel EEG data (Ossadtchi, et al., 2004). In this approach, ICA is applied to spatio-temporal data and components resembling abnormal epileptic activities selected by visual inspection and then interpreted by a neurophysiologist (Hesse & James, 2007; Ossadtchi, et al., 2004). Kobayashi and coworkers (2002) used ICA decomposition together with the RAP-MUSIC source localization approach (Mosher, Baillet, & Leahy, 1999; Mosher & Leahy, 1998; Mosher, Leahy, & Lewis, 1999) to detect potentially epileptogenic regions (Ossadtchi, et al., 2004). Rather than fitting a dipole to each independent component separately (Zhukov, Weinstein, & Johnson, 2000), Kobayashi and coauthors (2002) followed a multidimensional ICA paradigm and defined an inter-ictal subspace spanned by the columns of the estimated mixing matrix visually identified as corresponding to epileptic components (Ossadtchi, et al., 2004).

f. Methods based on artificial neural networks

In the sixth category belong approaches built upon artificial neural networks (ANNs) which simulate the behavior of a collection of neurons (Tzallas, Karvelis, et al., 2006). ANNs have been trained using either raw data (Ko & Chung, 2000; Ozdamar & Kalayci, 1998; Pang, Upton, Shine, & Kamath, 2003; Webber, et al., 1994) or select features (Acir, Oztura, Kuntalp, Baklan, & Guzelis, 2005; Castellaro et al., 2002; Gabor & Seyal, 1992; Liu, Zhang, & Yang, 2002; Pang, et al., 2003; Tzallas, Karvelis, et al., 2006; Webber, et al., 1994) to detect spikes. In the first case, windows of raw EEG data are fed into an ANN. In the second case, two types of features are used: (1) waveforms features such as duration, slope, sharpness, and amplitude, which are extracted from spikes and (2) context features, such as EEG variance and baseline crossings, which are extracted from the EEG activity surrounding the spikes.

g. Methods based on clustering techniques

Clustering techniques in the field of automated spike detection analysis has also been addressed. Hierarchical agglomerative methods and self organizing maps have been used for clustering EEG segments (Sommer & Golz, 2001). The nearest mean (NM) algorithm (Wahlberg & Salomonsson, 1996), the ant *K*-mean algorithm (Shen, Kuo, & Hsin, 2009) and the fuzzy *C*-means (FCM) algorithm (Inan & Kuntalp, 2007; Wahlberg & Lantz, 2000) have been employed in order to cluster spikes. In addition, the *K*-means algorithm has been used in order to cluster spikes and other types of transient waveforms (Exarchos, et al., 2006; Tzallas, Karvelis, et al., 2006).

h. Methods employed data mining and other classification techniques

Data mining (DM) techniques are also used to build automatic spike detection models, (Exarchos, et al., 2006; Valenti et al., 2006). In DM, the identification of spikes does not need a clear definition of spike morphology. In addition, other classification schemes such as support vector machines (SVMs) have also been applied to spike detection (Acir & Guzelis, 2004; Acir, et al., 2005; Tzallas et al., 2005). The main idea was to adjust the position of the separator (line, plane, hyperplane) between spike and non-spike patterns based on the distance from misclassified outliers.

i. Methods utilizing knowledge-based rules

The majority of the methods, mainly those belonging to the first four categories (mimetic, morphological, template matching and parametric) deals with the single EEG channel data only. Knowledge-based reasoning in addition to the aforementioned methods is widely used (Tzallas, Karvelis, et al., 2006). This arises from the need to incorporate knowledge of neurophysiologists that adopt spatial and temporal rules (Acir, et al., 2005; Dingle, Jones, Carroll, & Fright, 1993; Edwards, James, Coakley, & Brown, 1976; Glover, Raghavan, Ktonas, & Frost, 1989; James, 1997; James, et al., 1999; Liu, et al., 2002; Ozdamar, Yaylali, Jayakar, & Lopez, 1991; Tzallas, Karvelis, et al., 2006; Webber, et al., 1994). More specifically, Glover and coauthors (1989), Dingle and coauthors (1993), and Liu and coauthors (2002) used a knowledge-based system with a high degree of success, taking advantage of both spatial and temporal information. Ozdamar and his coworkers (1991) made use of spatial information by integrating the outputs of individual channel spike detection ANNs, from four channels into a single ANN module trained to recognize the common spatial distributions of spikes. Webber and coauthors (1994) used four channels simultaneously, while including spatial contextual information of a 1 sec long window around the spike, in the training of their ANN. James and coworkers (1999) have employed a spatial-combiner stage with the outputs of a self-organizing ANN, using a fuzzy logic approach, in order to incorporate spatial information in the multichannel EEG recordings. In a similar way, Acir and coauthors (2005) and Tzallas and coauthors (2006), in the final stage of their spike detection method, combined the outputs of the classification stage (ANN or SVM) in such a way as to confirm the presence of spike across two or more channels of the EEG recordings.

Based on the foregoing, it is apparent that when deciding on a method capable of the detection of spikes in the multichannel EEG recordings, a few number of important questions need to be answered. Fig.4 illustrates the questions and some of the possible answers (James, 1997). To sum up, these are:

- Should raw EEG recordings be used for the classification or should features be extracted first and the classification performed in the new feature space?
- If features are to be extracted, what features adequately describe spikes for the classification purposes?
- Once the decision made on raw vs. features, which machine learning algorithm should be used?

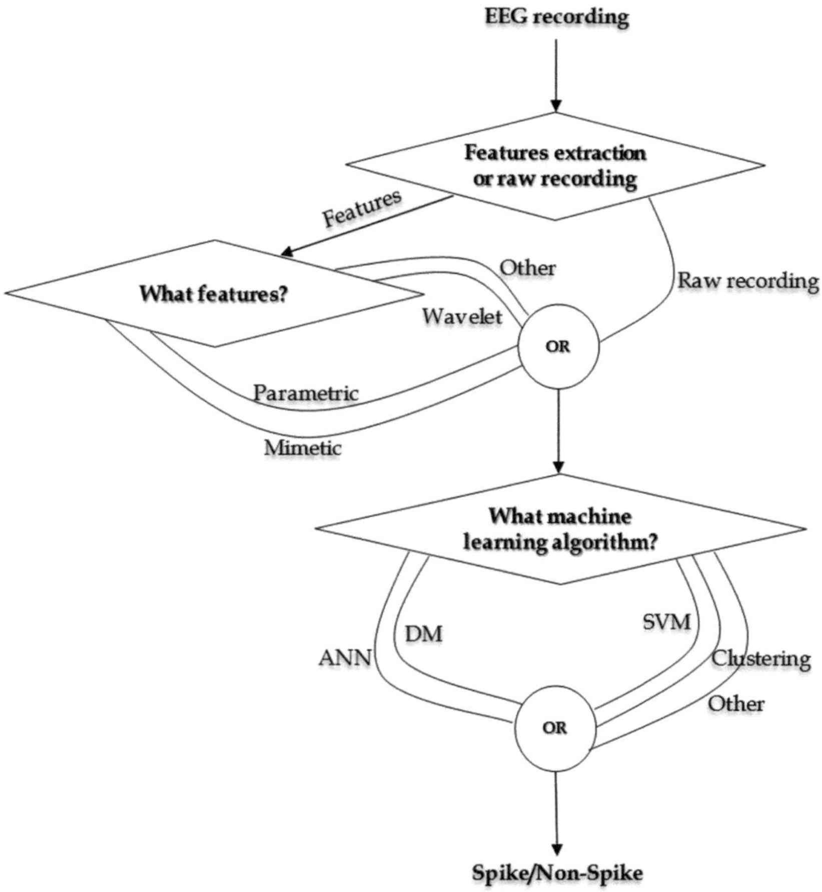

Fig. 4. Questions to be answered in choosing the best spike detection criterion. Once the method for spike detection has been established, it is important to keep in mind the need to incorporate spatial and temporal information (James, 1997).

2.1.1 Spike enhancement before spike detection analysis

From the preceding discussion, in the spike detection problem, a balance must be obtained between having high sensitivity and high selectivity. It is relatively easy to adjust method parameters to obtain performance where all spikes are found in a given patient, but this would usually be accompanied by an unacceptably large number of false detections (James, 1997; James, Hagan, Jones, Bones, & Carroll, 1997; Oikonomou, et al., 2007). Alternatively, it is relatively easy to have a method with very low false detection rates, but this would be accompanied by an unacceptably large number of missed events. Many researchers argue that it is better to have a high sensitivity, to minimize missed events, and to have more false detections that can be checked by a neurophysiologist, rather than missing the events altogether (James, 1997; Oikonomou, et al., 2007). If we look at the method from the point of view of minimizing the number of false detections then the number of missed events will increase. However, if spikes can be enhanced prior to the use of a spike detection criterion, it should be possible to increase the sensitivity minimizing missed events, while maintaining the selectivity at a satisfactory level. Thus, a spike enhancement stage would not be a detection stage, but it would simply aim to enhance anything vaguely spike like, is needed. This means that actual spikes, as well as spike like artefacts and background will be enhanced, i.e. a large number of unwanted waveforms will be enhanced along with real spikes. This is quite acceptable as long as the spike detection method has high selectivity. To our knowledge, there few methods that explicitly addressed the spike enhancement problem in epileptic EEG recordings (James, et al., 1997; Lopes da Silva, et al., 1975; Oikonomou, et al., 2007). Lopes da Silva and co-authors (1975) used the method of modelling the background EEG with an autoregressive prediction filter and detecting transient waveforms by examining the prediction error. The autoregressive filter was calculated from a segment of the background EEG which is assumed to be stationary. James and coworkers (1997) made use of the multireference adaptive noise cancelling (MRANC) in which the background EEG on adjacent channels in the multichannel EEG recording is used to adaptively cancel the background EEG on the channel under investigation. Oikonomou and coauthors (2007) have presented a method for spike enhancement in EEG recordings, based on time-varying autoregressive model in order to take advantage of the nonstationarity nature of the EEG signal. More specifically, the method was based on the assumption that EEG consists of an underlying background activity, which was assumed stationary, and superimposed transient nonstationarities such spikes and artifacts. The method used a time-varying autoregressive model for the accentuation of spikes and other transient waveforms that are similar to spikes. The parameters of the model were estimated by Kalman filter.

After that, a complete spike detection scheme can be thought as a two-stage process: enhancement and detection (Fig. 5).

The purpose of the enhancement stage is to make the spike samples stand out from the rest of the data, thereby simplifying the subsequent task of detection. Depending on the nature of the enhancement strategy, several EEG spike detection schemes have been proposed categorized into three broad classes: (i) time domain techniques (Kim & Kim, 2000; Malarvili, Hassanpour, Mesbah, & Boashash, 2005; Mukhopadhyay & Ray, 1998) (ii) signal modeling approaches (Dandapat & Ray, 1997; James, et al., 1997; Tzallas, Oikonomou, et al., 2006), and (iii) transform domain methods (Durka, 2004; Hassanpour, Mesbah, & Boashash, 2004).

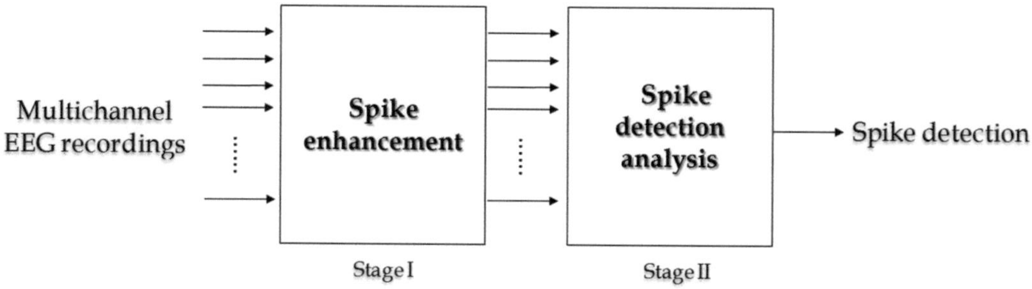

Fig. 5. Complete spike detection methods consists of two stages: (I) spike enhancement and (II) spike detection analysis. The spike enhancement stage processes an EEG recording by attenuating the background EEG, thus primarily leaving only transients waveforms -which are then classified as spikes or non-spikes by following stage II (spike detection analysis which is analytically described in the section 2.1). The main goal of the spike enhancement stage is to increase the sensitivity of the overall method to candidate spikes, while maximizing selectivity (minimizing the number of candidate spikes which are not epileptic passed onto the next stage) (Tzallas, Oikonomou, et al., 2006).

2.2 Automated epileptic seizure analysis

Automated epileptic seizure analysis (Fig. 6) refers collectively to methods for:

- epileptic seizures detection,
- epileptic seizures prediction, and
- epileptic seizures origin localization.

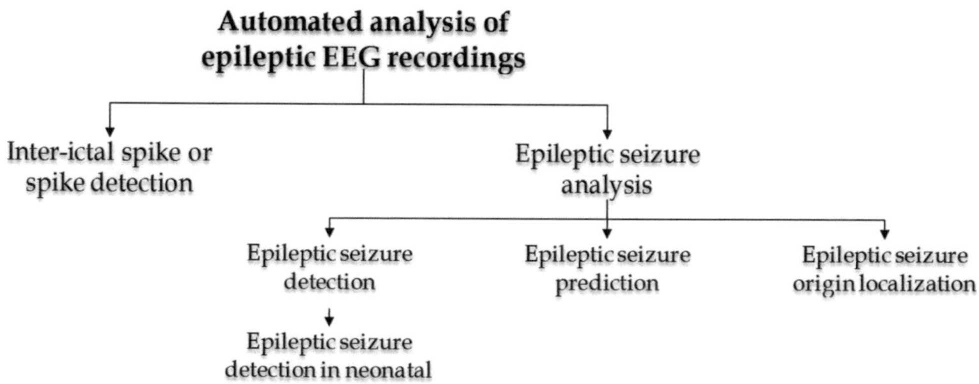

Fig. 6. Automated analysis of epileptic EEG recordings addresses two major problems: 1) inter-ictal spike detection or spike detection (section 2.1) and 2) epileptic seizure analysis. In addition, methods for automated epileptic seizure analysis can be divided into three categories: (i) epileptic seizure detection, (ii) epileptic seizure prediction, and (iii) epileptic seizure origin localization (Tzallas, et al., 2007a, 2007b, 2009).

In the literature, many algorithms for epileptic seizures detection have been proposed using classical signal processing methods (Gotman, 1999; McSharry, He, Smith, & Tarassenko, 2002). All suggested signal processing's methods aim to detect various patterns in EEG recordings that are the manifestation of an epileptic seizure. The entire process of methods developed for automated epileptic seizure detection can be generally subdivided into two main stages: (i) feature extraction, and (iii) classification (Fig. 7).

The selection of discriminative features is the basis of almost all epileptic seizure detection methods. Sometimes the choice for certain features is based on the physiological phenomena that need to be detected. Some authors referred to the fact that during an epileptic seizure many neurons fire synchronously (Gotman, 1999). To get a measure or this "synchronicity" they determined features such as the autocorrelation function (Liu, et al., 2002), the synchronization likelihood (Altenburg, Vermeulen, Strijers, Fetter, & Stam, 2003), or the nearest neighbour phase synchronization (van Putten, 2003). Other authors based their feature choice on morphological characteristics of epileptic EEG recordings. Epileptic seizures are often visible in EEG recordings as rhythmic discharges or multiple spikes. For spike detection, Gotman (1982) developed an algorithm that first breaks down the EEG signal into half-waves. Then morphological characteristics of these half-waves, such as amplitude and duration, were used to determine whether they are part of an epileptic seizure or not (Gotman, 1982, 1999).

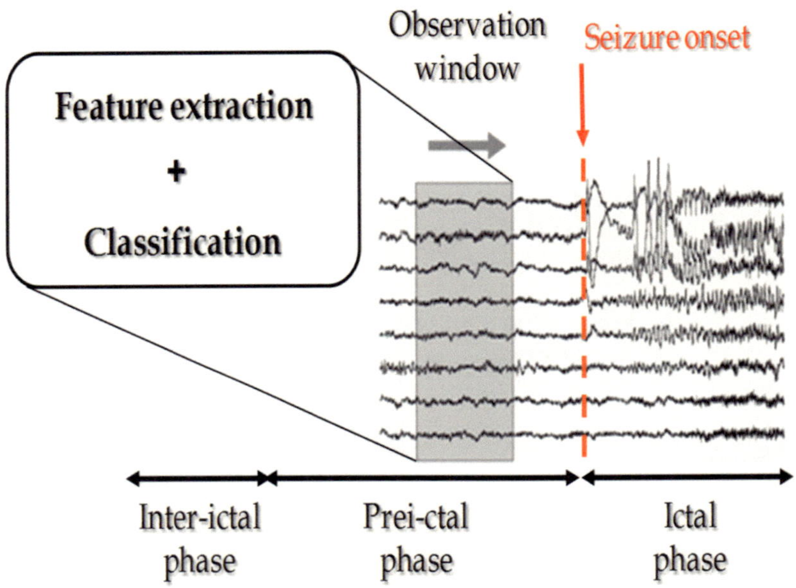

Fig. 7. Most of the automated epileptic seizure detection methods share certain common stages: (i) feature extraction and (ii) classification. By means of a moving-window analysis, features are calculated which is intended to characterise the multichannel EEG recordings. Then, the classification stage is employed to decide, from the calculated features, whether this EEG represents an epileptic seizure or not.

For rhythmic discharges, fast Fourier transform based (Polat & Gunes, 2007, 2008a, 2008b), frequency domain (Alkan, Koklukaya, & Subasi, 2005; Chua, Chandran, Acharya, & Lim, 2008; Gabor, 1998; Iscan, Dokur, & Tamer, 2011; Mousavi, Niknazar, & Vahdat, 2008; Murro et al., 1991; Nigam & Graupe, 2004; Sadati, Mohseni, & Magshoudi, 2006; Srinivasan, Eswaran, & Sriraam, 2005; Ubeyli, 2010a), time-frequency based (Martinez-Vargas, Avendano-Valencia, Giraldo, & Castellanos-Dominguez, 2011; Subasi & Gursoy, 2010; Tzallas, et al., 2007a, 2007b, 2009) or wavelet based features (Adeli, Ghosh-Dastidar, &

Dadmehr, 2007; Guler & Ubeyli, 2005, 2007; Guo, Rivero, Dorado, Rabunal, & Pazos, 2010; Guo, Rivero, & Pazos, 2010; Guo, Rivero, Seoane, & Pazos, 2009; Kiymik, Subasi, & Ozcalik, 2004; Lima, Coelho, & Eisencraft, 2010; H. Ocak, 2008; H. Ocak, 2009; Orhan, Hekim, & Ozer, 2011; Polat & Gunes, 2008b; Sadati, et al., 2006; Subasi, 2007a, 2007b; Subasi, Alkan, Koklukaya, & Kiymik, 2005; Subasi & Gursoy, 2010; Ubeyli, 2008c, 2009b, 2009c; Wang, Miao, & Xie, 2011) were often used. Some studies did not use prior information and just used large sets of various features. Aarabi and coauthors (2006) evaluated a large feature set containing various feature types. Their results showed that the most discriminative features for neonatal seizure detection[1] are morphological based features, such as amplitude, shape and duration of waveforms. In addition, time domain features such as statistical features (Adjouadi et al., 2005), Hjorth's descriptors (Hjorth, 1970), nonlinear features (Kannathal, Acharya, Lim, & Sadasivan, 2005; McSharry, et al., 2002)- correlation dimension (Elger & Lehnertz, 1998), Lyapunov exponent (Guler & Ubeyli, 2007; Guler, Ubeyli, & Guler, 2005; Ubeyli, 2006; Ubeyli, 2010b) and other features obtained from convolution kernels (Adjouadi et al., 2004), eigenvector methods (Naghsh-Nilchi & Aghashahi, 2010 ; Ubeyli, 2008a, 2008b, 2009a; Ubeyli & Guler, 2007), principal component analysis (PCA) (Ghosh-Dastidar, Adeli, & Dadmehr, 2008; Hesse & James, 2007; James & Hesse, 2005; Polat & Gunes, 2008a; Subasi & Gursoy, 2010), ICA (Hesse & James, 2007; James & Hesse, 2005; Subasi & Gursoy, 2010), crosscorrelation function (Chandaka, Chatterjee, & Munshi, 2009; Iscan, et al., 2011), and entropy (Guo, Rivero, Dorado, et al., 2010; Guo, Rivero, & Pazos, 2010; Kannathal, Acharya, et al., 2005; Kannathal, Choo, Acharya, & Sadasivan, 2005; Liang, Wang, & Chang, 2010; Naghsh-Nilchi & Aghashahi, 2010 ; H. Ocak, 2009; Srinivasan, Eswaran, & Sriraam, 2007; Wang, et al., 2011) have been proposed to characterize the EEG signal. It is also possible to select features using genetic programming (Firpi, Goodman, & Echauz, 2005; Guo, Rivero, Dorado, Munteanu, & Pazos, 2011). In this way, various features were extracted that were able to detect epileptic seizures, but these features did not have a physiological meaning.

Once a set of features has been obtained to characterise a section of EEG, it is necessary to apply a classification method in order to decide whether this section of EEG is taken from an epileptic seizure or not. Just as a wide variety of features has been used, an equally varied set of classification methods can be found in the literature. The classification methods varied from simple threshold (Altunay, Telatar, & Erogul, 2010; Martinez-Vargas, et al., 2011), rule based decisions (Gotman, 1990, 1999), or linear classifiers (Ghosh-Dastidar, Adeli, & Dadmehr, 2007; Iscan, et al., 2011; Liang, et al., 2010; Subasi & Gursoy, 2010) to ANNs (Ghosh-Dastidar, et al., 2007, 2008; Guler, et al., 2005; Mousavi, et al., 2008; Nigam & Graupe, 2004; Srinivasan, et al., 2005, 2007; Tzallas, et al., 2007a, 2007b, 2009; Ubeyli, 2006, 2009c; Ubeyli, 2010b) that have a complex shaped decision boundary. Other classification methods have been used using SVMs (Chandaka, et al., 2009; Guler & Ubeyli, 2007; Iscan, et al., 2011; Liang, et al., 2010; Lima, et al., 2010; Subasi & Gursoy, 2010; Ubeyli, 2008a; Ubeyli, 2010a), *k*-nearest neighbour classifiers (Guo, et al., 2011; Iscan, et al., 2011; Liang, et al., 2010; Orhan, et al., 2011; Tzallas, et al., 2009), quadratic analysis (Iscan, et al., 2011), logistic regression (Alkan, et al., 2005; Tzallas, et al., 2009), naive Bayes classifiers (Iscan, et al., 2011;

[1] The detection of epileptic seizures in neonates is quite different from that in adults: the discharges are often much slower (down to 0.5 Hz), epileptic seizure onset can be gradual and epileptic seizures can last several minutes, the waveforms of epileptic seizures and the inter-ictal background show a high level of variability.

Tzallas, et al., 2009), decision trees (Iscan, et al., 2011; Polat & Gunes, 2007; Tzallas, et al., 2009), Gaussian mixture model (Chua, et al., 2008; Lima & Coelho, 2011), mixture of expert model (Subasi, 2007b; Ubeyli, 2007, 2008c; Ubeyli & Guler, 2007) and adaptive neurofuzzy inference systems (Guler & Ubeyli, 2005; Kannathal, Choo, et al., 2005).

In addition to epileptic seizure detection methods, prediction methods have become increasingly valuable since detection of seizures at an early stage can warn a patient that a seizure is about to occur. Additionally, these methods can alert medical staff, and allow them to perform behavioural testing to further assess which specific functions may be impaired because of an epileptic seizure and help them in localizing the source of the epileptic seizure activity. Methods used to predict epileptic seizures include time-domain analysis (Lange, Lieb, Engel, & Crandall, 1983), frequency-based methods (Schiff et al., 2000), nonlinear dynamics and chaos (Lehnertz et al., 2001), methods of delays (Le Van Quyen et al., 2001), and intelligent approaches (Geva & Kerem, 1998). Advances in seizure prediction promise to give rise to implantable devices able to warn of impending seizures and to trigger therapy to prevent clinical epileptic attacks (Litt & Echauz, 2002; McSharry, Smith, & Tarassenko, 2003). Treatments such as electrical stimulation or focal drug infusion could be given on demand and might eliminate side effects in some patients taking antiepileptic drugs.

On the other hand, if drug control of epileptic seizures is not successful and if the epileptic seizures are serious enough, then a further option for treatment is surgery. Epilepsy surgery outcome strongly depends on the epileptic seizure origin localization. The analysis of ictal EEG recordings (scalp or intracranial) is a gold standard for definition of localization of sn epileptic seizure origin. Several linear (Parra, Spence, Gerson, & Sajda, 2005) and nonlinear methods (Acar, Aykut-Bingol, Bingol, Bro, & Yener, 2007) for analysis of epileptic EEG recordings as well as multi-way arrays models (Miwakeichi et al., 2004) have been used to understand the complex structure of epileptic seizure and localize seizure origin.

Table 1 shows a number of automated epileptic seizure detection methods found in the literature which is evaluated using the same dataset (Andrzejak et al., 2001). In Table 1, all methods are listed with their methodological standards (detection method, dataset, and classification accuracy). The dataset described in (Andrzejak, et al., 2001) is used for training and evaluation of these methods. This dataset includes five subsets five sets (denoted as *Z*, *O*, *N*, *F* and *S*), each containing 100 single-channel EEG segments of 23.6 sec duration, with sampling rate of 173.6 Hz. These segments were selected and cut out from continuous multi-channel EEG recordings after visual inspection for artifacts, e.g., due to muscle activity or eye movements. Sets *Z* and *O* consisted of segments taken from surface EEG recordings that were carried out on five healthy volunteers using a standardized electrode placement scheme. Volunteers were relaxed in an awake state with eyes open *(Z)* and eyes closed *(O)*, respectively. Sets *N*, *F*, and *S* originated from an EEG archive of presurgical diagnosis. Segments in set *F* were recorded from the epileptogenic zone, and those in set *N* from the hippocampal formation of the opposite hemisphere of the brain. While sets *N* and *F* contained only activity measured during seizure-free intervals, set S only contained epileptic seizure activity. All EEG signals were recorded with the same 128-channel amplifier system, using an average common reference. The data were digitized at 173.61 samples per second using a 12-bit resolution and they have the spectral bandwidth of the acquisition system, which varies from 0.5 to 85 Hz.

Author(s)	Year	Epileptic Seizure Detection Method		Dataset	Accuracy (%)
		Feature Extraction	Classification		
Nigam & Graupe	2004	Nonlinear pre-processing filter	Diagnostic neural network	*Z-S*	97.2
Scrinivasan et al	2005	Time & Frequency domain analysis	Recurrent neural network	*Z-S*	99.6
Kannathal et al	2005a	Chaotic measures	Surrogate data analysis	*Z-S*	90
Kannathal et al	2005b	Entropy measures	Adaptive neuro-fuzzy inference system	*Z-S*	92.22
Güler & Übeyli	2005	Wavelet transform	Adaptive neuro-fuzzy inference system	*Z-O-N-F-S*	98.68
Güler et al	2005	Lyapunov exponents	Recurrent neural network	*Z-F-S*	96.79
Übeyli	2006	Lyapunov exponents	Artificial neural networks	*Z-F-S*	95
Sadati et al	2006	Wavelet transform	Adaptive neuro-fuzzy network	*Z-F-S*	85.9
Adeli et al	2007	Wavelet transform & Chaos analysis	Statistical analysis (Analysis of variance-ANOVA)	*Z-F-S*	-
Dastidar et al	2007	Wavelet transform & Chaos analysis	*K*-means clustering, Discriminant analysis, Artificial neural networks	*Z-F-S*	96.7
Güler & Übeyli	2007	Wavelet transform & Lyapunov exponents	Support vector machines	*Z-O-N-F-S*	99.28
Polat & Günes	2007	Fast Fourier transform	Decision Tree	*Z-S*	98.72
Subasi	2007	Wavelet transform	Mixture of expert model	*Z-S*	95
Tzallas et al	2007a	Time-frequency analysis	Artificial neural networks	*Z-S* *Z-F-S* *ZONF-S* *ZO-NF-S*	100 99.28 97.73 97.72
Tzallas et al	2007b	Time-frequency analysis	Artificial neural networks	*Z-S* *Z-F-S* *Z-O-N-F-S*	100 100 89
Übeyli & Güler	2007	Eigenvector methods	Mixture of expert model	*Z- O-N-F-S*	98.60
Scrinivasan et al	2007	Approximate Entropy	Artificial neural networks	*Z-S*	100
Polat & Günes	2008a	Principal Component Analysis & Fast Fourier transform	Artificial immune recognition system	*Z-S*	100

Author(s)	Year	Epileptic Seizure Detection Method		Dataset	Accuracy (%)
		Feature Extraction	Classification		
Polat & Günes	2008b	Wavelet transform & Fast Fourier transform & Autoregressive model	Decision Tree	*Z-S*	99.32
Chua et al	2008	Higher order spectra & Power spectral density	Gaussian mixture model	*Z-F-S*	93.11
Dastidar et al	2008	Principal Component Analysis	Artificial neural networks	*Z-F-S*	99.3
Ocak	2008	Wavelet transform & Approximate entropy & Genetic algorithm	Learning vector quantization	*ZNF-S*	98
Mousavi et al	2008	Wavelet transform & Autoregressive model	Artificial neural networks	*Z-N-S*	96
Übeyli	2008a	Eigenvector methods	Support vector machines	*Z-O-N-F-S*	99.30
Übeyli	2008b	Wavelet transform	Mixture of expert model	*Z-F-S*	93.17
Chandaka et al	2009	Cross correlation	Support vector machines	*Z-S*	95.95
Ocak	2009	Wavelet transform & Approximate entropy	Surrogate data analysis	*ZNF-S*	96.65
Guo et al	2009	Wavelet transform & Relative wavelet energy	Artificial neural networks	*Z-S*	95.2
Tzallas et al	2009	Time-frequency analysis	Naives Bayes, k-nearest neighbor classifier, Logistic regression, Decision trees, Artificial neural networks	*Z-S* *Z-F-S* *Z-O-N-F-S*	100 100 89
Übeyli	2009a	Eigenvector methods	Recurrent neural network	*Z-O-N-F-S*	98.15
Übeyli	2009b	Wavelet transform	Probabilistic neural networks	*Z-O-N-F-S*	97.63
Guo et al	2010a	Wavelet transform & Approximate entropy	Artificial neural networks	*Z-S* *ZONF-S*	99.85 98.27
Guo et al	2010b	Wavelet transform & Line length feature	Artificial neural networks	*Z-S* *ZONF-S* *ZO-NF-S*	99.6 97.75 97.77
Liang et al	2010	Time frequency analysis & Approximate entropy	Linear least squares, Linear discriminant analysis, Support vector machines	*ZONF-S* *Z-F-S* *Z-O-N-F-S*	98.51 98.67 85.9

Author(s)	Year	Epileptic Seizure Detection Method		Dataset	Accuracy (%)
		Feature Extraction	Classification		
Subasi &Gursoy	2010	Wavelet transform & Principal component analysis & Independent component analysis & Linear discriminant analysis	Support vector machines	*Z-S*	100
Naghsh-Nilchi & Aghashahi	2010	Eigenvector methods & MUSIC	Artificial neural networks	*Z- N- S*	97.5
Altunay et al	2010	Linear prediction filter	Thresholding function	*Z- O- N- F- S*	93.6
Übeyli	2010a	Burg autoregressive & Moving average & Least squares modified Yule–Walker autoregressive moving average	Support vector machines	*Z- S*	99.56
Übeyli	2010b	Lyapunov exponents	Probabilistic neural networks	*Z- O- N-F-S*	98.05
Lima et al	2010	Wavelet transform	Support vector machines	*Z- S*	100
Guo et al	2011	Genetic programming based feature extraction system	*K*-nearest neighbor classifier	*Z- S* *Z- F-S*	99 93.5
Wang et al	2011	Wavelet transform & Shannon entropy	*K*-nearest neighbor classifier	*Z-S*	100
Iscan et al	2011	Cross correlation & Power spectral density	Support vector machines, Decision tree, Quadratic analysis, linear discriminant analysis, Naïve Bayes, *K*-nearest neighbor classifier	*Z- S*	100
Martinez-Vargas et al	2011	Time-frequency analysis	Thresholding function	*ZO- NF-S*	99%
Orhan et al	2011	Wavelet transform	*K*-nearest neighbor classifier, Artificial Neural Network	*Z-S* *Z-F-S* *ZO-NF-S* *ZO-NFS* *ZONF-S*	100 96.67 95.60 98.80 100

Z: (Healthy) Relaxed in an awake state with eyes open, *O:* (Healthy) Relaxed in an awake state with eyes closed, *N:* Recorded from the hippocampal formation of the opposite hemisphere of the brain (seizure-free), *F:* Recorded from within the epileptogenic zone (seizure free), *S:* During seizure activity

Table 1. Classification accuracies (in percent) obtained by automated epileptic seizure methods which are evaluated using a publicly available dataset (Andrzejak, et al., 2001).

3. Conclusion

Locating epileptic activity in the form of epileptic seizures or inter-ictal spikes in EEG recordings (usually lasting days or weeks in case of long-term recordings) is a demanding, time-consuming task because this activity constitutes a small percentage of the entire recording. This difficulty has motivated the development of automated methods that scan, identify, and then present to a neurophysiologist epochs containing epileptic events. Two types of automated methods for analysis of epileptic EEG recordings have been reported in the literature: those aimed at inter-ictal spike detection, and those aimed at epileptic seizure analysis and characterization of abnormal EEG activities in long-term recordings. In this chapter, a literature survey of the significant and recent studies that are concerned with effective detection of spike and epileptic seizures using EEG signals are presented. The main goal behind this review is to assist the researchers in the field of EEG signal analysis to understand the available methods and adopt the same for the detection of neurological disorders associated with EEG recordings.

4. References

Aarabi, A., Wallois, F., & Grebe, R. (2006). Automated neonatal seizure detection: a multistage classification system through feature selection based on relevance and redundancy analysis. *Clin Neurophysiol, 117* 328-340.

Acar, E., Aykut-Bingol, C., Bingol, H., Bro, R., & Yener, B. (2007). Multiway analysis of epilepsy tensors. *Bioinformatics, 23*(13), i10-18.

Acir, N., & Guzelis, C. (2004). Automatic spike detection in EEG by a two-stage procedure based on support vector machines. *Comput Biol Med, 34*(7), 561-575.

Acir, N., Oztura, I., Kuntalp, M., Baklan, B., & Guzelis, C. (2005). Automatic detection of epileptiform events in EEG by a three-stage procedure based on artificial neural networks. *IEEE Trans Biomed Eng, 52*(1), 30-40.

Adeli, H., Ghosh-Dastidar, S., & Dadmehr, N. (2007). A wavelet-chaos methodology for analysis of EEGs and EEG subbands to detect seizure and epilepsy. *IEEE Trans Biomed Eng, 54*(2), 205-211.

Adjouadi, M., Cabrerizo, M., Ayala, M., Sanchez, D., Yaylali, I., Jayakar, P., et al. (2005). Detection of interictal spikes and artifactual data through orthogonal transformations. *J Clin Neurophysiol, 22*(1), 53-64.

Adjouadi, M., Sanchez, D., Cabrerizo, M., Ayala, M., Jayakar, P., Yaylali, I., et al. (2004). Interictal spike detection using the Walsh transform. *IEEE Trans Biomed Eng, 51*(5), 868-872.

Alkan, A., Koklukaya, E., & Subasi, A. (2005). Automatic seizure detection in EEG using logistic regression and artificial neural network. *J Neurosci Methods, 148*(2), 167-176.

Altenburg, J., Vermeulen, R. J., Strijers, R. L., Fetter, W. P., & Stam, C. J. (2003). Seizure detection in the neonatal EEG with synchronization likelihood. *Clin Neurophysiol, 114*(1), 50-55.

Altunay, S., Telatar, Z., & Erogul, O. (2010). Epileptic EEG detection using the linear prediction error energy. *Expert Systems with Applications*(37), 5661-5665.

Andrzejak, R. G., Lehnertz, K., Mormann, F., Rieke, C., David, P., & Elger, C. E. (2001). Indications of nonlinear deterministic and finite-dimensional structures in time series of brain electrical activity: Dependence on recording region and brain state. *Physical Review E, 64*(6), 061907.

Birkemeier, W. P., Fontaine, A. B., Celesia, G. G., & Ma, K. M. (1978). Pattern recognition techniques for the detection of epileptic transients in EEG. *IEEE Trans Biomed Eng, 25*(3), 213-217.

Casson, A., Yates, D., Smith, S., Duncan, J., & Rodriguez-Villegas, E. (2010). Wearable electroencephalography. What is it, why is it needed, and what does it entail? *IEEE Eng Med Biol Mag, 29*(3), 44-56.

Castellaro, C., Favaro, G., Castellaro, A., Casagrande, A., Castellaro, S., Puthenparampil, D. V., et al. (2002). An artificial intelligence approach to classify and analyse EEG traces. *Neurophysiol Clin, 32*(3), 193-214.

Chandaka, S., Chatterjee, A., & Munshi, S. (2009). Cross-correlation aided support vector machine classifier for classification of EEG signals. *Expert Systems with Applications 36*(2), 1329-1336.

Chua, K. C., Chandran, V., Acharya, R., & Lim, C. M. (2008). Automatic identification of epilepsy by HOS and power spectrum parameters using EEG signals: a comparative study. *Conf Proc IEEE Eng Med Biol Soc, 2008*, 3824-3827.

Dandapat, S., & Ray, G. C. (1997). Spike detection in biomedical signals using midprediction filter. *Med Biol Eng Comput, 35*(4), 354-360.

Davey, B. L., Fright, W. R., Carroll, G. J., & Jones, R. D. (1989). Expert system approach to detection of epileptiform activity in the EEG. *Med Biol Eng Comput, 27*(4), 365-370.

Diambra, L., & Malta, C. P. (1999). Nonlinear models for detecting epileptic spikes. *Phys. Lett. A, 241*, 61-66.

Dingle, A. A., Jones, R. D., Carroll, G. J., & Fright, W. R. (1993). A multistage system to detect epileptiform activity in the EEG. *IEEE Trans Biomed Eng, 40*(12), 1260-1268.

Durka, P. J. (2004). Adaptive time-frequency parametrization of epileptic spikes. *Phys Rev E Stat Nonlin Soft Matter Phys, 69*(5 Pt 1), 051914.

Edwards, J. H., James, C. J., Coakley, W. T., & Brown, R. C. (1976). The effect of ultrasonic cavitation on protein antigenicity. *J Acoust Soc Am, 59*(6), 1513-1514.

El-Gohary, M., McNames, J., & Elsas, S. (2008). User-guided interictal spike detection. *Conf Proc IEEE Eng Med Biol Soc, 2008*, 821-824.

Elger, C. E., & Lehnertz, K. (1998). Seizure prediction by non-linear time series analysis of brain electrical activity. *Eur J Neurosci, 10*(2), 786-789.

Exarchos, T. P., Tzallas, A. T., Fotiadis, D. I., Konitsiotis, S., & Giannopoulos, S. (2006). EEG transient event detection and classification using association rules. *IEEE Trans Inf Technol Biomed, 10*(3), 451-457.

Faure, C. (1985). Attributed strings for recognition of epileptic transients in EEG. *Int J Biomed Comput, 16*(3-4), 217-229.

Firpi, H., Goodman, E., & Echauz, J. (2005). Genetic programming artificial features with applications to epileptic seizure prediction. *Conf Proc IEEE Eng Med Biol Soc, 5*, 4510-4513.

Gabor, A. J. (1998). Seizure detection using a self-organizing neural network: validation and comparison with other detection strategies. *Electroencephalogr Clin Neurophysiol, 107*(1), 27-32.

Gabor, A. J., & Seyal, M. (1992). Automated interictal EEG spike detection using artificial neural networks. *Electroencephalogr Clin Neurophysiol, 83*(5), 271-280.

Geva, A. B., & Kerem, D. H. (1998). Forecasting generalized epileptic seizures from the EEG signal by wavelet analysis and dynamic unsupervised fuzzy clustering. *IEEE Trans Biomed Eng, 45*(10), 1205-1216.

Ghosh-Dastidar, S., Adeli, H., & Dadmehr, N. (2007). Mixed-band wavelet-chaos-neural network methodology for epilepsy and epileptic seizure detection. *IEEE Trans Biomed Eng, 54*(9), 1545-1551.

Ghosh-Dastidar, S., Adeli, H., & Dadmehr, N. (2008). Principal component analysis-enhanced cosine radial basis function neural network for robust epilepsy and seizure detection. *IEEE Trans Biomed Eng, 55*(2 Pt 1), 512-518.

Glover, J. R., Jr., Raghavan, N., Ktonas, P. Y., & Frost, J. D., Jr. (1989). Context-based automated detection of epileptogenic sharp transients in the EEG: elimination of false positives. *IEEE Trans Biomed Eng, 36*(5), 519-527.

Goelz, H., Jones, R. D., & Bones, P. J. (2000). Wavelet analysis of transient biomedical signals and its application to detection of epileptiform activity in the EEG. *Clin Electroencephalogr, 31*(4), 181-191.

Gotman, J. (1982). Automatic recognition of epileptic seizures in the EEG. *Electroencephalogr Clin Neurophysiol, 54*(5), 530-540.

Gotman, J. (1990). Automatic seizure detection: improvements and evaluation. *Electroencephalogr Clin Neurophysiol, 76*(4), 317-324.

Gotman, J. (1999). Automatic detection of seizures and spikes. *J Clin Neurophysiol, 16*(2), 130-140.

Gotman, J. (2003). Noninvasive methods for evaluating the localization and propagation of epileptic activity. *Epilepsia, 44 Suppl 12*, 21-29.

Gotman, J., & Gloor, P. (1976). Automatic recognition and quantification of interictal epileptic activity in the human scalp EEG. *Electroencephalogr Clin Neurophysiol, 41*(5), 513-529.

Guedes de Oliveira, P. Q., C., Lopes da Silva, F. (1983). Spike detection based on a pattern recognition approach using a microcomputer. *Electroencephalogr Clin Neurophysiol. , 56*(1), 97-103.

Guler, I., & Ubeyli, E. D. (2005). Adaptive neuro-fuzzy inference system for classification of EEG signals using wavelet coefficients. *J Neurosci Methods, 148*(2), 113-121.

Guler, I., & Ubeyli, E. D. (2007). Multiclass support vector machines for EEG-signals classification. *IEEE Trans Inf Technol Biomed, 11*(2), 117-126.

Guler, I., Ubeyli, E. D., & Guler, I. (2005). Reccurent neural networks employing Lyapunon exponents in EEG recordings. *Expert Systems with Applications, 29*(3), 2005.

Guo, L., Rivero, D., Dorado, J., Munteanu, C. R., & Pazos, A. (2011). Automatic feature extraction using genetic programming: An application to epileptic EEG classification. *Expert Systems with Applications, 38*, 10425-10436.

Guo, L., Rivero, D., Dorado, J., Rabunal, J. R., & Pazos, A. (2010). Automatic epileptic seizure detection in EEGs based on line length feature and artificial neural networks. *J Neurosci Methods, 191*(1), 101-109.

Guo, L., Rivero, D., & Pazos, A. (2010). Epileptic seizure detection using multiwavelet transform based approximate entropy and artificial neural networks. *J Neurosci Methods, 193*(1), 156-163.

Guo, L., Rivero, D., Seoane, J. A., & Pazos, A. (2009). *Classification of EEG signals using relative wavelet energy and artificial neural networks*. Paper presented at the Conf Proc of the first ACM/SIGEVO Summit on Genetic and Evolutionary Computation

Hassanpour, H., Mesbah, M., & Boashash, B. (2004). *EEG Spike Detection Using Time-Frequency Signal Analysis.* Paper presented at the Proc. of IEEE Int. Conf. Ac., Speech. & Sign. Proc.

Hesse, C. W., & James, C. J. (2007). Tracking and detection of epileptiform activity in multichannel ictal EEG using signal subspace correlation of seizure source scalp topographies. *Med Biol Eng Comput, 45*(10), 909-916.

Hjorth, B. (1970). EEG analysis based on time domain properties. *Electroencephalogr Clin Neurophysiol, 29*(3), 306-310.

Inan, Z. H., & Kuntalp, M. (2007). A study on fuzzy C-means clustering-based systems in automatic spike detection. *Comput Biol Med, 37*(8), 1160-1166.

Iscan, Z., Dokur, Z., & Tamer, D. (2011). Classification of electroencephalogram signals with combined time and frequency features. *Expert Systems with Applications, 38*, 10499-10505.

James, C. J. (1997). *Detection of epileptiform activity in the electroencephalogram using the electroencephalogram using artificial neural networks.* University of Canterbury, Christchurch.

James, C. J., Hagan, M. T., Jones, R. D., Bones, P. J., & Carroll, G. J. (1997). Multireference adaptive noise canceling applied to the EEG. *IEEE Trans Biomed Eng, 44*(8), 775-779.

James, C. J., & Hesse, C. W. (2005). Independent component analysis for biomedical signals. *Physiol Meas, 26*(1), R15-39.

James, C. J., Jones, R. D., Bones, P. J., & Carroll, G. J. (1999). Detection of epileptiform discharges in the EEG by a hybrid system comprising mimetic, self-organized artificial neural network, and fuzzy logic stages. *Clin Neurophysiol, 110*(12), 2049-2063.

Kannathal, N., Acharya, U. R., Lim, C. M., & Sadasivan, P. K. (2005). Characterization of EEG--a comparative study. *Comput Methods Programs Biomed, 80*(1), 17-23.

Kannathal, N., Choo, M. L., Acharya, U. R., & Sadasivan, P. K. (2005). Entropies for detection of epilepsy in EEG. *Comput Methods Programs Biomed, 80*(3), 187-194.

Kim, K. H., & Kim, S. J. (2000). Neural spike sorting under nearly 0-dB signal-to-noise ratio using nonlinear energy operator and artificial neural-network classifier. *IEEE Trans Biomed Eng, 47*(10), 1406-1411.

Kiymik, M. K., Subasi, A., & Ozcalik, H. R. (2004). Neural networks with periodogram and autoregressive spectral analysis methods in detection of epileptic seizure. *J Med Syst, 28*(6), 511-522.

Ko, C. W., & Chung, H. W. (2000). Automatic spike detection via an artificial neural network using raw EEG data: effects of data preparation and implications in the limitations of online recognition. *Clin Neurophysiol, 111*(3), 477-481.

Kobayashi, K., Akiyama, T., Nakahori, T., Yoshinaga, H., & Gotman, J. (2002a). Systematic source estimation of spikes by a combination of independent component analysis and RAP-MUSIC. I: Principles and simulation study. *Clin Neurophysiol, 113*(5), 713-724.

Kobayashi, K., Akiyama, T., Nakahori, T., Yoshinaga, H., & Gotman, J. (2002b). Systematic source estimation of spikes by a combination of independent component analysis and RAP-MUSIC. II: Preliminary clinical application. *Clin Neurophysiol, 113*(5), 725-734.

Kobayashi, K., James, C. J., Nakahori, T., Akiyama, T., & Gotman, J. (1999). Isolation of epileptiform discharges from unaveraged EEG by independent component analysis. *Clin Neurophysiol, 110*(10), 1755-1763.

Ktonas, P. Y. (1983). Automated analysis of abnormal electroencephalograms. *Crit Rev Biomed Eng, 9*(1), 39-97.

Ktonas, P. Y., Luoh, W. M., Kejariwal, M. L., Reilly, E. L., & Seward, M. A. (1981). Computer-aided quantification of EEG spike and sharp wave characteristics. *Electroencephalogr Clin Neurophysiol, 51*(3), 237-243.

Lange, H. H., Lieb, J. P., Engel, J., Jr., & Crandall, P. H. (1983). Temporo-spatial patterns of pre-ictal spike activity in human temporal lobe epilepsy. *Electroencephalogr Clin Neurophysiol, 56*(6), 543-555.

Le Van Quyen, M., Martinerie, J., Navarro, V., Boon, P., D'Have, M., Adam, C., et al. (2001). Anticipation of epileptic seizures from standard EEG recordings. *Lancet, 357*(9251), 183-188.

Lehnertz, K., Andrzejak, R. G., Arnhold, J., Kreuz, T., Mormann, F., Rieke, C., et al. (2001). Nonlinear EEG analysis in epilepsy: its possible use for interictal focus localization, seizure anticipation, and prevention. *J Clin Neurophysiol, 18*(3), 209-222.

Liang, S. F., Wang, H. C., & Chang, W. L. (2010). Combination of EEG Complexity and Spectral Analysis for Epilepsy Diagnosis and Seizure Detection. *EURASIP Journal on Advances in Signal Processing, 2010*, 853434.

Lima, C. A., & Coelho, A. L. (2011). Kernel machines for epilepsy diagnosis via EEG signal classification: A comparative study. *Artif Intell Med.*

Lima, C. A., Coelho, A. L., & Eisencraft, M. (2010). Tackling EEG signal classification with least squares support vector machines: a sensitivity analysis study. *Comput Biol Med, 40*(8), 705-714.

Litt, B., & Echauz, J. (2002). Prediction of epileptic seizures. *Lancet Neurol, 1*(1), 22-30.

Liu, H. S., Zhang, T., & Yang, F. S. (2002). A multistage, multimethod approach for automatic detection and classification of epileptiform EEG. *IEEE Trans Biomed Eng, 49*(12 Pt 2), 1557-1566.

Lopes da Silva, F. H., A., D., & H., S. (Eds.). (1975). *Detection of nonstationarities in EEGs using the autoregressive model – an application to EEGs of epileptics* (Kunkel, H. Dolce, G. ed.). Stuttgart: Gustav Fischer Verlag.

Malarvili, M. B., Hassanpour, H., Mesbah, M., & Boashash, B. (2005). *A Histogram-Based Electroencephalogram Spike Detection.* Paper presented at the Proc. 8th Intern. Symp. on Sig.l Proc. & Its Applic.

Martinez-Vargas, J. D., Avendano-Valencia, L. D., Giraldo, E., & Castellanos-Dominguez, G. (2011). *Comparative analysis of Time Frequency Representations for discrimination of epileptic activity in EEG Signals.* Paper presented at the Conf Proc of the 5th International IEEE EMBS Conference on Neural Engineering.

McGrogan, N. (2001). *Neural Network Detection of Epileptic Seizures in the Electroencephalogram.* Oxford University, Oxford.

McSharry, P. E., He, T., Smith, L. A., & Tarassenko, L. (2002). Linear and non-linear methods for automatic seizure detection in scalp electro-encephalogram recordings. *Med Biol Eng Comput, 40*(4), 447-461.

McSharry, P. E., Smith, L. A., & Tarassenko, L. (2003). Prediction of epileptic seizures: are nonlinear methods relevant? *Nat Med, 9*(3), 241-242; author reply 242.

Michel, C. M., Seeck, M., & Landis, T. (1999). Spatiotemporal Dynamics of Human Cognition. *News Physiol Sci, 14*, 206-214.

Miwakeichi, F., Martinez-Montes, E., Valdes-Sosa, P. A., Nishiyama, N., Mizuhara, H., & Yamaguchi, Y. (2004). Decomposing EEG data into space-time-frequency components using Parallel Factor Analysis. *Neuroimage, 22*(3), 1035-1045.

Mormann, F., Andrzejak, R. G., Elger, C. E., & Lehnertz, K. (2007). Seizure prediction: the long and winding road. *Brain, 130*(Pt 2), 314-333.

Mosher, J. C., Baillet, S., & Leahy, R. M. (1999). EEG source localization and imaging using multiple signal classification approaches. *J Clin Neurophysiol, 16*(3), 225-238.

Mosher, J. C., & Leahy, R. M. (1998). Recursive MUSIC: a framework for EEG and MEG source localization. *IEEE Trans Biomed Eng, 45*(11), 1342-1354.

Mosher, J. C., Leahy, R. M., & Lewis, P. S. (1999). EEG and MEG: forward solutions for inverse methods. *IEEE Trans Biomed Eng, 46*(3), 245-259.

Mousavi, S. R., Niknazar, M., & Vahdat, B. V. (2008). *Epileptic seizure detection using AR model on EEG signals.* Paper presented at the Conf Proceedings of Cairo International Biomedical Engineering Conference (CIBEC '08), Cairo, Egypt.

Mukhopadhyay, S., & Ray, G. C. (1998). A new interpretation of nonlinear energy operator and its efficacy in spike detection. *IEEE Trans Biomed Eng, 45*(2), 180-187.

Murro, A. M., King, D. W., Smith, J. R., Gallagher, B. B., Flanigin, H. F., & Meador, K. (1991). Computerized seizure detection of complex partial seizures. *Electroencephalogr Clin Neurophysiol, 79*(4), 330-333.

Naghsh-Nilchi, A. R., & Aghashahi, M. (2010). Epilepsy seizure detection using eigen-system spectral estimation and Multiple Layer Perceptron neural network. *Biomedical Signal Processing and Control 5* 147–157.

Nigam, V. P., & Graupe, D. (2004). A neural-network-based detection of epilepsy. *Neurol Res, 26*(1), 55-60.

Nishida, S., Nakamura, M., Ikeda, A., & Shibasaki, H. (1999). Signal separation of background EEG and spike by using morphological filter. *Med Eng Phys, 21*(9), 601-608.

Ocak, H. (2008). Optimal classification of epileptic seizures in EEG using wavelet analysis and genetic algorithm. *Signal Processing, 88*(7), 1858–1867.

Ocak, H. (2009). Automatic detection of epileptic seizures in EEG using discrete wavelet transform and approximate entropy. *Expert Systems with Applications, 36*(2), 2027–2036.

Oikonomou, V. P., Tzallas, A. T., & Fotiadis, D. I. (2007). A Kalman filter based methodology for EEG spike enhancement. *Comput Methods Programs Biomed, 85*(2), 101-108.

Orhan, U., Hekim, M., & Ozer, M. (2011). EEG signals classification using the K-means clustering and a multilayer perceptron neural network model. *Expert Systems with Applications* (38), 13475–13481.

Ossadtchi, A., Baillet, S., Mosher, J. C., Thyerlei, D., Sutherling, W., & Leahy, R. M. (2004). Automated interictal spike detection and source localization in magnetoencephalography using independent components analysis and spatio-temporal clustering. *Clin Neurophysiol, 115*(3), 508-522.

Ozdamar, O., & Kalayci, T. (1998). Detection of spikes with artificial neural networks using raw EEG. *Comput Biomed Res, 31*(2), 122-142.

Ozdamar, O., Yaylali, I., Jayakar, P., & Lopez, C. (1991). *Multilevel neural network system for EEG spike detection.* Paper presented at the Conf Proc 4th IEEE symp IEEE Computer society press Washington.

Pang, C. C., Upton, A. R., Shine, G., & Kamath, M. V. (2003). A comparison of algorithms for detection of spikes in the electroencephalogram. *IEEE Trans Biomed Eng, 50*(4), 521-526.

Parra, L. C., Spence, C. D., Gerson, A. D., & Sajda, P. (2005). Recipes for the linear analysis of EEG. *Neuroimage, 28*(2), 326-341.

Polat, K., & Gunes, S. (2007). Classification of epileptiform EEG using a hybrid system based on decision tree classifier and fast Fourier transform. *Applied Mathematics and Computation, 187*(2), 1017–1026.

Polat, K., & Gunes, S. (2008a). Artificial immune recognition system with fuzzy resource allocation mechanism classifier, principal component analysis and FFT method based new hybrid automated identification system for classification of EEG signals. *Expert Systems with Applications, 34*(3), 2039-2048.

Polat, K., & Gunes, S. (2008b). A novel data reduction method: Distance based data reduction and its application to classification of epileptiform EEG signals. *Applied Mathematics and Computation, 200*(1), 10-27.

Pon, L. S., Sun, M., & Robert, J. S. (2002). *The bi-directional spike detection in EEG using mathematical morphology and wavelet transform.* Paper presented at the 6th International Conference on Signal Processing.

Sadati, N., Mohseni, H. R., & Magshoudi, A. (2006, 16-21 July). *Epileptic Seizure Detection Using Neural Fuzzy Networks.* Paper presented at the Proc. of the IEEE Intern. Conf. on Fuzzy Syst., Canada.

Sankar, R., & Natour, J. (1992). Automatic computer analysis of transients in EEG. *Comput Biol Med, 22*(6), 407-422.

Schiff, S. J., Aldroubi, A., Unser, M., & Sato, S. (1994). Fast wavelet transformation of EEG. *Electroencephalogr Clin Neurophysiol, 91*(6), 442-455.

Schiff, S. J., Colella, D., Jacyna, G. M., Hughes, E., Creekmore, J. W., Marshall, A., et al. (2000). Brain chirps: spectrographic signatures of epileptic seizures. *Clin Neurophysiol, 111*(6), 953-958.

Senhadji, L., Dillenseger, J. L., Wendling, F., Rocha, C., & Kinie, A. (1995). Wavelet analysis of EEG for three-dimensional mapping of epileptic events. *Ann Biomed Eng, 23*(5), 543-552.

Senhadji, L., & Wendling, F. (2002). Epileptic transient detection: wavelets and time-frequency approaches. *Neurophysiol Clin, 32*(3), 175-192.

Shen, T. W., Kuo, X., & Hsin, Y. L. (2009). *Ant K-Means Clustering Method on Epileptic Spike Detection.* Paper presented at the Fifth International Conference on Natural Computation.

Sommer, D., & Golz, M. (2001). *Clustering EEG-segments using hierarchical agglomerative methods and self-organising maps.* Paper presented at the Annual conference of the European Neural Network Society.

Srinivasan, V., Eswaran, C., & Sriraam, N. (2005). Artificial neural network based epileptic detection using time-domain and frequency-domain features. *J Med Syst, 29*(6), 647-660.

Srinivasan, V., Eswaran, C., & Sriraam, N. (2007). Approximate entropy-based epileptic EEG detection using artificial neural networks. *IEEE Trans Inf Technol Biomed, 11*(3), 288-295.

Subasi, A. (2007a). Application of adaptive neuro-fuzzy inference system for epileptic seizure detection using wavelet feature extraction. *Comput Biol Med, 37*(2), 227-244.

Subasi, A. (2007b). Signal Classification using wavelet feature extraction and a mixture of expert model. *Expert Systems with Applications, 32*(4), 1084-1093.

Subasi, A., Alkan, A., Koklukaya, E., & Kiymik, M. K. (2005). Wavelet neural network classification of EEG signals by using AR model with MLE preprocessing. *Neural Netw, 18*(7), 985-997.

Subasi, A., & Gursoy, I. (2010). EEG signal classification using PCA, ICA, LDA and support vector machines. *Expert Systems with Applications 37*, 8659–8666.

Tzallas, A. T., Karvelis, P. S., Katsis, C. D., Fotiadis, D. I., Giannopoulos, S., & Konitsiotis, S. (2006). A method for classification of transient events in EEG recordings: application to epilepsy diagnosis. *Methods Inf Med, 45*(6), 610-621.

Tzallas, A. T., Katsis, C. D., Karvelis, P. S., Fotiadis, D. I., Giannopoulos, S., & Konitsiotis, S. (2005, 20-25 November). *Classification of Transient Events in EEG recordings using Support Vector Machines* Paper presented at the Conf Proc of the 3rd European Medical & Biological Engineering Conference, Prague.

Tzallas, A. T., Oikonomou, V. P., & Fotiadis, D. I. (2006). Epileptic spike detection using a Kalman filter based approach. *Conf Proc IEEE Eng Med Biol Soc, 1*, 501-504.

Tzallas, A. T., Tsipouras, M. G., & Fotiadis, D. I. (2007a). Automatic seizure detection based on time-frequency analysis and artificial neural networks. *Comput Intell Neurosci*, 80510.

Tzallas, A. T., Tsipouras, M. G., & Fotiadis, D. I. (2007b). The use of time-frequency distributions for epileptic seizure detection in EEG recordings. *Conf Proc IEEE Eng Med Biol Soc, 2007*, 3-6.

Tzallas, A. T., Tsipouras, M. G., & Fotiadis, D. I. (2009). Epileptic seizure detection in EEGs using time-frequency analysis. *IEEE Trans Inf Technol Biomed, 13*(5), 703-710.

Ubeyli, E. D. (2006). Analysis fo EEG signals using Luapunov exponents. *Neural Network World, 16*(3), 257-273.

Ubeyli, E. D. (2007). Modified mixture of experts for analysis of EEG signals. *Conf Proc IEEE Eng Med Biol Soc, 2007*, 1546-1549.

Ubeyli, E. D. (2008a). Analysis of EEG signals by combining eigenvector method and multiclass support vector machines. *Comput Biol Med, 38*(1), 14-22.

Ubeyli, E. D. (2008b). Implementing eigenvector methods/probabilistic neural networks for analysis of EEG signals. *Neural Netw, 21*(9), 1410-1417.

Ubeyli, E. D. (2008c). Wavelet/mixture of experts network structure for EEG classification. *Expert Systems with Applications, 37*, 1954-1962.

Ubeyli, E. D. (2009a). Analysis of EEG signals by implementing eigenvector methods/recurrent neural networks. *Digital Signal Processing 1, 9* 134–143.

Ubeyli, E. D. (2009b). Combined neural network model employing wavelet coefficients for EEG signals classification. *Digital Signal Processing, 19*, 297-308.

Ubeyli, E. D. (2009c). Probabilistic neural networks combined with wavelet coefficients for analysis of EEG signals. *Expert systems, 26*(2), 147-159.

Ubeyli, E. D. (2010a). Least squares support vector machine employing model-based methods coefficients for analysis of EEG signals. *Expert Systems with Applications, 37*, 233–239.

Ubeyli, E. D. (2010b). Lyapunov exponents/probabilistic neural networks for analysis of EEG signals. *Expert Systems with Applications 37* 985–992.

Ubeyli, E. D., & Guler, I. (2007). Features extracted by eigenvector methods for detection variability of EEG signals. *Pattern Recognition Letters, 28*(5), 592-603.

Valenti, P., Cazamajou, E., Scarpettini, M., Aizemberg, A., Silva, W., & Kochen, S. (2006). Automatic detection of interictal spikes using data mining models. *J Neurosci Methods, 150*(1), 105-110.

van Putten, M. J. (2003). Nearest neighbor phase synchronization as a measure to detect seizure activity from scalp EEG recordings. *J Clin Neurophysiol, 20*(5), 320-325.

Wahlberg, P., & Lantz, G. (2000). Methods for robust clustering of epileptic EEG spikes. *IEEE Trans Biomed Eng, 47*(7), 857-868.

Wahlberg, P., & Salomonsson, G. (1996). Feature extraction and clustering of EEG epileptic spikes. *Comput Biomed Res, 29*(5), 382-394.

Wang, D., Miao, D., & Xie, C. (2011). Best basis-based wavelet packet entropy feature extraction and hierarchical EEG classification for epileptic detection. *Expert Systems with Applications*(8), 14314–14320.

Waterhouse, E. (2003). New horizons in ambulatory electroencephalography. *IEEE Eng Med Biol Mag, 22*(3), 74-80.

Webber, W. R., Litt, B., Wilson, K., & Lesser, R. P. (1994). Practical detection of epileptiform discharges (EDs) in the EEG using an artificial neural network: a comparison of raw and parameterized EEG data. *Electroencephalogr Clin Neurophysiol, 91*(3), 194-204.

Wilson, S. B., & Emerson, R. (2002). Spike detection: a review and comparison of algorithms. *Clin Neurophysiol, 113*(12), 1873-1881.

Xu, G., Wang, J., Zhang, Q., Zhang, S., & Zhu, J. (2007). A spike detection method in EEG based on improved morphological filter. *Comput Biol Med, 37*(11), 1647-1652.

Zhukov, L., Weinstein, D., & Johnson, C. (2000). Independent component analysis for EEG source localization. *IEEE Eng Med Biol Mag, 19*(3), 87-96.

5

Automated Non-Invasive Identification and Localization of Focal Epileptic Activity by Exploiting Information Derived from Surface EEG Recordings

Amir Geva[1], Merav Ben-Asher[1], Dan Kerem[2], Mayer Aladjem[1] and Alon Friedman[3]

[1]*Dept. Electrical Engineering, Ben Gurion University in the Negev, Be'er Sheva*
[2]*School for Marine Sciences, University of Haifa, Haifa*
[3]*Departments of Physiology and Neurobiology, Faculty of Health Sciences, Ben Gurion University in the Negev, Be'er Sheva, Israel*

1. Introduction

Non-invasive epileptic source localization techniques, relying on interictal activity in the EEG and mainly motivated by the need for presurgical localization of the epileptic focus in intractable focal epilepsy, have been widely studied and validated (Blume et al., 2001; Ebersole, 2000; Gavaret et al., 2009; Lantz et al., 2003; Michel *et al.*, 2004; Zumsteg et al., 2005, 2006). Source localization methods most often utilize electric source imaging (ESI), a technique which applies various mathematical algorithms and constraining neurophysiological hypotheses on scalp electroencephalographic (EEG) recordings in order to reconstruct the three dimensional brain electrical activity (reviewed in Michel et al., 2004). Correlation of such modeling with actual intra-cerebral recordings is not always straightforward and some focal deep sources do not seem to have surface representation sufficient for stable modeling (Merlet & Gotman, 2001). Advantage may also be taken of extracranial magnetic fields (MEG) which like scalp electric potentials are due to current density distribution that arises from post-synaptic processes of underlying neuronal populations. The MEG forms the basis for the various versions of low resolution brain electromagnetic tomography (LORETA), which have been advanced to tackle the solution of the inverse problem (the 3D reconstruction of electric neuronal activity based on extracranial measurements) (Pascual-Marqui, 2002).

Metabolic imaging investigations are considered not suitable to study the temporal dynamics of seizures due to poor temporal resolution. Yet, a new and rapidly developing non-invasive imaging technique of EEG-correlated functional magnetic resonance imaging (EEG-fMRI) which maps regional changes in cerebral oxygenation and blood flow that are time-locked to interictal epileptiform discharges (IED) identified on the simultaneously recorded EEG, can identify metabolic activations correlated to IEDs (Vulliemoz et al., 2009).

Due to its high temporal resolution, EEG is the most frequently used non-invasive method for the diagnosis of epilepsy and for localizing its source. However, the cortical potentials are severely attenuated by bone and soft tissues as they spread to the scalp, requiring amplification that facilitates contamination by muscular and external potentials with a low signal-to-noise ratio. The scalp recording is hence a mixture of neural signals, artifacts and noise, causing the identification of epileptic activity and its origin to be a challenging quest, even for experienced clinicians. Thus, there is still a great need for efficient signal processing methods in order to clean the signal and localize its neural sources.

During the last decade, blind source separation (BSS) has been successfully used for various EEG-processing applications, including artifact rejection (Bouzida et al., 2005; Frank & Frishko, 2007; James & Gibson, 2003; Shoker et al., 2005; Xue et al., 2006), discrimination of mental tasks in healthy individuals (Chiappa & Barber, 2005; Liu et al., 2004; Makeig et al., 2004; Sutherlanda & Tanga, 2006), localization of abnormal brain signals (Latif & Sanei, 2005; Richards, 2004; Zhukov et al., 2000) and diagnosis of Alzheimer's disease (Cichocki et al., 2005).

Several attempts have been made to extract epileptic signals using BSS. It has been demonstrated that removal of artifacts using BSS improves seizure detection (Liu et al., 2004). While many studies addressed the problem of seizure recognition, identifying the epileptic source out of separated components was in most cases done subjectively by visual inspection (Hesse & James, 2005; James & Lowe, 2000; Kobayashi et al., 1999). Other workers, e.g. Jing & Sanei (2007), employed clinical information concerning the location of the epileptic focus and prior assumptions regarding its frequency. Faul and colleagues (2005) and Jarchi and colleagues (2009) developed a method for seizure source localization based on a complexity measure and singular value fraction. These studies relied on the different temporal characteristics of normal EEG activity and pre-ictal/ictal signals, the latter consisting of slower, rhythmic activity, spikes or spike and wave complexes. The flexibility of fuzzy clustering algorithms (Bezdek, 1981) enables them to uncover hidden states without forcing a priori constraints on cluster features. This has been used by researchers to identify brain states based on EEG activity, following previous feature extraction. Geva and Kerem (1998) used a fuzzy clustering algorithm (Gath & Geva, 1989) following wavelets analysis on pre-ictal recordings to classify brain states and forecast epileptic seizures. Chang and Po (2005) used wavelet analysis followed by fuzzy k-means clustering to track brain states of meditation from EEG recordings. Others had since used a similar clustering procedure for classifying polymorphic seizures according to their similarity in feature space (Meier et al., 2008).

With the rising popularity of BSS for EEG analysis, its ability to extract features for further clustering was highlighted. In fact, BSS is able to distinguish between overlapping clusters, a feat not possible otherwise. Hallez and colleagues (2007) have shown that the use of BSS to filter noise and artifacts prior to dipole estimation and clustering of the dipoles improved the localization. Makeig et al (2004) clustered scalp maps and activity power spectra of independent components drawn from EEGs of subjects performing visual tasks. They were able to identify clusters with functionally distinct activity patterns such as eye movements and muscle artifacts.

Since the epileptic activity patterns are likely to change over time, attempting to identify them by their temporal/spectral structures alone may prove problematic. Imposing constraints on the nature of the epileptic activity may also reduce the likelihood of identifying it. Thus, there is a call for an unsupervised method that will blindly extract and localize the epileptic signals. Since in patients with partial/focal epilepsy there is a single (or a few) focus (i) which triggers an epileptic seizure, it was decided to search for sources with a similar spatial distribution on the scalp, but not necessarily with similar temporal characteristics. For this purpose, a method is here proposed which combines BSS and cluster analysis procedures. The BSS procedure is used to separate the temporal components of the epileptic source from noise and background neural activity. Then a cluster analysis procedure is implemented on topographic maps of the BSS components. The motivation for the latter is the fact that focal epileptic activity tends to originate from the same small brain region (or neuronal population), which is expected to produce a cluster of spatially restricted (not scattered) maps of the epileptic BSS components on the scalp. Even if and when the pre-ictal activity is projected from the source to other distinct and restricted brain locations such as the contra-lateral lobe (Baumgartner et al., 1995), it is still expected that member maps of the 'epileptic-source cluster' would be the least expanded. The method is tested on twelve patients with source-localized focal epilepsy who were video-monitored as part of their pre-surgical evaluation. The obtained results show that the method is a promising tool for accurate detection of focal activity on the scalp.

In Section 2, the method is derived and in Section 3 results of its application on the EEG of epileptic patients are validated against the clinical source localization. Conclusions are provided in Section 4.

2. A method for scalp localization of the epileptic zone

In Section 2.1, the data acquisition and preprocessing steps are described then in Section 2.2, topological maps of the BSS components are constructed for EEG time windows, which are likely to contain information on the epileptic focus. In Section 2.3, a clustering procedure of the maps is developed and finally, by employing general clinical knowledge on an epileptic focus, a procedure is derived for the identification and localization of the latter in a given patient.

2.1 Data acquisition and EEG preprocessing

Data acquisition and noise reduction preprocessing of the EEG signals have crucial influence on the efficiency of the BSS and cluster analysis procedures. From each of 12 patients several (2-7) EEG records of variable lengths (0.8-35.5 min) were obtained, up to but excluding the seizure (see Table 1). EEG was recorded from 27 electrodes placed according to the 10-20 system, using a 128-channel digital acquisition unit (Bio-Logic, USA). The sampling rate was set to 256Hz and an input bandpass filter with the range of 0.3-100Hz was applied. All electrodes were referenced to a common electrode (FCz).

EEG recordings from the scalp are severely contaminated with noise and artifacts, since the high resistance of the skull attenuates the neural signals, resulting in low signal-to-noise ratios. Moreover, there are many potential localized non-neural sources of electrical activity on the scalp, such as eye, eyelid, tongue or jaw movements, scalp muscle activity and cardiac activity, that produce scalp potentials as large as, or even larger than the EEG in general and the epileptic activity in particular (Nunez & Srinivasan, 2006). At very low frequencies, muscle movement artifacts mainly overpower the neurological signals. Since epileptic activity is known to occur at relatively high frequencies, a 1.5-40 Hz finite impulse response bandpass filter (Proakis & Manolakis, 2007) was also applied on the EEG signals.

Brain activity is constantly changing, resulting in changes of the statistical properties of the EEG signal (Sanei & Chambers, 2007). In order to attain signal stationarity which is required by the BSS methods and in order to minimize loss of information, the EEG processing was replicated on 50% overlapping windows (Delorme & Makeig, 2004) of 60 seconds duration, a length found suitable in a preliminary study.

Fig. 1 presents pre-processed EEG during the final five seconds of a 60 s window of Patient #2 (Tables 1 and 2). Pre-seizure epileptic activity is observed in temporo-occipital electrodes bilaterally. The suspected epileptic focus in this patient was the left temporal lobe (predominantly recorded in electrode T5). This was also validated using 64 electrode sub-dural grids and sub temporal strip (8x2) electrodes, implanted prior to resection of the temporal lobe.

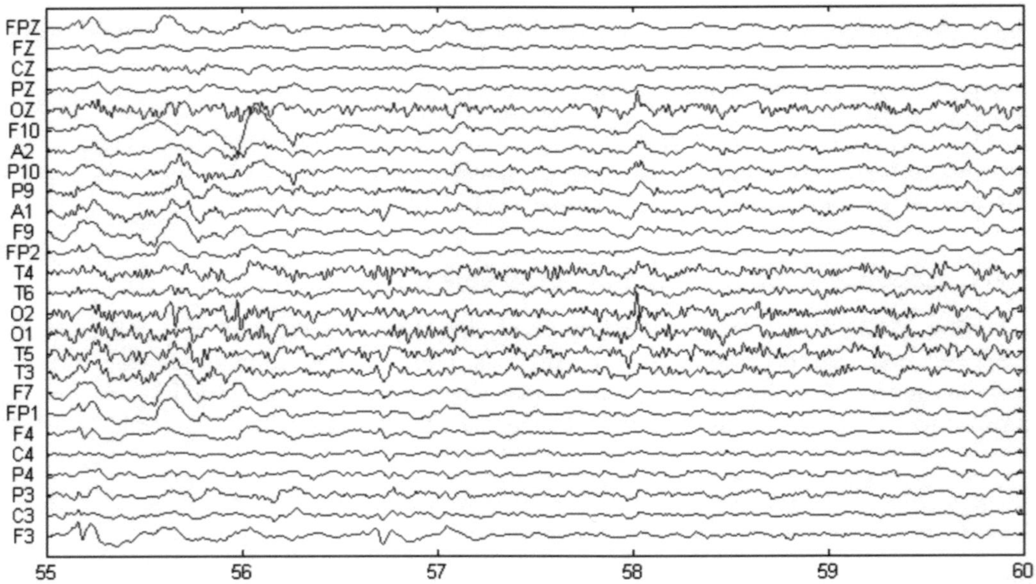

Fig. 1. Preprocessed EEG signals for patient # 2.

2.2 Topological maps of BSS components

In the first stage of the presented method, BSS was employed for the reconstruction of neural (true EEG) and extra-neural components in the filtered electrical recordings. The input for the BSS procedure was the pooled EEG recordings (all pre-seizure records) of each patient. As mentioned above, pre-ictal and inter-ictal scalp recordings are a mixture of normal neural activity, epileptic neural activity, non-neural potentials (e.g. scalp and eye muscles) and external artifacts. Using the linear BSS method, it was assumed that the measured activity is a weighted sum of multiple independent sources. Delorme & Makeig (2004) have shown that the linear model is a good approximation of the source mixing into the EEG recordings. In addition, it was concluded that the electrical fields of the recorded EEG are practically independent because neural and extra-neural activities are anatomically and physiologically separate processes (Hyvärinen & Oja, 2000), which motivated the use of independent component techniques (Hyvärinen et al., 2001) for BSS in this work.

For the present case, the linear BSS model can be expressed as follows in matrix notations:

$$\hat{s}(t) = \boldsymbol{W}\boldsymbol{x}(t), \tag{1}$$

where vector $\boldsymbol{x}(t)$ comprises the recorded EEG signals $\{x_j(t)\}_{j=1}^{n}$ from each of n electrodes, (n=27 in this application) of a given 60 sec time window of a given patient, vector $\hat{\boldsymbol{s}}(t)$ with elements $\{\hat{s}_i(t)\}_{i=1}^{n}$, which gives the restorations of the separated sources and $\boldsymbol{W}$ is a demixing matrix computed by a BSS procedure. After trying out in a preliminary study the most popular linear BSS procedures (Bell & Sejnowski, 1995; Belouchrani et al., 1997; Cardoso, 1999; Hyvärinen, 1999; Ziehe & Muller,1998; Ziehe et al., 2004) for their ability to localize the epileptic foci by our method, Infomax (Bell & Sejnowski, 1995) was chosen as being the most suitable. The Infomax method was used to compute $\boldsymbol{W}$ for each 60 s time window and then a topographic map of the restorations $\{\hat{s}_i(t)\}_{i=1}^{n}$ (1) was created by a back-projecting technique (Makeig et al., 1997):

$$\mathbf{x}_{bp}^{i}(t) = \left(\mathbf{W}_i^{(-1)}\right)^T \hat{s}_i(t), \tag{2}$$

where T denotes transpose operation, $\boldsymbol{W}_i^{(-1)}$ is the ith column of $\boldsymbol{W}^{-1}$ being the inverse of the demixing matrix $\boldsymbol{W}$ and $\mathbf{x}_{bp}^{i}(t)$ is a back-projected vector comprising activities $\mathbf{x}_{bpj}^{i}(t)$ at locations j=1,2,...,n on the scalp implied by the restoration $\hat{s}_i(t)$.

Following Jung et al (2001), for each EEG time window and for each $\hat{s}_i(t)$, a unique time instance is computed:

$$t_i{}^* = \arg\max_t \hat{V}ar_{bp}^{i}(t), \tag{3}$$

$$\hat{V}ar_{bp}^{i}(t) = \frac{1}{n}\sum_{j1=1}^{n}\left(x_{bp_{j1}}^{i}(t) - \frac{1}{n}\sum_{j2=1}^{n} x_{bp_{j2}}^{i}(t)\right)^2, \tag{4}$$

where $\hat{V}ar_{bp}^{i}(t)$ denotes sample variance of the back projected activities $\mathbf{x}_{bpj}^{i}(t)$ $j = 1,2,...,n$. Following this, a vector is constructed:

$$B_{bp}^{i} = \left[x_{bp_1}^{i}(t_i{}^*),\ldots,x_{bpn}^{i}(t_i{}^*)\right]^T \tag{5}$$

comprising the back projected activities at time $t_i{}^*$ for each EEG time window and each $\hat{s}_i(t)$.

The topographic maps can be conceptually grasped as depicting the spatial manifestation of a given separated source. Inter-ictal epileptic activity in focal epilepsy is thought to be restricted to a small localized brain volume. Therefore, it is assumed that the back projected quantities $\{x_{bpj}^{i}(t_i{}^*)\}_{j=1}^{n}$ at the locations $j = 1,2,...,n$ on the scalp are not likely to represent epileptic activity if 'hot' regions are relatively widely distributed. Hence, $Pe(i)$ is defined:

$$Pe(i) = \frac{\max_j \left(x_{bpj}^i (t^*) \right)}{\sum_{j=1}^{n} x_{bpj}^i (t^*)} \times 100 \tag{6}$$

as a measure of the likelihood of B_{bp}^i to represent epileptic activity and it is proposed to define and reject B_{bp}^i (5) as being non epileptic, if $Pe(i)$ <10%. For illustration, in Figure 2a the restorations $\{\hat{s}_i(t)\}_{i=1}^n$ for the $\{x_j(t)\}_{j=1}^n$ shown in Figure 1 are presented. An expert EEG analyst would recognize epileptic-like activity in $\hat{s}_i(t)$ restorations: i = 5-8, 13-15 and 19, most of which indeed display highly non uniform heat maps of the values of $\{x_{bpj}^i(t_i^*)\}_{j=1}^n$ (5) (Figure 2b). The latter are reflected by large values of their $Pe(i)$ (6) (in the range of 15-61%), which indicate a strong orientation to a specific location on the scalp. It is again noteworthy that such orientation in this patient involves diverse scalp locations (temporal, and occipital, bilaterally). It should also be noted that localized heat maps and high $Pe(i)$ s can also be seen in restorations such as i = 17 and 21, that are seemingly not associated with epileptic activity. The latter may be artifacts or other brain signals that are not expected to reoccur systematically and with a fixed spatial representation throughout the pre-ictal period, and as such will not be grouped by the clustering of the BSS topographic maps, described in the next section.

2.3 Clustering of the BSS topographic maps

After computing B_{bp}^i s (5) for all overlapping 60 s windows obtained from several pre-seizure time interval recordings for a single patient, all B_{bp}^i s having $Pe(i) > 10\%$ are collated as being potentially epileptic. Similar B_{bp}^i s are then grouped by a cluster analysis method. In preliminary trials, an attractive accuracy in focus localization was found by using unsupervised fuzzy clustering analysis, employing a correlation distance (Gath & Geva, 1989):

$$d\left(B_{bp}^*, B_{bp}^{**}\right) = 1 - \frac{B_{bp}^{*\,T} B_{bp}^{**}}{\left\|B_{bp}^*\right\| \left\|B_{bp}^{**}\right\|} \tag{7}$$

where B_{bp}^* and B_{bp}^{**} denote arbitrary vectors of B_{bp}^i (5) and $\|\bullet\|$ is the L2 norm of the vectors. The efficacy of $d\left(B_{bp}^*, B_{bp}^{**}\right)$ in this case may stem from its invariance to rotation and dilation (Duda et al., 2000), which seems to be critical for the grouping of epileptic B_{bp}^i s.

The cluster analysis procedure results in K groups of B_{bp}^i s (K being set by a validation criterion explained in Gath & Geva (1989)), for each of which the following statistics are computed:

1. The averaged sample variance

$$\overline{V}^{(k)} = \frac{1}{n} trace\left(\hat{\Sigma}_{bp}^{(k)}\right), \ k = 1,2,...,K \tag{8}$$

where $\hat{\Sigma}_{bp}^{(k)}$ is the sample covariance matrix of B_{bp}^{i}'s in group k.

2. The average $Pe(i)$ value (6) of B_{bp}^{i} s in group k, denoted by $\overline{Pe}^{(k)}$, $k = 1,...,K$.

The values of these two statistics for m members of each group are arranged in descending order, after which $Score\overline{V}^{(k)}$ and $Score\overline{Pe}^{(k)}$ are defined, assigning score 1 for the largest value and K for the smallest.

The two scores are joined in an index:

$$I_{e.a}^{(k)} = \frac{K + Score\overline{Pe}^{(k)} - Score\overline{V}^{(k)}}{2(K-1)+1} \tag{9}$$

having a range of $0 \leq I_{e.a}^{(k)} \leq 1$, which is normalized for each patient:

$$\tilde{I}_{e.a}^{(k)} = \frac{I_{e.a}^{(k)} - I_{e.a}^{\min}}{I_{e.a}^{\max} - I_{e.a}^{\min}} \tag{10}$$

with $I_{e.a}^{\min} = \min\limits_{k}\left\{I_{e.a}^{(k)}\right\}$, $I_{e.a}^{\max} = \max\limits_{k}\left\{I_{e.a}^{(k)}\right\}$. The normalized index $\tilde{I}_{e.a}^{(k)}$ has similar extrimum values of 0 and 1 for all patients. The expression $I_{e.a}^{(k)}$ (9) reflects the accepted medical knowledge for the pre-ictal activity. For a group (cluster) of B_{bp}^{i} s which is most likely to represent epileptic activity (henceforth: 'epileptic group'), $I_{e.a}^{(k)}$ tends to have a small value, implying a low value of $\overline{V}^{(k)}$ (8) and/or a high value of $\overline{Pe}^{(k)}$.

The means of the B_{bp}^{i} s (5) within the groups are then computed, which are denoted by:

$$\overline{B}_{bp}^{(k)} = \left[\overline{x}_{bp_1}^{(k)}, ..., \overline{x}_{bp}^{(k)}\right]^T, \ k = 1,2,...,K . \tag{11}$$

Finally, a number $J^{(k)}$ is calculated:

$$J^{(k)} = \arg\max_{j}\left\{\overline{x}_{bpj}^{(k)}\right\}_{j=1}^{n}, \tag{12}$$

being the EEG scalp electrode number (1-27) having the highest mean activity value $\overline{x}_{bpj}^{(k)}$ (5) within group k. The number $J^{(k)}$ for the 'epileptic group' is likely to indicate the EEG electrode with the closest orientation to the epileptic brain focus and thus, the most appropriate to represent its reflection upon the surface of the scalp.

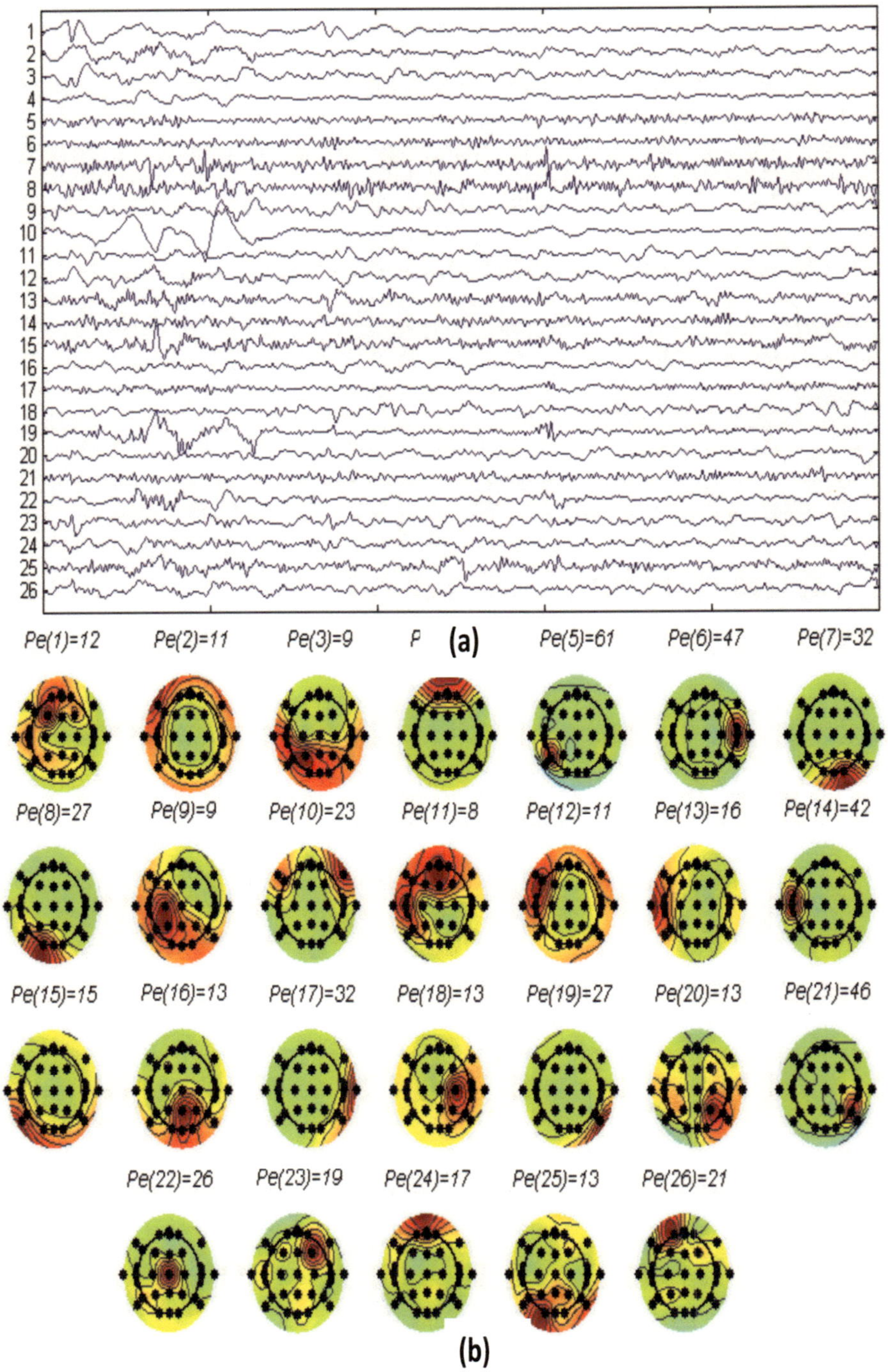

Fig. 2. BSS results for patient # 2. (a) BSS restorations $\hat{s}_i(t)$, $i = 1,...,n$ obtained from $\{x_j(t)\}_{j=1}^{n}$ in Figure 1. (b) Heat maps of the values $x_{bpj}^{i}(t_i{}^*)$ (5) at points $j = 1,2,...,n$ on the scalp. $Pe(i)$ values (6) are shown above each map.

The model's results for patient #2 are shown in Figure 3. Figure 3a presents the heat maps of $\overline{B}_{bp}^{(k)}$ (11) for the three of seven groups identified by the cluster analysis, having the smallest values of the index $I_{e.a}^{(k)}$ (10), given in parentheses over the maps. They are reminiscent of restorations 5,6 and 7 of Figure 2, the latter being members in clusters K=1, 2 and 3, respectively. For completeness, Figures 3b,c present the values of $\overline{V}^{(k)}$ and $\overline{Pe}^{(k)}$, respectively, for all seven groups. The first group (K=1), with $I_{e.a}^{(k)}$=0, is the "epileptic group". The number $J^{(1)}$ computed by (12) suggests that the epileptic focus is mostly reflected in electrode T5. Groups 2 and 3 are projections to the contra-lateral hemisphere, which interestingly scored higher than the map centered over T3, yet trailed the 'epileptic group' considerably. Non-epileptic restorations in this patient (e.g. numbers 17 and 21 in Figure 2) either did not form clusters or formed clusters with high $I_{e.a}^{(k)}$ values.

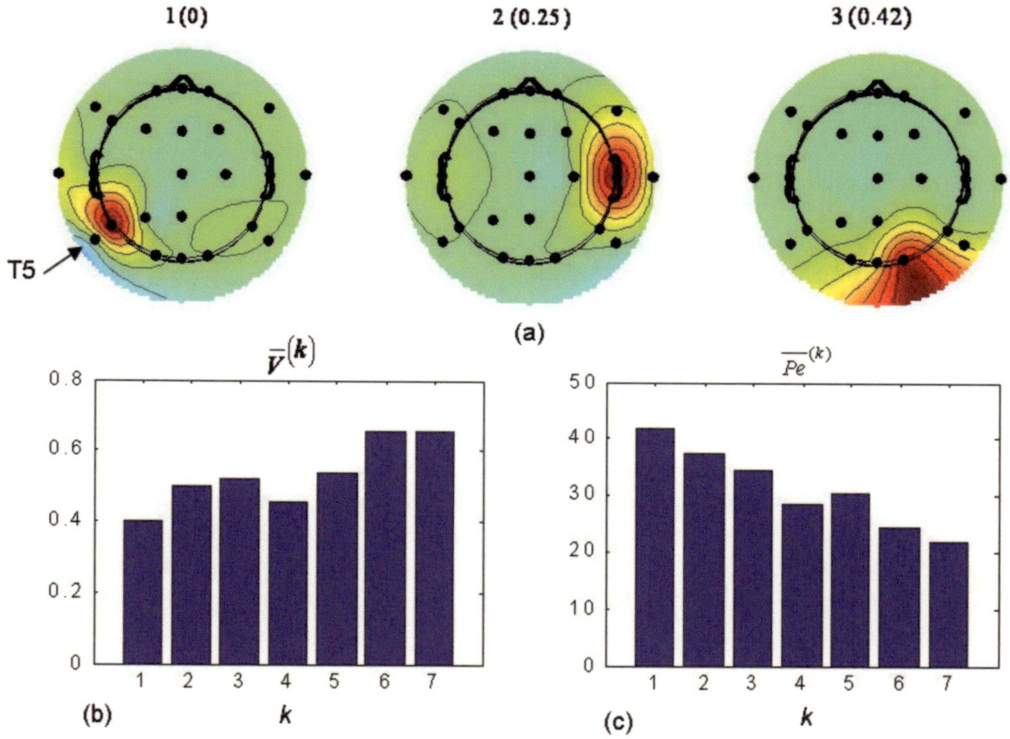

Fig. 3. Localization of the epileptic focus for patient #2. (a) Heat maps of $\overline{B}_{bp}^{(k)}$ for the three groups having the smallest values of $\tilde{I}_{e.a}^{(k)}$ (10) (shown in parentheses above maps); (b) and (c) The averaged sample variance $\overline{V}^{(k)}$ (8) and the average $Pe(i)$ value (6) of B_{bp}^{i} s in group k, $\overline{Pe}^{(k)}$, respectively, for all seven groups identified for this patient. Map 1 is that of the 'epileptic group' and location T5 (arrow) came out as closest to the source.

3. Experimental results

The results obtained by the proposed method on 12 patients are now presented. Table 1 lists the relevant clinical information.

Patient №	Age (y)	Gender	Age at disease onset (y)	Seizure frequency*	Validation method for epileptic focus localization	Number of seizures analyzed	Range (mean) of pre-ictal period (minutes)
1	17	F	9	6/m	IR	3	4.5-6.4 (5.3)
2	8	M	3	2/w	IR	5	7.4-35.3 (17.5)
3	8.5	M	7	15/d	MRI	2	3.8-9.8 (6.8)
4	12	F	1	10/d	IR	2	10-10.2 (10.1)
5	16	F	6	1/d	MRI	5	0.8-5 (2.3)
6	20	M	0.7	2/d	MRI, IR	2	2.2-9.3 (5.7)
7	26	F	19	2/d	MRI	7	5.7-12.9 (7.8)
8	19	F	6	2/d	SPECT	2	6.7-9.6 (8.2)
9	54	F	0.5	3 /w	MRI	5	5.5-26.6 (10.7)
10	22	F	2	6/w	MRI, IR	3	5.1-13.8 (9.8)
11	34	M	4	2/d	MRI, IR	7	2-5.5 (3.6)
12	54	M	50	2/w	MRI	2	9.4-11.5 (10.5)

Table 1. Patients' clinical details. * At the time of video monitoring (per day (d), week (w), month (m)). IR - Invasive Recordings.

The clinical material contains five cases (patients 1-4, 7) with focal lateral temporal lobe epilepsy, two cases (patients 8, 9) with mesial temporal lobe epilepsy, three (patients 10-12) with frontal lobe epilepsy and two (patients 5, 6) with a focus in the parietal lobe. The location of the suspected epileptic brain region was determined by the clinical neurophysiologist and correlated with brain MRI images or inter-ictal SPECT. For six cases (# 1, 2, 4, 6, 10, 11) additional validation for the localization of the epileptic region was obtained using invasive recordings (IR) by intracranial (mostly sub-dural) electrodes implanted prior to surgical resection of the epileptic brain region. Two to seven scalp EEG strips were obtained from each patient (Table 1), all ending at the seizure's onset. The clinical onset time of seizures was determined by the clinical neurophysiologist based on video-EEG recorded data (seizure onset was determined as the first sign of a seizure - either clinically or neurophysiologicaly). In figure 4, the IR verification method results are shown for patient #2.

Figure 5a illustrates an MRI scan of patient #5 showing a suspected dysplastic cortical lesion in the left parietal region (arrow). Cortical dysplasia is a developmental cortical lesion in which neurons fail to migrate properly and is frequently associated with epilepsy. The epileptic region in many cases is in close proximity to the dysplastic brain region. In Figure 5b, the results for focal localization (heat map of $\overline{B}_{bp}^{(k)}$ (11) corresponding to the smallest $\tilde{I}_{e.a}^{(k)}$ (10)) are shown, pointing on the same left parietal origin.

In Table 2 the result obtained for the 12 study patients are summarized. In all patients, the epileptic source was lateralized correctly. This achievement is not trivial when being reminded that it is not based on any temporal relations but only on average spatial spread. In 9 cases (patients №s 1-5, 7,9,10 and 12) the smallest value of $\tilde{I}_{e.a}^{(k)}$ (10) for the heat map of $\overline{B}_{bp}^{(k)}$ was obtained with correct identification of $J^{(1)}$ as being closest to the focus.

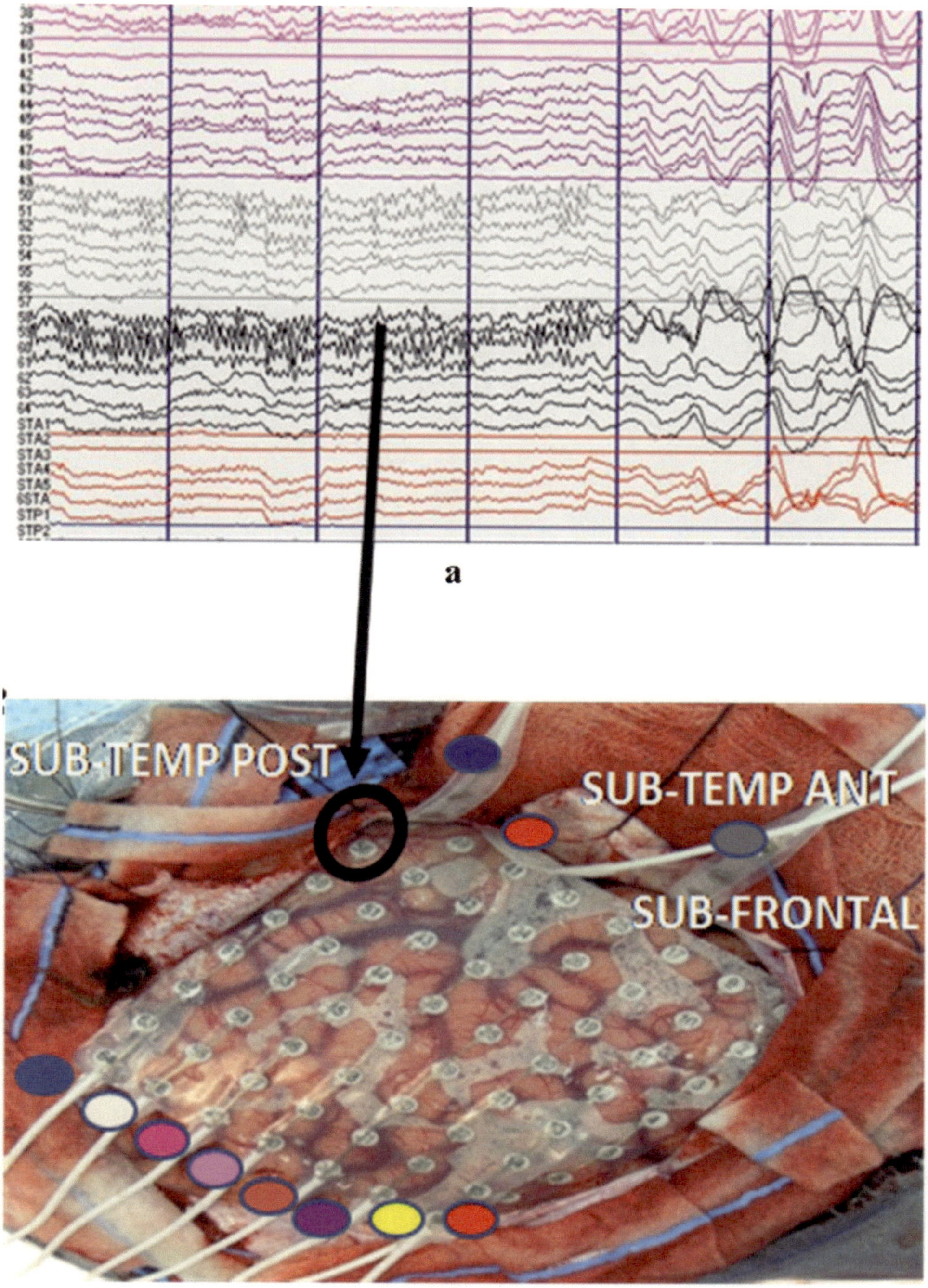

a

b

Fig. 4. Verification of focus localization for patient #2. Invasive recordings (a) from a grid of sub-dural electrodes implanted over the left anterior lateral temporal lobe of the patient (b). This focus is closest to the T5 scalp electrode, the same one identified by the proposed method (see Fig. 3a).

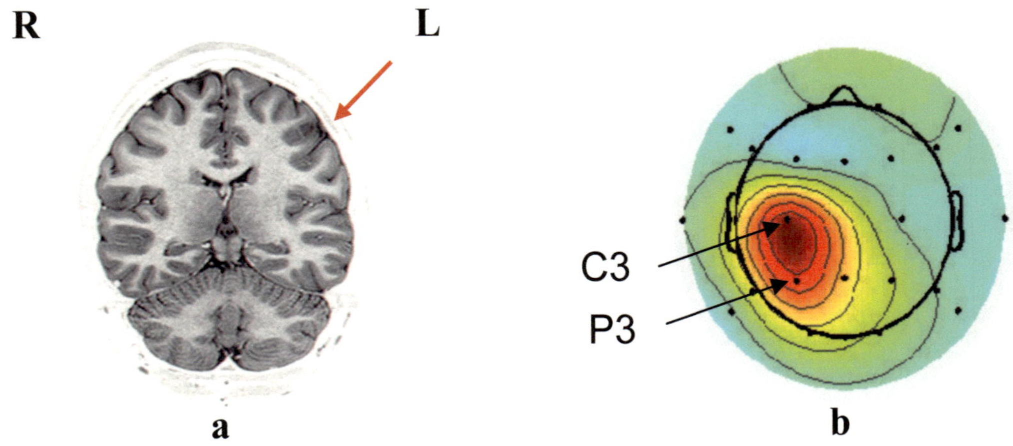

Fig. 5. Focus localization for patient #5. (a) MRI brain scan. The arrow points on the left parietal region, under surface electrodes P3,C3. (b) Heat map of $\overline{B}_{bp}^{(k)}$ (11) of the group with the lowest value of $\tilde{I}_{e.a}^{(k)}$ (10), pointing to the same left parietal origin and two surface electrodes.

For the other 3 patients (marked by *) questionable consistency was obtained with clinical observation. In patient 6, the center of the detected epileptic cluster was not focused around one specific electrode but was equally distant from three surface electrodes and hence the $\tilde{I}_{e.a}^{(k)}$ of the actual closest scalp electrode was not the lowest (although having a relatively small value). Patient 8 had mesial temporal lobe epilepsy (MTLE) which was difficult to localize clinically. In patients with this infliction, interictal epileptiform discharges arising from restricted and deep cortical areas such as the hippocampus may be hardly distinguishable from background EEG activity or may even remain undetected on scalp EEG recordings (i.e. Nayak et al., 2004). Hence, the results of the method concerning this patient were defined as questionable. More cases with MTLE need be analyzed in order to assess the method's applicability to this major variant of focal epilepsy. Patient #11 had a frontal focus (closest to FP1) and hence it was harder to detect due to blinking artifacts affecting that region. This problem has recently been successfully resolved using a constrained BSS procedure (Shoker et al., 2005).

Present findings		clinical localization	Patient №
Epileptic heat maps having smallest $\tilde{I}_{e.a}^{(k)}$	$J^{(1)}$		
	T3	Left Temporal (T3-T5)	1
	T5	Left temporal (T5)	2
	T4	Right temporal (T4-T6)	3
	A1	Left temporal (A1-T3)	4
	C3	Left centro- parietal (C3-P3)	5
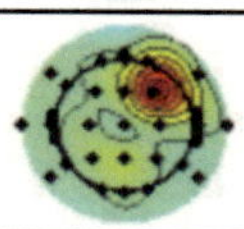	F4	Right centro- parietal (C4-P4)	6*
	T3	Left temporal lobe (T3)	7
	T3	Left mesial temporal lobe	8*
	T3	Left temporal (T3)	9
	F7	Left fronto- parietal (F7)	10
	T3	Left fronto-temporal (F9)	11*
	F8	Right frontal (F4-F8)	12

Table 2. Patients' focus localization

4. Summary and conclusions

In this chapter a method is introduced for localizing an epileptic source, exploiting information from scalp electroencephalographic (EEG) recordings. Blind source separation (BSS) and cluster analysis are combined with generally known pathological features of the epileptic focus. The combination of these unsupervised methods allows us to detect hidden epileptic patterns, at times, unrevealed by other means. The proposed method may help minimize the number and extent of invasive electrodes used in ambiguous cases, when surface EEG recordings do not offer a definite diagnosis. It should be stressed that the method is not intended to replace the expert clinical evaluation; rather, it has been designed to focus the clinician's attention on regions which are more likely to harbor the source of the epileptic activity and which may merit a more careful anatomical (using MRI) or neurophysiological evaluation.

In the derivation of the method, we first describe the data acquisition and preprocessing of the EEG signals, steps that have crucial influence on the following steps. BSS is then employed for restoration of independent neural and extra-neural components in the EEG signals. The seizure in focal epilepsy is thought to arise in a small brain volume. Based on this assumption, a measure $Pe(i)$ (6) is derived for the likelihood of the BSS back-projected quantities B_{bp}^{i} (5) to represent epileptic activity. Then, similar B_{bp}^{i} s with above-threshold $Pe(i)$ are grouped by a cluster analysis method. Finally, an index $\bar{I}_{e.a}^{(k)}$ (10) is derived, which indicates the group of B_{bp}^{i} s most likely to represent epileptic activity. By identifying this group and its dominant electrode we actually localize the epileptic focus.

We then present results of the method, as applied on twelve patients having prior invasive neuro-physiological verification of the actual localization of the epileptic focus. Correct lateralization was achieved for all 12 patients. For nine of the twelve patients the method pinpointed the scalp location closest to epileptic focus. For the other three patients, questionable consistency with clinical observation was obtained. Being automated and unsupervised, the method is a promising tool for pre-surgical evaluation.

5. References

Baumgartner, C. Lindinger, G. Ebner, A. Aull, S. Serles, W. Olbrich, A. Lurger, S. Czech, T. Burgess, R. & Lüders, H. (1995). Propagation of interictal epileptic activity in temporal lobe epilepsy. *Neurology*, Vol.45, pp. 118-22

Bell, AJ. & Sejnowski, TJ. (1995). An information-maximization approach to blind separation and blind deconvolution. *Neural Computation*, Vol.7, pp. 1129-1159

Belouchrani, A. Abed-Meraim, K. Cardoso, JF. & Moulines, E. (1997). A blind source separation technique. *IEEE Transactions on Signal Processing*, Vol.45, pp. 434-444

Bezdek, JC. (1981). *Pattern recognition with fuzzy objective function algorithms*, Plenum, New York, USA

Blume, WT. Holloway, GM. & Wiebe, S. (2001). Temporal epileptogenesis: localizing value of scalp and subdural interictal EEG data. *Epilepsia*, Vol.42, pp. 508-514

Bouzida, N. Peyrodie, L. & Vasseur, C. (2005). ICA and a gauge of filter for the automatic filtering of an EEG signal *Proceedings of theIEEE International Joint Conference on Neural Networks*, pp. 2508-2513, Montreal, Canada, July 31-August 4, 2005

Cardoso, JF. (1999). High-order contrasts for independent component analysis. *Neural computation*, Vol.11, pp. 157-192

Chang, KM. & Po, PC. (2005). Meditation EEG interpretation based on novel fuzzy-merging strategies and wavelet features. *Biomedical Engineering Applications*, Vol.17, pp. 167-175

Chiappa, S. Barber D (2005) Generative independent component analysis for EEG classification. *Proceedings of the European Symposium on Artificial Neural Networks, ESANN' 2005*, Burges, pp. 297-302.

Cichocki, A. Shishkina SL, Mushac T, Leonowicz Z (2005) EEG filtering based on blind source separation (BSS) for early detection of Alzheimer's disease. *Clinical Neurophysiology* 116:729–737.

Corsini, J. Shoker L, Sanei S, Alarcòn G (2006) Epileptic seizure predictability from scalp EEG incorporating constrained blind source separation. *IEEE Transactions on Biomedical Engineering* 53:790-799.

Delorme, A. Makeig S (2004) EEGLAB: an open source toolbox for analysis of single-trial EEG dynamics including independent component analysis. *Journal of Neuroscience Methods* 134:9-21.

Duda, RO. Hart PE, Stork DG (2000) Pattern Classification, 2d Ed. Wiley, New York.

Ebersole, JS. (2000) Noninvasive localization of epileptogenic foci by EEG source modeling. *Epilepsia* 41:24–33.

Faul, S. Marnane L, Lightbody G, Boylan G, Connolly S (2005) A method for the blind separation of sources for use as the first stage of a neonatal seizure detection system. *Proceedings of IEEE International Conference on Acoustics, Speech, and Signal Processing ICASSP '05*, Philadelphia pp. 409–412.

Frank, RM. Frishko GA (2007) Automated protocol for evaluation of electromagnetic component separation (APECS): application of a framework for evaluating statistical methods of blink extraction from multichannel EEG. *Clinical Neurophysiology* 118:80-97.

Gath, I. & Geva, AB. (1989). Unsupervised optimal fuzzy clustering. *IEEE Transactions on Pattern Analysis and Machine Intelligence*, Vol.7, pp. 773-781

Gavaret, M. Trébuchon, A. Bartolomei, F. Marquis, P. McGonigal, A. Wendling, F. Regis, J. Badier, JM. Chauvel, P. (2009). Source localization of scalp-EEG interictal spikes in posterior cortex epilepsies investigated by HR-EEG and SEEG. *Epilepsia*, Vol.50, pp. 276-289

Geva, AB. Kerem, DH. (1998). Forecasting generalized epileptic seizures from the EEG signal by wavelet analysis and dynamic unsupervised fuzzy clustering. *IEEE Transactions on Biomedical Engineering*, Vol.45, pp. 1205-1216

Hallez, H. Vanrumste, B. Grech, R. Muscat, J. De Clercq, W. Vergult. A. D'Asseler, Y. Camilleri, KP. Fabri, SG.Van Huffel, S. & Lemahieu, I. (2007). Review on solving the forward problem in EEG source analysis. *Journal of Neuroengineering and Rehabilitation*, Vol.4, p. 46 (29 pp.)

Hesse, CW. & James, CJ. (2005). Tracking epileptiform activity in the multichannel ictal EEG using spatially constrained independent component analysis, *Proceedings of the 27th Annual International Conference of the IEEE Engineering in Medicine and Biology Society*, pp. 2067-2070, Shanghai, China, September 1-3, 2005

Hyvärinen, A. (1999). Fast and robust fixed-point algorithms for independent component analysis. *IEEE Transactions on Neural Networks*, Vol.10, pp. 626-634

Hyvärinen, A. & Oja, E. (2000). Independent component analysis: Algorithms and applications. *Neural Networks*, Vol.13, pp. 411-430

Hyvärinen, A. Karhunen, J. Oja, E. (2001). *Independent component analysis*, Wiley, New York, USA

James, CJ. & Lowe, D. (2000). Using independent component analysis & dynamical embedding to isolate seizure activity in the EEG. *Proceedings of the 22nd annual international conference of the IEEE Engineering in Medicine and Biology Society*, pp. 1329-1332, Chicago, Illinois, USA, July 23-28, 2000

James, CJ. & Gibson, OJ. (2003). Temporally constrained ICA: an application to artifact rejection in electromagnetic brain signal analysis. *IEEE Transactions on Biomedical Engineering*, Vol.50, pp. 1108-1116

Jarchi, D. Boostani, R. Taheri, M. & Sanei, S. (2009). Seizure source localization using hybrid second order blind identification and extended rival penalized competitive learning algorithm. *Biomedical Signal Processing and Control*, Vol.4, pp. 108-117

Jing, M. & Sanei, S. (2007). A novel constrained topographic independent component analysis for separation of epileptic seizure signals. *Computational Intelligence and Neuroscience*, Vol.1, pp. 1-7

Jung, TP. Makeig, S. Mckeown, MJ. Bell, AJ. Lee, TW. & Sejnowski, TJ. (2001). Imaging brain dynamics using independent component analysis. *Proceedings of the IEEE*, Vol.897, pp. 1107-1122

Kobayashi, K. Nakahori, CJ. Akiyama, T. & Gotman, J. (1999). Isolation of epileptiform discharges from unaveraged EEG by independent component analysis. *Clinical neurophysiology*, Vol.110, pp. 1755-1763

Lantz, G. Spinelli, L. Seeck, M. Grave de Peralta, R. Sottas, C. & Michel, C. (2003). Propagation of interictal epileptiform activity can lead to erroneous source localizations: a 128-channel EEG mapping study. *Journal of Clinical Neurophysiology* Vol.20, pp. 311-319

Latif, MA. & Sanei, S. (2005). Localization of brain abnormal signal sources using blind source separation, *Proceedings of the 27th Annual International Conference of the IEEE Engineering in Medicine and Biology Society*, pp. 1170-1172, Shanghai, China, September 1-3, 2005

Liu, H. Hild, KE. Gao, JB. Erdogmus, D. Príncipe, JC. Chris, J. (2004). Evaluation of a BSS algorithm for artifacts rejection in epileptic seizure detection, *Proceedings of the 26th Annual International Conference of the IEEE Engineering in Medicine and Biology Society*, pp. 91-94, San Francisco, California, USA, September 1-5, 2004

Makeig, S. Jung, TP. Bell, AJ. Mckeown, MJ. Ghahremani, D. & Sejnowski, TJ. (1997). Blind separation of auditory event-related brain responses into independent components. *Proceedings of the National Academy of Sciences*, Vol.94, pp. 10979-10984

Makeig, S. Westerfield, M. Jung, TP. Covington, J. Townsend, J. Sejnowski, TJ. & Courchesne, E. (1999). Functionally independent components of the late positive event-related potential during visual spatial attention. *The Journal of Neuroscience*, Vol.19, pp. 2665-2680

Makeig, S. Delorme, A. Westerfield, M. Jung, TP., Townsend, J. Courchesne, E. & Sejnowski, TJ. (2004). Electroencephalographic brain dynamics following manually responded visual targets. *Public Library of Science Biology*, Vol. 2, pp. 747-762

Meier, R. Dittrich, H. Schulze-Bonhage, A. Aertsen, A. (2008). Detecting epileptic seizures in long-term human EEG: a new approach to automatic online and real-time detection and classification of polymorphic seizure patterns. *Journal of Clinical Neurophysiology*, Vol.25, pp. 1-13

Merlet, I. & Gotman, J. (2001). Dipole modeling of scalp electroencephalogram epileptic discharges: correlation with intracerebral fields. *Clinical Neurophysiology*, Vol.112, pp. 414-430

Michel, C. Lantz, G. Spinelli, L. Grave de Peralta, R. Landis, T. & Seeck, M. (2004). 128-channel EEG source imaging in epilepsy: clinical yield and localization precision. *Journal of Clinical Neurophysiology*, Vol.21, pp. 71-83

Nunez, PL. & Srinivasan, R. (2006). *Electric fields of the brain: The Neurophysics of EEG*, Oxford University Press, New York, USA

Nayak, D. Valentin, A. Alarcon, G. Garcia Seoane, JJ. Brunnhuber, F. Juler, J. Polkey, CE. & Binnie, CD. (2004). Characteristics of scalp electrical fields associated with deep medial temporal epileptiform discharges. *Clinical Neurophysiology*, Vol.115, pp. 1423-35

Pascual-Marqui. RD. (2002). Standardized low resolution brain electromagnetic tomography (sLORETA): technical details. *Methods and Findings in Experimental & Clinical Pharmacology*, Vol.24D, pp. 5-12

Proakis, JG. & Manolakis, DG. (2007). *Digital signal processing*, Pearson Prentice-Hall, New Jersey, USA

Richards, JE. (2004). Recovering dipole sources from scalp-recorded event-related-potentials using component analysis: principal component analysis and independent component analysis. *International Journal of Psychophysiology*, Vol.54, pp. 201-220

Sanei, S. & Chambers, JA. (2007). *EEG Signal Processing*, Wiley, New York, USA

Sepulveda, F. Meckes, M. & Conway, BA. (2004). Cluster separation index suggests usefulness of non-motor EEG channels in detecting wrist movement direction intention. *Proceedings of the IEEE conference on cybernetics and intelligent systems*, pp. 943-947, Singapore, December 1-3, 2004

Shoker. L, Sanei, S. Wang, W. & Chambers, JA. (2005). Removal of eye blinking artifact from the electro-encephalogram, incorporating a new constrained blind source separation algorithm. *Medical & Biological Engineering & Computing*, Vol. 43 pp. 290-295

Sutherlanda, MT. & Tanga AC. (2006). Reliable detection of bilateral activation in human primary somatosensory cortex by unilateral median nerve stimulation. *Neuroimage*, Vol.33, pp. 1042-1054

Vulliemoz, S. Thornton, R. Rodionov, R. Carmichael, DW. Guye, M. Lhatoo, S. McEvoy, AW., Spinelli, L. Michel, CM. Duncan, JS. & Lemieux, L. (2009). The spatio-temporal mapping of epileptic networks: Combination of EEG–fMRI and EEG source imaging . *Neuroimage*, Vol.46, pp. 834-843

Xue, Z. Li, J. Li, S. & Wan, B. (2006) Using ICA to remove eye blink and power line artifacts in EEG. *Proceedings 1st international conference on innovative computing, information and control*, ICICIC'06, Beijing pp. 107-110

Zhukov, L. Weinstein, D. & Johnson, C. (2000). Independent component analysis for EEG source localization. *IEEE Engineering in Medicine and Biology*, Vol.19, pp. 87-96

Ziehe, A. & Muller, KR. (1998). TDSEP - an efficient algorithm for blind separation using time structure. *Proceedings of conf. on artificial neural networks*, ICANN'98, Skovde, pp. 675-680

Ziehe, A. Lasko, P. Nolte, G. & Muller, KR. (2004). A fast algorithm for joint diagonalization with non-orthogonal transformations and its application to blind source separation. *Journal of Machine Learning Research*, Vol.5, pp. 777-800

Zumsteg, D. Friedman, A. Wennberg, RA. & Wieser, HG. (2005). Source localization of mesial temporal interictal epileptiform discharges: correlation with intracranial foramen ovale electrode recordings. *Clinical Neurophysiology*, Vol.116, pp. 2810-2818

Zumsteg, D. Friedman, A. Wieser, HG. & Wennberg, RA. (2006). Source localization of interictal epileptiform discharges: comparison of three different techniques to improve signal to noise ratio. *Clinical neurophysiology*, Vol. 117, pp. 562-571

6

Hyper-Synchronization, De-Synchronization, Synchronization and Seizures

Jesús Pastor, Rafael García de Sola and Guillermo J. Ortega
Instituto de Investigación Sanitaria Hospital de la Princesa, Madrid
Spain

1. Introduction

Ranging from its most basic mechanisms to the clinical symptoms, epilepsy is tightly associated with the word "synchronization". In fact, synchronization phenomena underlying epilepsy are described in several mechanisms at different temporal and spatial scales. At the lowest spatial scale, hippocampal and neocortical interictal spikes appear as the result of synchronized activity of pyramidal cells. At a larger spatial scale, epileptic seizures are usually described as a state of "hypersynchrony" encompassing extended cortical areas. Synchronization and epilepsy are so associated one to each other that lack of synchronization, or desynchronization, has been highlighted in recent years as a key aspect of the underlying dynamic in this pathology. The word synchronization comes from two Greek words, χρονοζ (chronos) and συν (same), which means "sharing the same time"; therefore, a synchronized event is always composed by the temporal coincidence of two or more actions. However, it is usually understood the existence of an underlying mechanism that cause the synchronization itself. In this sense, synchronization is assumed differently from chance, because no deterministic causal effect exists in the last one. On the other hand, the use of the word synchronization is generally associated with a mechanism, known or not, that makes possible the temporal coincidence. In the above sense, and from the very beginning of epilepsy research, the word synchronization is found in many aspects of this pathology. Every time we found the word synchronization in epilepsy, one is tempted to think in a pathological substrate that would make it possible. However, it seems that synchronization in epilepsy has suffered from bifurcating routes since the beginnings of the quantitative descriptions of epileptic phenomena. The very clever and insightful descriptions made by the epilepsy researches in the late 40's and 50's (Penfield & Jasper, 1954) were plagued by the words synchronization and hypersynchronization, and today they are still used almost in the same fashion that were originally used. However, since the first description of the synchronization phenomena by Christian Huygens in 1673 to now-days, there has been a profound revision and enlargement of the concept of synchronization especially in the last years (Pikovsky et al., 2001). Chaos theory, complex networks methodologies and nonlinear time series analysis have dug into the traditional concept of synchronization with the net result of a completely new proposal of what synchronization actually is. Today, in fact, there is no more a single synchronization phenomena, but instead, the traditional term has been split in several, more specific terms to characterize the numerous forms of the underlying mechanism but also, of the different kind of synchronized objects. It seems that the new understanding we have today of synchronization is far from being adopted by the epileptologists, especially by the physician community. New terms as lag or full synchronization are rarely seen in clinical works, which is rather frustrating, because synchronization research in the last years has opened up new and powerful techniques, which would very useful in the improvement of diagnostic or therapeutic techniques.

This chapter is intended to review some of the new advances in synchronization in general and specifically in epilepsy research. A very brief mathematical introduction will be presented in order to fully understand the whole range of synchronization concepts presented in the chapter, either in the theoretical or empirical fields. Our aim thus is to show only the most basic methods and applications of synchronization and epilepsy.

2. Contemporary concepts of synchronization

In general terms, *synchronization between two systems is defined as the adjustment of their internal rhythms due to an existing (weak) interaction between them* (Pikovsky et al., 2001). Therefore, the following ingredients are essential for synchronization:

a. There must exist two or more self-sustained oscillators, i.e., systems capable of generating their own rhythms,
b. The systems adjust their own rhythms due to a (weak) interaction between them, and
c. The adjustment of rhythms occurs in a certain range of systems' mismatch; in particular, if the frequency of one oscillator is slowly varied, the second system follows this variation.

The presence of self-sustained oscillators is needed for synchronization. This requirement is fundamental at the time of differentiating actual synchronization from other phenomena, for instance resonance. Self-sustained oscillators typically are represented mathematically by nonlinear differential equations, as for example Van der Pol oscillator. This kind of oscillator is able to oscillate with its own rhythm without external driving. Moreover, self-sustained oscillators can adjust their frequency, which is the key concept in synchronization. Two isolated different self-sustained oscillators, with different intrinsic frequencies can oscillate at the same frequency when they interact with each other. Note that synchronization refers to a dynamical process instead of a stationary state. Synchronization is a process by which two or more systems adjust their rhythms in the course of time. This means that the interaction allows (generally small) variations of the intrinsic rhythms, but always try to reach the common frequency. One aspect of the formal synchronization definition worth of mentioning is point b), which states that the rhythms are adjusted by a weak interaction. Although this is the very general concept, adjustment of rhythms through the weak interaction allows differentiating activity raised due to a true synchronization between two or more systems from, the activity derived from a compound system in which several subsystems are tightly connected. Note lastly that the synchronization definition used encompass the case of a unidirectional synchronization known as synchronization by driving. We will review below basically the most important concepts of oscillators and synchronization.

2.1 Linear, nonlinear, and chaotic oscillations

The simplest of all oscillators is the harmonic oscillator, described mathematically by:

$$\frac{d^2z}{dt^2} + \omega^2 z = 0 \tag{1}$$

The Equation (1), which is a single second-order differential equation, can also be written as a system of two first-order differential equations, that is:

$$\begin{aligned} \frac{dx_1}{dt} &= x_2 \\ \frac{dx_2}{dt} &= -\omega^2 x_1 \end{aligned} \tag{2}$$

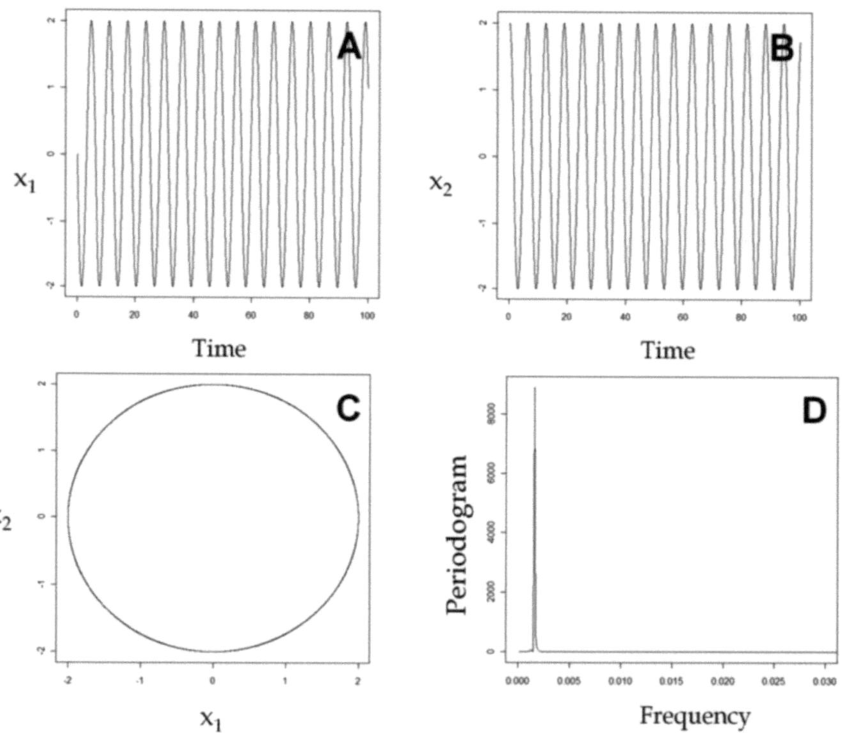

Fig. 1. Harmonic oscillator, solution of Equation 2.

The solution of both equations, (1) and (2), is the well-known $z(t) = A_0 \sin(\omega t + \varphi)$ a periodic function with a fixed frequency $f = \omega / 2\pi$, amplitude A_0, phase $(\omega t + \varphi)$, and an initial phase φ. This kind of oscillator has an intrinsic frequency that cannot be varied. In Figure 1 is displayed the time series from x_1 (panel A) and x_2 (panel B). Also displayed in the phase portrait, that is, the evolution of the system in the phase space of variables, x_1 and x_2 (panel C). In panel C, Figure 1, is depicted the power spectrum of this system, which obviously consists of only one frequency. As we have stated above, two or more oscillators of this type cannot be synchronized because they cannot adjust its intrinsic frequency *f*. It turns out that equation (1) needs a nonlinear term in order to achieve self-sustained oscillations. Perhaps the best-known nonlinear oscillator is the Van der Pol oscillator with parameter μ,

$$\frac{d^2z}{dt^2} + \mu(z^2 - 1)\frac{dz}{dt} + z = 0 \tag{3}$$

or

$$\begin{aligned} \frac{dx_1}{dt} &= x_2 \\ \frac{dx_2}{dt} &= -x_1 - \mu(x_1{}^2 - 1)x_2 \end{aligned} \tag{4}$$

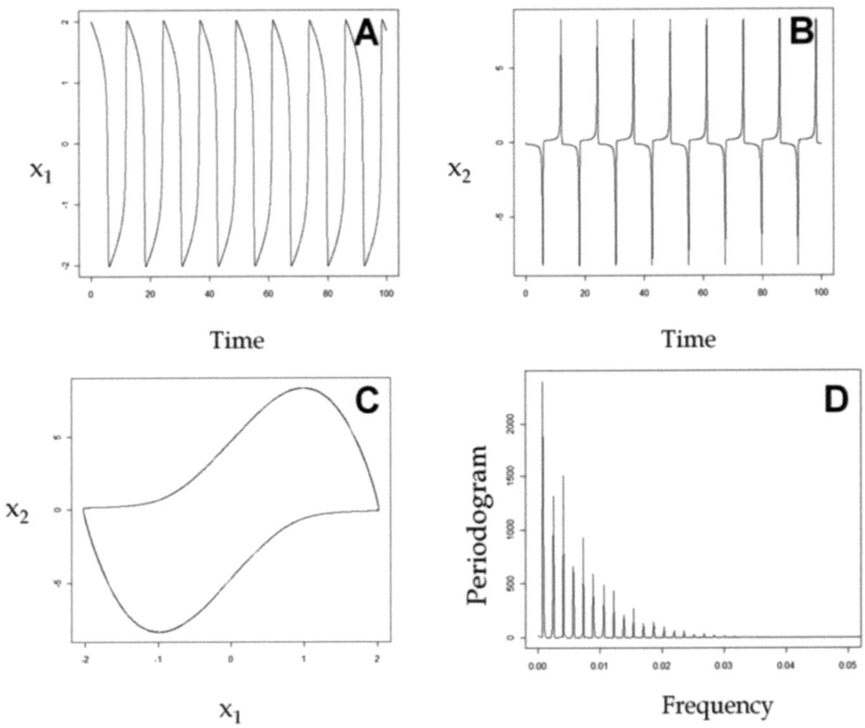

Fig. 2. Van der Pol oscillator for a value of μ=5.5.

In this case, an analytical solution cannot be obtained, so numerical solutions must be calculated. The Van der Pol oscillators, like many other nonlinear oscillators, posses an intrinsic frequency that can be adjusted or entrained by an external driver. In this sense, these kinds of oscillators are fundamentally different from the harmonic one. In Figure 2 is displayed the time series from both coordinates, x_1 (panel A) and x_2 (panel B). Note the clear deviation from a pure sine solution of Equation (2). This fact is also evident en panel C of Figure 2, displaying the characteristic limit cycle. Departure from a pure sine is more evident in the power spectrum, panel D, Figure 2. In this case, the main frequency, corresponding to the fundamental period of the oscillator is accompanied with several harmonic frequencies, due to the non-harmonic character of the time series. Note that in this case it is still possible to define a phase of the oscillator, as in the case of the harmonic oscillator, but in rotating from 0 to 2π the amplitude will change (Figure 3C).

Furthermore, there exists another kind of oscillators, chaotic oscillators, which are fundamentally different from the linear and nonlinear ones, mainly because its behavior does not possess anymore a fundamental frequency. Instead, a chaotic oscillator typically displays a behavior with many different frequencies. Perhaps the best known of all of chaotic oscillators is the Lorenz system, which takes the following form:

$$\begin{aligned} \frac{dx_1}{dt} &= \sigma(x_2 - x_1) \\ \frac{dx_2}{dt} &= x_1(\rho - x_3) - x_2 \\ \frac{dx_3}{dt} &= x_1 x_2 - \beta x_3 \end{aligned} \tag{5}$$

where typically σ=10, ρ=28 and β=8/3,. With these parameters values, Equation (5) displays chaotic behavior. Note that in this case, three first order differential equations are needed to describe the complex dynamic of the oscillator, an essential difference needed for a chaotic regime.

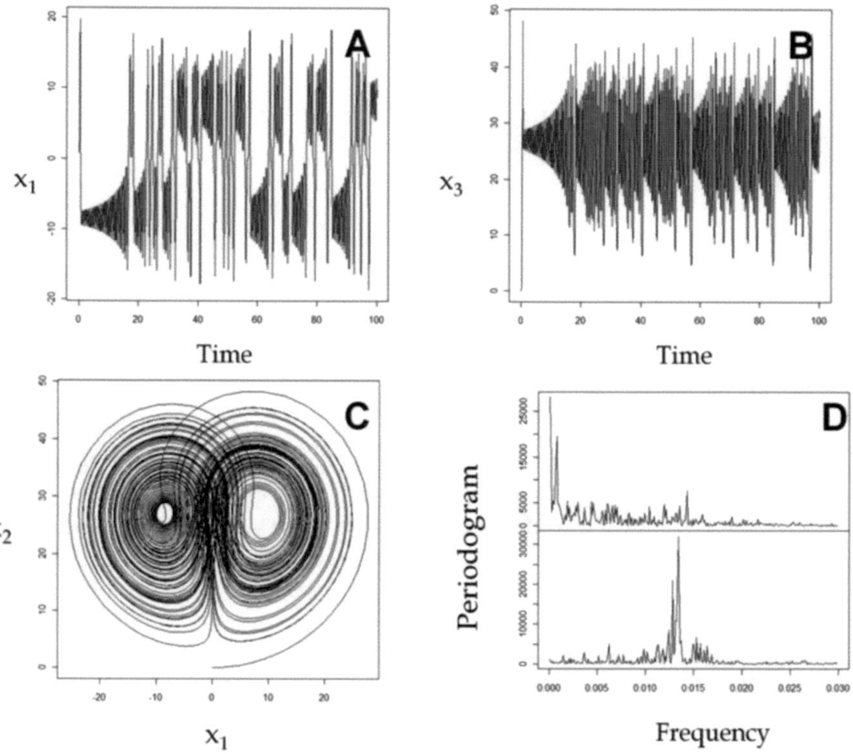

Fig. 3. Lorenz system

In Figure 3, time series of variables x_1 (panel A) and x_3 (panel B) are displayed, drastically different from the linear and nonlinear oscillators. The phase portrait in panel C shows complex oscillations in a bounded region, called attractor, which are very different from the typical limit cycle of nonlinear oscillators, whereas Panel D shows the power spectra of both coordinates, x_1 and x_3. Note the great difference between the power spectrum of variable x_1 (upper) from the power spectrum of variable x_3 (lower). In this last case, the system may be practically considered with a unique, or several but similar frequencies. This is an important point in chaotic systems because, unlike regular oscillators, it is sometimes impossible to define an instantaneous frequency and, in this case, an average frequency is defined. This point will be clear at the time to differentiate frequency from phase synchronization.

Considering the three cases, the linear oscillator (2), the nonlinear oscillator (4) and the chaotic one (5), it is always possible to write these equations in a more general form:

$$\frac{d\mathbf{x}}{dt} = \mathbf{F}(\mathbf{x}) \tag{6}$$

where $\mathbf{x}=(x_1, x_2)$ for (2) and (4), and $\mathbf{x}=(x_1,x_2,x_3)$ for the Lorenz equations (5). For a system like (6), $\mathbf{x}$ is call the state variable and completely determines the state of the system at each time t. Note that experimentally one usually can only access one or a few variables of the system through the experimental time series recorded.

Now, we can express the general form of coupling between two systems, **F** with state variable **x** and **G** with state variable **y**, in the form:

$$\begin{aligned} \frac{d\mathbf{x}}{dt} &= \mathbf{F}(\mathbf{x}) + \mathbf{c}_1(\mathbf{x},\mathbf{y}) \\ \frac{d\mathbf{y}}{dt} &= \mathbf{G}(\mathbf{y}) + \mathbf{c}_2(\mathbf{x},\mathbf{y}) \end{aligned} \tag{7}$$

Systems **F** and **G** for example could be a Van der Pol oscillator and a Lorenz system, respectively, where both systems would interact each one the other through the coupling terms $\mathbf{c}_1(\mathbf{x},\mathbf{y})$ and $c_2(\mathbf{x},\mathbf{y})$. The existence of both terms implies a bidirectional coupling. In the case of a unidirectional coupling, for example from **F** to **G** only $c_2(\mathbf{x},\mathbf{y})$ will remain.

2.2 Nonlinear systems and experimental records

Up to now, theoretical considerations have been made regarding oscillators and coupling between them. However, when dealing with neurophysiological records from patients with epilepsy, one cannot access the underlying interacting systems, like those of Equation 7. Instead, experimental time series from the various types of neurophysiological records are readily obtained in most epilepsy centers. These time series are in fact a "window" to the underlying dynamic, which formally can be considered as function of it. For instance, given a neurophysiological time series eeg(t), it is always possible to consider that record as generated by an underlying system, as the one of Equation (6). In this sense, the experimental record eeg(t) = $\Phi[x(t)]$, where Φ is a function which maps the, generally unknown, multivariate system state **x** onto de univariate time series eeg(t). The simplest of all Φ would be the one which gives a coordinate of the system, eeg(t) = $x_1(t)$. In Figure 4 is sketched the process in the upper curved arrow, which maps state variable **x** in the underlying system to the experimental record eeg(t).

The opposite process is nevertheless possible, due to the famous "embedding theorem" of Takens (Stam, 2005), which is at the heart of the modern approach of time series analysis.

The embedding theorem allows reconstructing an equivalent dynamical system to the original one. If we have a univariate neurophysiological record eeg(t), it is possible to reconstruct an equivalent state variables **x'**(t) of the original **x**(t), in such a way that x'(t)=ψ[egg(t)] by means of a function ψ. There exist many ways to choose ψ, but the simplest and most common way to go from the time series to the reconstructed dynamics is by means of the time delays methods (Stam, 2005). In Figure 4 is sketched the process of dynamics' reconstruction from the experimental time series eeg(t).

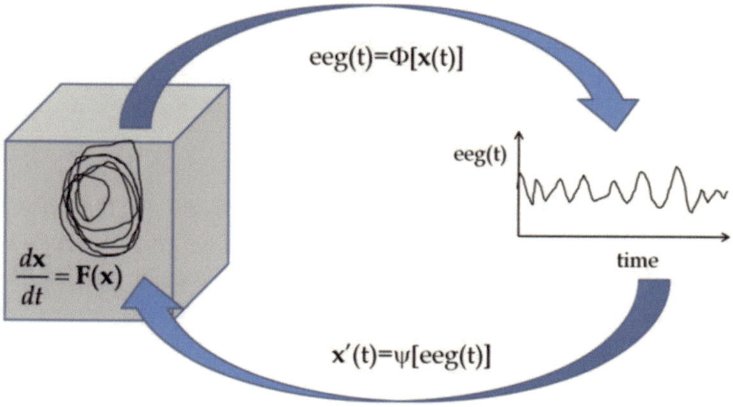

Fig. 4. Go and back from time series to dynamical system.

The above considerations are of special importance at the time to address the synchronization issue. Synchronization of time series was the main objective in the past, whether in time or frequency domain. However, with the advent of nonlinear and chaos theory, the concept of synchronization has been extended to the underlying dynamic. Therefore, is important to take into account what kind of synchronization one is dealing with. It may happen that one kind of synchronization implies another, as we will see below, but in other cases, this is not.

2.3 Synchronization

2.3.1 Frequency synchronization

This is perhaps the most intuitive form of synchronization in which two systems, for instance oscillators like Van der Pol, adjust their rhythms through their mutual interaction in order to share the same frequency or period. One important point to highlight is that frequency synchronization does not need that exactly the same frequency be shared by two synchronized systems with state variables **x** and **y** respectively. Instead, the following relation is valid: $n_x \omega_x - n_y \omega_y = 0$, where n_x and n_y are integers and ω_x and ω_y are the frequencies of systems **x** and **y** respectively.

Frequency synchronization can also be achieved in chaotic systems. For instance, Lorenz systems can synchronize another oscillator through its strong frequency displayed in the x_3 component (Figure 3D, lower panel). This frequency, which corresponds to the alternations between the two lobes in the "attractor" (Figure 3C), may be used to entrain a regular oscillator with periodic motion. However, when looking at the power spectrum of this variable or at the variable itself, it is clear that the frequency is not as well defined as in the case of the Van der Pol oscillator, due to its chaotic dynamic. Therefore, it is usual to define an average frequency in these systems and synchronization through frequency does not need an exact instantaneous frequency. In this last sense, frequency synchronization, at least in chaotic systems seems to be less restrictive than other types of synchronization (phase or identical), because instantaneous values of variables may be different.

2.3.2 Phase synchronization

When considering non-chaotic systems, this is a very intuitive notion of synchronization. Two oscillators are phase synchronized when their phases are entrained, that is:

$$\left|n_x\varphi_x(t) - n_y\varphi_y(t)\right| \leq const$$

where $\varphi_x(t)$ and $\varphi_y(t)$ are the phases of the oscillators. As in the case of frequency synchronization, n_x and n_y stand for possible integers, but we can eliminate them, yielding $\left|\varphi_x(t) - \varphi_y(t)\right| \leq const.$ That is, the difference between the phases in the oscillations remain constant through time. In particular, if const=0, we get $\varphi_x(t) = \varphi_y(t)$, both systems have the same phase for every time. This is called in-phase synchronization. One important point to highlight is that phase synchronization is irrespective of amplitude, thus, two phase-synchronized systems may have different trajectories with their amplitudes totally uncorrelated. In the case of chaotic systems, however, the phase is a more complex concept. For the existence of phase synchronization however, frequency synchronization of needed and therefore is more restricted.

2.3.3 Identical synchronization

In this case the state variables of the systems become identical, that is, $\mathbf{x}(t) = \mathbf{y}(t)$. However, in order to achieve a complete synchronization between two systems **F** and **G**, both must be identical, in form and parameters values. This is therefore the most restrictive form of synchronization. Identical synchronization implies phase and frequency synchronization.

2.3.4 Lag synchronization

This type of synchronization is a generalization of the identical synchronization in which the state variables of a system $\mathbf{x}(t)$ are identical to the state variables of another $\mathbf{y}(t)$, but with lag difference of τ,

$$\mathbf{x}(t) = \mathbf{y}(t+\tau)$$

Note that for $\tau = 0$, identical synchronization is recovered.

2.3.5 Generalized synchronization

The concept of generalized synchronization arises in the case of unidirectional coupling between two systems that establish a relation known as "master-slave", when one system (slave) obeys the dynamics of the other (master). In Equation (7) for example making $c_2(\mathbf{x},\mathbf{y})=0$ we will have a unidirectional coupling from $\mathbf{G}(\mathbf{y})$ to $\mathbf{F}(\mathbf{x})$. In this case, we will expect that the state variables **x** will be determined from the interplay between its own dynamic (**F**) and the coupling $c_1(\mathbf{x},\mathbf{y})$. However, when a generalized synchronization is established between both systems, the state variable **x** will be completely determined from the state variables from the other system, **y**. In this case, it is possible to write the master-slave relation in the following way:

$$\mathbf{x}(t)=\boldsymbol{\Gamma}[\mathbf{y}(t)] \tag{8}$$

where Γ is a general (vectorial) function which maps the state variable **y**, of system **G** to state variable **x** of system **F**. Because Γ is unknown, this kind of synchronization is difficult to detect, especially because the experimental series coming from the system, one of the state variables for example, do not must follow one each other. However, by using nonlinear time series techniques, such as the embedding methodology (Stam, 2005) it is possible to reconstruct state variables from a single time series and therefore to assess the existence of a functional relation between both systems.

2.3.6 Full (complete) synchronization

Up to now, the aforementioned synchronization types where described in the case of two interacting systems, which can obviously could be generalized to the case of more. However, when dealing with a large number of interacting oscillators it may be more practical to quantify the average synchronization in the whole set rather than specifying synchronization between every pair of them. The study and characterization of synchronization in a large set of oscillators is an active field of research today, which has been pioneered by the works of Winfree (1967) and Kuramoto (1975). In the case of a population of interacting systems, Equation 7 should be generalized to include N oscillators instead of two, the kind of coupling between the systems, the underlying topology of the network of oscillators, etc. Trying to characterize such a system from a dynamical point of view is a colossal task, so in order to simplify as much as possible the problem, but capturing the essential phenomenology, the following system is studied:

$$\frac{d\vartheta_i}{dt}=\omega_i+\frac{K}{N}\sum_{j=1}^{N}\sin(\vartheta_i-\vartheta_j) \quad (i,j=1,\ldots,N) \tag{9}$$

where K is a normalization factor. Every oscillator is now replaced by its instantaneous phase θ_i and its natural or intrinsic frequency ω_i, as in the case of nonlinear oscillators. Doing this way, the model does not consider the amplitude evolution. The system therefore is composed of N limit cycle oscillators mutually coupled each other through the terms $\sin(\theta_i - \theta_j)$. However, Equation 9 does not take into account the spatial distribution of the oscillators, a relevant information that should be considered and which open a whole world of possibilities (Arenas et al., 2008).

$$\frac{d\vartheta_i}{dt}=\omega_i+\sum_{j=1}^{N}\sigma_{ij}a_{ij}\sin(\vartheta_i-\vartheta_j) \quad (i,j=1,\ldots,N) \tag{10}$$

where σ_{ij} is the coupling strength between adjacent oscillators in the underlying network and a_{ij} are the elements of the connectivity matrix, that is, 1's or 0's whether the links exist or not and which determine much of the network topology.

Finally, the degree of synchronization in the whole network is evaluated by using the *order parameter*,

$$r(t)e^{i\phi(t)} = \frac{1}{N}\sum_{i=1}^{N} e^{i\vartheta_i(t)} \tag{11}$$

where the modulus $0 \le r(t) \le 1$ measures the average phase coherence of the oscillators and the degree of synchronization. When $r \sim 1$, a full or complete synchronization is obtained, and the oscillators are said to be phase locked.

2.4 Synchronization measures

Synchronization has been traditionally detected by several numerical methods, the most common of all the cross-correlation (Press et al., 2007) in the time domain. In electroencephalographic (EEG) analysis the cross-spectrum, which is nothing else but a cross-correlation in the frequency domain, dominates the field for several years. The search of brain *functional connectivity* and *effective connectivity* (Friston et al., 1993) boosts many researchers to introduce the use of non-traditional techniques. Functional connectivity is defined as "a temporal correlation between spatially remote neurophysiological events", whereas effective connectivity is defined as "the influence that one neural system exerts over another either directly or indirectly". This distinction between both concepts allows to clearly dividing the existing methods in two main classes. On the one hand, methods to quantify functional connectivity overpass the causality issue, focusing only in the existence of a deterministic relation between two systems. In this sense, synchrony detection is generally thought as an indication of a functional connectivity between two distinct brain regions. On the other hand, effective connectivity involves temporal causation, so appropriate methods must be used in this case. Traditionally, cross-correlation has been also used, fundamentally at the cellular level, to detect effective connectivity between, for instance, two neurons, by looking the time lag which maximizes the cross-correlation estimate. New and powerful methods have been developed in the last years to infer the effective connectivity, as the Granger causality (Granger, 1969) and transfer entropy (Schreiber, 2000) to mention a few.

We will briefly describe some of the most used methods to calculate synchronization in neurophysiological records, in particular those coming from epileptic patients. Cross-correlation was one of the first methods used to calculate synchronization. Being a linear method and with the increasing use of nonlinear time series analysis techniques (Stam, 2005), it was replaced by other, more subtle methods. However, in the last years, cross-correlation has suffered a revival in its use to quantify functional connectivity between cortical areas, powered mostly by developments in complex networks techniques. Dynamics and synchronization in complex networks (Arenas, 2008) from neurophysiological systems need fast and reliable estimates of synchronization between huge quantities of pairs, in order to compare estimates of functional connectivity against large databases of anatomical connectivity from human data (Hilgetag & Kaiser, 2004). Cross-correlation offers both properties. Moreover, the discovery that cross-correlation, although a linear statistic, perform as well as nonlinear synchronization measures when applied to neurophysiological data (Netoff & Schiff, 2002; Quian Quiroga et al., 2002: Ortega et al., 2008) justifies in most cases its first use, at least as an initial exploratory task. As is well-known in the field of time series analysis, it is always mandatory to employ a full battery of methods to estimate an underlying quantity. In the case of synchronization, several methods would bring different aspect of the true functional connectivity.

There exists many excellent books and reviews of time series and synchronization methods, just to mention few (Pereda et al. 2005; Lehnertz et al., 2009).

We will not get into details about statistical validation of the results obtained in each of the following methods, a topic which is of paramount importance. Many times, synchronization calculations generate significant but otherwise false indications of connectivity between two systems, due to the limited quantity of analyzed data or its lack of stationarity. Statistical testing of the results should always be performed, by the classical methods or with the use of surrogate methodology. We will only describe here the basic concept underlying each of them and it use in epilepsy research.

Synchronization is a bivariate measure, and most often measures of synchronization between more than two systems is needed. Almost every modern neurophysiological equipment produce multivariate time series, recorded from different brain areas, so special techniques and methods should be developed to deal with these kinds of spatially extended data. Moreover, the recent developments in complex networks (Boccaletti et al., 2006) and synchronization over them (Arenas et al., 2008) have made researchers to discover new methods or rediscover traditional ones. This is the case for example the searching of community structures in networks i.e., the organization in groups of tightly connected members. New (Girvan & Newman, 2002) and traditional clustering analysis (Boccaletti et al., 2006) methods are used with the aim to describe in the most succinct and comprehensive fashion the synchronization pattern in a complex network. In the last part of this section we will show only one classical method, hierarchical clustering, which allows to organize the synchronized activity in more than two interacting systems.

Whatever kind of neurophysiological record we wish to analyze, typically it will consist of a multivariate data set $\mathbf{x}$ with the recorded electrical activity of N_{chan} channels, and N_{dat} data points in each channel:

$\mathbf{x} = x_i(k), i = 1, N_{chan}$ *and* $k = 1, N_{win}$ where index *i* represents the channel number and *k* is the discretized time.

2.4.1 Linear correlation

Cross-correlation analysis is perhaps the most used method to estimate synchronization between two variables. It was the favourite choice during the 50's and 60's of the past century at the time of measuring correlation between spikes discharges in microelectrodes studies. However, in the mid-60's with the advent of the Cooley-Tukey method (Press et al., 2007) to quickly calculate the Fourier Transform, the fast Fourier transform algorithm, moves synchronization search to the frequency domain in many fields. Given two continuous signals *x(t)* and *y(t)*, the cross-correlation function between both signals is defined as:

$$\rho_{xy}(\tau) = \frac{\mathrm{cov}\left[x(t), y(t+\tau)\right]}{\sqrt{\mathrm{var}\left[x(t)\right]}\sqrt{\mathrm{var}\left[y(t+\tau)\right]}} \tag{12}$$

where *cov* stands for covariance and *var* for variance. Cross correlation is a function of τ, the lag between both signals. In looking for synchronization at the same time thus τ=0. One can easily estimate the cross-correlation for two discretized time series, x_i and x_j, at times k, with the Pearson correlation coefficient (Press et al., 2007),

$$\rho_{ij}(0) = \frac{\sum_{k=1}^{N_{win}} (x_i(k) - \overline{x_i})(x_j(k) - \overline{x_j})}{\sqrt{\sum_{k=1}^{N_{win}} (x_i(k) - \overline{x_i})^2 \sum_{k=1}^{N_{win}} (x_j(k) - \overline{x_j})^2}} \qquad (13)$$

Where $\rho_{ij}(0)$ implies that no lag between x_i and x_j exists. Correlation coefficient range in $-1 \le \rho_{ij} \le 1$, where a value of -1 implies a perfect inverse linear correlation between both time series, and a value of 1 implies a perfect linear relation. The linear character of (13) resides in the fact it is a fit to a straight line and ρ_{ij} gives its slope. The case of zero correlation however, implies only the *inexistence of linear correlation,* but a nonlinear interaction between both signals may be present. Cross-correlation is essentially an amplitude method in the sense that it quantifies co-movements in two time series by "comparing" amplitudes in the signals. When the signals are similar Equation (13) gives robust results, which is not the case for example when both signals have very different amplitudes. As many methods, it also requires the stationarity of the time series, both in its means and variance. While many modern techniques specifically designed to deal with nonlinear time series flourish in the epilepsy literature, cross-correlation analysis is still being used to detect synchronization, mostly for its intuitive interpretation, ease of implementation and statistical evaluation.

While (12) or (13) evaluate correlation in the time domain, there exists and equivalent way to do the same thing in the frequency domain, through the cross-spectrum, and in particular by using the coherence function:

$$Coh_{xy}(f) = \frac{|P_{xy}(f)|^2}{P_{xx}(f)P_{yy}(f)} \qquad (14)$$

where $P_{xy}(f)$ is the cross-spectrum of variables x and y, and P_{xx} and P_{yy} correspond to the power spectrum of x and y respectively. Unfortunately, coherence is a measure that does not separate the effects of amplitude and phase in the interrelations between the signals and, as in the case of time correlation; it can be applied only to stationary signals.

2.4.2 Phase synchronization

The concept of phase synchronization was introduced by Rosenblum *et al.* (1996) in relation with chaotic oscillators. It has been also extended to the case of noisy oscillators. The power of the method resides in that it measures the phase relationship, independently on the signal amplitude. In order to evaluate differences between phases in two signals, one must firstly define the *instantaneous phase* of the signal, by means of the analytical signal concept. For a continuous signal $x_i(t)$ the associated analytical or complex signal is defined as:

$$z_i(t) = x_i(t) + i\tilde{x}_i(t) = A_i(t)e^{i\varphi_i(t)}$$

where $\tilde{x}_i(t)$ is the Hilbert transform of $x_i(t)$

$$\tilde{x}_i(t) = \frac{1}{\pi} p.v. \int_{-\infty}^{\infty} \frac{x(t')}{t-t'} dt' \tag{15}$$

where *p.v.* stands for (Cauchy) Principal Value. The instantaneous phase is thus,

$$\phi_i(t) = \arctan \frac{\tilde{x}_i(t)}{x_i(t)} \tag{16}$$

Therefore, in the case of two signals $x_i(t)$ and $x_j(t)$, phase difference between both can be calculated as

$$\phi_i(t) - \phi_j(t) = \arctan \frac{\tilde{x}_i(t)x_j(t) - x_i(t)\tilde{x}_j(t)}{\tilde{x}_i(t)x_j(t) + x_i(t)\tilde{x}_j(t)} \tag{17}$$

This gives the instantaneous phase difference between both signals.

In order to implement numerically the above definition over two time series $x_i(k)$ and $x_j(k)$, the mean phase coherence R_{ij} was introduced (Mormann et al., 2000):

$$R_{ij} = \left| \frac{1}{N_{win}} \sum_{k=1}^{N_{win}} e^{i\Delta\alpha_{ij}(k)} \right| \tag{18}$$

calculated in the time window N_{win}, where $\Delta\alpha_{ij}(k) = \phi_i(k) - \phi_j(k)$ is the instantaneous phase difference at the discretized time k. It is clear from Equation (18) that R_{ij} follows $0 \le R_{ij} \le 1$. The literature (Pikovski et al., 2000) gives useful hints for the numerical calculation of the Hilbert Transform of a time series, i.e. Equation (15).

2.4.3 Mutual information

A very different kind of approach to evaluate association between two variables is through the information theory approach (Cover & Thomas, 2006). For a single time series, x_i one can estimate its probability distribution, $P(x_i)$ by partitioning the entire range of values taken by x_i in N_{bins} bins and then, count the number of points n_i falling in each bin *l*. In this way, the relative occurrence n_l/N_{win} estimate $p_i(l)$ the probability that a point in the time series *i* fall in the bin *l*. Then, the Shannon entropy is defined as:

For a single time series, x_i of length N_{win} one can estimate its probability distribution, $P(x_i)$ by partitioning the entire range of values taken by x_i in N_{bins} bins and then, count the number of points n_l falling in each bin *l*. In this way, the relative occurrence n_l/N_{win} estimate $p_i(l)$ the probability that a point in the time series *i* fall in the bin *l*. Then, the Shannon entropy is defined as:

$$H[P(x_i)] = H(x_i) = -\sum_{l=1}^{N_{bins}} p_i(l) \log_2 p_i(l) \text{ for time series } x_i \text{, and likewise} \tag{19}$$

$$H\left[P(x_j)\right] = H(x_j) = -\sum_{l=1}^{N_{bins}} p_j(l)\log_2 p_j(l) \text{ for time series } x_j \tag{20}$$

Analogously, using the joint probability distribution $P(x_i, x_j)$, the joint entropy between x_i and x_j (or properly between $P(x_i)$ and $P(x_j)$) is

$$H\left[P(x_i, x_j\right] = H(x_i, x_j) = -\sum_{l=1}^{N_{bins}} \sum_{k=1}^{N_{bins}} p_{ij}(l,k)\log_2 p_{ij}(l,k) \tag{21}$$

Finally, the mutual information $MI(x_i, x_j)$ between x_i and x_j is

$$MI(x_i, x_j) = H(x_i) + H(x_j) - H(x_i, x_j) \tag{22}$$

which clearly shows that it is a symmetric function of x_i and x_j. *MI* is positive, being zero when -the probabilities distributions of x_i and x_j are independent. *MI* can be thought as a generalization of the linear correlation coefficient. It gives the reduction in the uncertainty in x_i due to the knowledge of x_j.

Although *MI* is bounded from below, giving *MI* = 0 when there is statistical independence between x_i and x_j, it is not bounded from above. In this context it is useful to define a mutual information statistic with the property of $0 \le R_{ij} \le 1$.

$$\lambda_{ij} = \sqrt{1 - e^{-2MI_{ij}}} \tag{23}$$

which now satisfies $0 \le \lambda_{ij} \le 1$.

2.4.4 Spatial synchronization

When dealing with spatially extended system, as it most the case with neurophysiological signals, especial methods should be used in order to organize synchronization between every pair of interacting systems. One traditional method is the hierarchical clustering (Boccaletti et al., 2006) that allows to arrange a group of, say N objects into smaller groups, such that objects belonging to these groups are more tightly linked with member of the same group than with the other groups. The net result is to obtain a set of groups, or clusters, such that members within each cluster are more closely related to one another than objects assigned to different clusters. Formally, it is better to work with "distances" instead of "correlations" between objects, such that a tight correlation between objects is equivalent to a closer distance, so this transformation is usually performed. The whole same procedure is able to be extended in the case of synchronization between time series. When dealing with multivariate records, firstly, a measure of synchronization is chosen, typically Pearson coefficient (Zhou et al.,2006), Equation (13), though other measures are also used (Ortega et al., 2008; Ortega et al., 2010) to check the results' reliability, and then a transformation to distances is performed. Then, a favorite clustering algorithm is used in order to create the hierarchical organization, which will uncover the synchronization clusters present in the data. See section 3.5 for a concrete example. This is only one of the many existing methods to organize synchronization in a set of time series.

3. Synchronization in epilepsy

3.1 Interictal epileptogenic discharges (IED)

Epilepsy is a synchronization/desynchronization pathology and the first indication of an abnormal synchronization process is in the most basic clinical sign of epilepsy, the appearance of IED, in the form of spikes and/or sharp waves in electrophysiological records. IED are usually present in epileptic patients and rarely in normal subject (Noatchar & Rémi, 2009; Walczak et al., 2008). In the case of temporal lobe epilepsy (TLE) epilepsy, spikes and sharp waves are frequently found in EEG records at both sides, even though strictly unilateral occurrence of IED has an excellent predictive value for a successful resective surgery. There is no difference in the diagnostic information between spikes and sharp waves. Although IED are difficult to define precisely, there exist widespread consensus about two or three criteria IED should meet, namely (Noatchar & Rémi, 2009; Walczak et al., 2008; Pastor et al., 2010):

a. IED should be clearly distinguishable from the background activity,
b. Duration must be <200 msec. The Committee on Terminology distinguishes between spikes, which have a duration <70 msec, and sharp waves, which have a duration between 70 and 200 msec, and
c. The IED must have a physiologic field. Practically, this means that the IED are recorded by more than one electrode and has a voltage gradient across the scalp. This requirement helps distinguish IEDs from artifacts.

IED are the mesoscopic manifestation of the cellular paroxysmal depolarizing shift (PDS) (McCormik & Contreras, 2001; Timofeev & Steriade, 2004; Speckmann et al., 2011), which occurs at the cellular level in the pyramidal cells. A single PSD is a sudden neuronal depolarization of large amplitude of 20-40 mV and long-lasting duration of 50-200 ms. During the depolarizing shift, a train of action potentials are triggered. Although intra as well extracellular recordings can detect a single PDS, it would be impossible to identify it by macroelectrodes, as those of EEG and/or electrocorticography (ECoG). The appearance of IED in these kinds of records is actually the overt activity in several pyramidal cells suffering PSD at the same time. The mechanisms underlying this synchronized activity are diverse, for instance:

a. Highly coupled large cortical network: interictal spikes appear to be generated through a brief period of runaway excitation that spreads rapidly through a large local network of neurons, lasting 80–200 ms and being terminated largely by the activation of inhibitory synaptic conductances and intrinsic K^+ currents (McCormik & Contreras, 2001).
b. Ephaptic (non-synaptic) interaction: This is another influence identified in the generation of IED. The close proximity of cell bodies and dendrites in the hippocampus results in a direct activation of neighbor cells by currents circulating in the extracellular space (an electrical field effect). The entry of positive charges into one neuron results in a negative charge in the extracellular space causing a decrease in the potential difference (e.g. depolarization) across the membrane of neighboring neurons. This apparent depolarization may then influence the timing of action potential generation in neighboring cells and therefore bring the network into synchrony
c. Change in the extracellular concentration of ions: Also a non-synaptic mechanism in which periods of intense activity, from brief synchronized bursts of action potentials in

a population of neurons to more prolonged discharges, result in significant increases in K^+ and decreases in Ca^{2+}. These changes in ion concentration may significantly increase neuronal excitability and promote epileptogenesis.

From the above three possible mechanism of IED generation, it is possible to identify three different classes of synchronization mechanisms. The first one clearly corresponds to the case of synchronization due to a strong coupling between oscillators. A strong coupling between pyramidal and interneurons, which in many cases is through gap junctions (Traub et al., 2004), would make a system of interconnected groups of oscillatory neurons behave as a unique system. In this case, the presence of IED in macroscopic records appears in fact as the overt activity of bigger cluster of neurons than the basic unit of oscillatory behavior does. Therefore, synchronized activity in the form of IED is mainly due to a structural characteristic of the underlying network. Numerical simulations in the dentate gyrus (Morgan & Soltesz, 2008) with 50000 granulate cells shows that hyperexcitablity in the network, which promotes de appearance of network synchronization is enhanced by the presence of a small number of highly and strongly interconnected "hubs" in the whole network ("scale-free" networks). The presence of these hubs in the network will promote de appearance of full synchronization in the whole network which will manifest as and IED in macroscopic records.

On the other side, the other two possible mechanisms for the generation of IED are, nor mediated by gaps junctions, but neither by chemical synapses. In these cases instead, synchronization of oscillatory activity at the cellular level is mediated by weak interaction through the extracellular space. However, weak interaction does not mean that full synchronization in the IED generating network will not be achieved. On the contrary, weak interaction may promote full synchronization in the network depending fundamentally of the underlying topology (Arenas et al., 2008).

Perhaps both properties in pyramidal-interneurons and granulate-interneurons networks are present whether in physiological and pathological situations. However, the interaction between both mechanisms of synchronization may be different in each situation, and the appearance of IED, as the result of full synchronization between small groups of neurons will be determined by a particular enhancement between them.

3.2 Hypersynchronization

The word hypersynchronization is traditionally used to describe two of the main characteristics of epilepsy. On the one hand, IED are described as "hypersynchronous" events due to excessive simultaneous neuronal discharge (Chang et al., 2011). As we have review in the above section, IED are produced by a full synchronization of PDS of neurons in a small cortical network, mediated by two synchronization mechanisms. In these sense, the term hypersynchronization should be replaced, we think, for a more appropriate *local full synchronization*, because IED are caused by a small spatial coordination of several neurons suffering PDS at the same time, due to both, strong and weak interactions.

On the other side, hypersynchronization is also used to describe global, high amplitude patterns in electroencephalographic recordings, that is epileptic seizures. Firstly, as we have stated above, neither synchronization nor hypersynchronization are "states", but dynamical process. In this sense, individual neurons adjust their rhythms during the ictal event, and the overt activity varies strongly in the elapsed time of the seizure. At the cellular level for instance, is has been shown that only 30 % of neurons change their firing frequency (Babb et al., 1987) during partial seizures. At larger scales, electrodecremental seizure' onset (Chang et al., 2011), which display a desynchronized neuronal activity at the seizure onset, as well as the whole correlation changes (Schindler et al. 2007) during the seizures. In this last case, synchronization, measured through zero-lag correlation, remain constant or even decrease during the first part of the seizure's development in the case of a secondary generalization, and increase only in the second half of the seizure.

3.3 Desynchronization

Although synchronization and hypersynchronization are at the heart of epilepsy, desynchronization has emerged in recent years as a property of neurophysiological signals, playing a central role in several issues. However, desynchronization in epilepsy may be traced back as far as 1963 (Gestaut et al., 1963) in the description of electrodecremental seizures (Arroyo et al., 1993; Chang et al., 2011; Tatum et al., 2008), described in both partial and generalized seizures. Electrodecremental seizures display an EEG pattern with low-voltage and fast activity which progress by increasing voltage and decreasing frequency. Due to the obvious lack of automatic or quantitative methods at that time (Gastaut et al., 1963), desynchronized activity was simply referred as an abolishment of fundamental rhythms at the onset of seizures. It is interestingly to note that some EEG seizure patterns were commonly described as desynchronization-synchronization or desynchronization-hypersynchronization processes, stressing the importance of the dynamical changes, and thus highlighting that synchronization is a process, not a state. Further research (Arroyo et al., 1993) denied the existence of desynchronization, favoring instead the existence of synchronized activity at higher frequencies. This fact could not be observed originally due to the formerly paper-based analysis performed, inadequate for a quantitative assessment.

More recently, desynchronization has entered into epileptology with new strength, not only at the macroscale, as in the cases of EEG patterns, but at the cellular level (Netoff & Schiff, 2002) too.

Perhaps the first methodological reference to desynchronization process in epileptic signals was that of Mormann *et al.* (Mormann et al., 2000), where intracranial EEG (iEEG) records from TLE patients were analyzed with phase synchronization methodology. Two kinds of findings were reported regarding des/synchronization activity, spatial and temporal results. Spatial findings showed that ipsilateral synchronization is greater than contralateral one in 14 out of 17 TLE patients during the interictal period. However, as those iEEG electrodes record activity coming from entorhinal cortex and hippocampal body, comparison between both temporal sides is difficult. As it was showed (Mormann et al., 2000) synchronization matrix is highly modular in each side due to the high synchronization within each structure, entorhinal cortex and hippocampus. Thus, the spatial average for the whole set of electrodes in each side is difficult to understand. The temporal findings, however, are easier to interpret. They found a clear difference between the synchronization behavior during the interictal and pre-ictal periods. There exists an abrupt drop in the mean phase coherence 15-20 minutes prior to the seizure onset, displaying therefore a very important desynchronization mark in the seizure process. The interpretation behind these findings is that the whole recorded area is in a state of increased susceptibility for pathological synchronization, that is, desynchronization, and an abrupt temporal drop in synchronization seems to favor the recruitment of neuronal tissue for a pathological synchronization. A second hypothesis underlying desynchronization prior to seizures seems to be favored by a second work (Mormann et al., 2003). In this case, the seizure preceding desynchronization appears to be caused by the existence of two different kinds of regions in the recording areas. One region is in the physiological levels of synchronization and the other one is already involved in the progressive pathological synchronization coming from the epileptic focus. In this way, phase synchronization between electrodes located at both areas will give abnormal lower values of synchronization. Worthy to note, synchronization drop prior to seizure allows anticipating seizure onset in one hour (on average). Similar results were presented by (Le Van Quyen et al., 2003), although only for the β-band (10-25 Hz).

These last results jointly with the earliest findings of the decremental, desynchronizing-synchronizing seizures seem to point out that a seizure by itself is a pure synchronizing process. Schindler et al. (Schindler et al., 2007) have dug into the focal seizure's dynamical structure, finding that desynchronization also play a key role in the seizure development. They analyzed the correlation matrix constructed with the zero-lag correlation coefficients (Equation 13) between every pair of electrodes (depth and subdural), during seizure with focal onset. Their results show that during the clinical seizure onset, synchronization structure remains unchanged, and then progressively increase before the seizure terminate. Interestingly, in the case of seizure with secondary generalization, desynchronization dominates during the first half of the seizure extent, a finding also observed by Wendling et al. (Wendling et al., 2003). How a localized synchronous activity inside the seizure onset zone may progress toward a more extended desynchronized activity during the first part of the seizure? The authors (Schindler et al., 2007) provide the answer by proposing that due to different conduction times, and perhaps traveling by different paths (Milton et al., 2007), synchronous activity at the seizure onset zone reach distant cortical areas with time delay relative to each other. In this way, the seizure correlation structure during the spreading will be mostly desynchronized. Because secondary generalization have more extensive spreading than complex partial seizures, this hypothesis explain why desynchronization is greater in the former. However, causality may also be reversed in the sense that an already present desynchronization may be the cause of the seizure spreading (Schindler et al., 2007). This interpretation is supported by recent findings (Ortega et al., 2010) in TLE, where there exists an intrinsic imbalance in the synchronized activity between both temporal lobes, being the ipsilateral lobe more desynchronized than the contralateral one.

The above mesoscopic findings seem to be reproduced at the cellular scale. In an *in vitro* experimental seizure-like events in hippocampal slices, Netoff and Schiff showed (2002) that desynchronization between pairs of neurons, is tightly associated to the seizure-like events. Comparing, by using a battery of linear and nonlinear methods, interictal and seizure-like periods, they were able to demonstrate the existence of desynchronization during the seizures. Therefore, desynchronization appears as essential for seizure initiation and maintenance, at least at the cellular level. Beyond the important results concerning desynchronization during seizures, the work of Netoff and Schiff also illuminates about two fundamental aspects of synchronization. The first one is related with the methodology employed to quantify synchronization. The finding reported shows that synchronization estimates strongly depends upon the numerical method employed and the underlying analyzed signal. A linear method, as the Pearson correlation used over nonlinear signals, like neuronal activity, may gives better results than a nonlinear method, like phase synchronization or mutual information. In any case, the morale is to employ a battery of synchronization methods, which examine different aspects of the underlying synchronization. The second issue raised by the work of Netoff and Schiff is related with the controversy of broad-band versus narrow-band analysis. As it was showed (Netoff & Schiff, 2002), only in the case of dominant-frequency signals, narrow band analysis seem to perform better than broad-band. On the contrary, when faced with compound signals, synchronization should be quantified by using the broad-band signal.

A more recent work (van Drongelen et al., 2005) combining modeling and electrophysiological experiments confirms and explains cellular desynchronization during the seizures. A network model of 656 cortical neurons can generate and sustain seizure-like activity if the excitatory coupling strength falls below a certain threshold. This fact certainly contradicts the common belief that strong excitatory coupling is needed to synchronize neurons.

3.4 Background interictal activity and IED synchronization

In order to identify IEDs, they must appear simultaneously in several neighboring electrodes, as it was mentioned in a previous section. However, this fact does not imply the existence of a true synchronization of IED ("conduction synchrony" in Lachaux et al., 1999), but merely it reproduces the field of the brain source, which is captured by several, neighboring electrodes at the same time. Characterization and tracking of IED have a long history in epilepsy due to its clinical importance (Bourien et al., 2005 and references therein). Source localization (Townsend & Ebersole, 2008), the search of IED generators, is perhaps the main electroencephalographic task in the pre-surgical studies of drug-resistant epileptic patients. Simultaneous occurrence of IED in distant areas, however, is considered as true synchronization phenomena and in fact, it was one of the earliest demonstrations of synchronization at a large scale (Spencer & Spencer, 1994) in epileptic patients, demonstrating hippocampal-entorhinal interaction. Since then, IED and synchronization have being side by side many times. This fact contrasts with the much more recent works which use instead the full interictal background signal to assess synchronization (Mormann et al., 2000). Furthermore, it is known that IED are present in records from epileptic patients, and for instance, they usually appear more frequently in the focal side than the contralateral side, in temporal lobe epilepsy. Thus, one question, to our best knowledge not being explored until very recently (Ortega et al., 2010) is the relation between interictal background signal and the IED content of the signal. It would be nice to know, for instance, if and how the presence of interictal spikes affects synchronization calculations in an EEG record.

One further issue that must be addressed before to compare IED synchronization against interictal background activity is the difference between the kind and quantity of IED present at the several neurophysiological records: EEG, foramen ovale electrodes (FOE), ECoG, depth electrodes, etc. An EEG IED (Walczak et al., 2008) must have a duration < 200 ms. In particular, a sharp wave has a temporal structure between 70 ms and 200 ms and spikes must have a duration < 70 ms. These values seem to be invariant along neurophysiological techniques, while the amount of IED present in these records may vary largely depending of the selected procedure. Although the EEG spikes' frequency varies largely, a typical epileptic record with frequent spikes has a spikes' frequency > 60 spike/hour or 1 spike/min, but it can be as high as 4 spikes/min or more. In FOE, a typical record shows approximately 3.5 spikes/min (Clemens et al., 2003: Ortega et al., 2010), but it can as high as 30 spikes/min. We can estimate roughly a proportion of 4 FOE/EEG, that is, there are 4 FOE spikes for a single EEG spike. This ratio is extreme in the case of ECoG. It has been showed (Tao et al., 2005) that most of intracranial spikes, recorded with subdural electrodes, are missing in the EEG record. Only 10% of cortical spikes, with a source < 10 cm^2 are also recorded in scalp EEG. We can therefore estimate a relation of 10 ECoG/EEG. Thus, it is safe to estimate a higher frequency of 40 spikes/min for a ECoG record. Depth electrodes seem to posses the highest spike rate (Bourien et al., 2005), 3322 spikes/hour, or equivalently, 55.3 spikes/min, with a maximum of approximately 6100 spikes/hour, or 100 spikes/min.

The above numbers allow us to estimate the "percentage of time" that spikes occupies in the background signal. Considering the maximum duration of a spike as 200 ms, or 0.2 seconds, we will get for each recording technique the following:

EEG: 0.2 sec x 4 /60 sec = 0.012

FOE: 0.2 sec x 30/ 60 sec = 0.1

ECoG: 0.2 sec x 40 / 60 sec = 0.133

Depth: 0.2 sec x 100 / 60 sec = 0.33

The above numbers yields the following conclusions: IED structures "occupy", at most 1.2% of the total time in an EEG record, at most 10% in a FOE record, at most 13,3% in an ECoG record and a maximum of 33.3 % for a depth electrode record. Note that our assumptions have been extremely conservatives by excess, taking maxima number of spikes in each case. With those percentages of time that spikes occupy in each type of neurohysiological record, we have implemented the following simulation, in order to evaluate the potential influence of IED in the full signal synchronization. A single IED structure typically last at most 200 ms. We then generate two Gaussian stochastic processes with a given value of correlation and we have replaced part of the records by a simulated IED. The simulated IED, represented by a sine wave cycle, is inserted at the same time in both time series. In this way we can study the influence of IED time, on the synchronization estimate by increasing the proportion of the IED time in the records, ranging from 0% (no IED) to 100% (a whole IED). Lastly we have plotted the ratio between ρ(stoch,stoch), that is, the correlation between both purely stochastic process, to ρ (stoch+IED,stoch+IED), the correlation between the processes with the inserted simulated IED. This ratio, ρ (stoch, stoch)/ ρ (stoch+IED, stoch+IED) allows to compare a given value of correlation between a pair of experimental signals against a pair of signals with theoretical IED inserted. This is showed in Figure 5. As can be seen, for small levels of correlation between both signals the influence of IED time is much higher (left part of the figure) than the case of high levels of correlation (right part), and also for high proportion of IED time (upper left part). Horizontal lines indicate the maximum percentage of IED times contained in the typical neurophysiological records, as calculated before. By knowing the calculated values of synchronization, in this case Pearson correlation, it is possible to know whether IED are affecting the synchronization estimate or not. For example, in the case of EEG, practically there are no correlation values in the experimental series which can be influenced by the IED contained. With as lower values as 1.2% of the EEG time series occupied by IED, the value of the correlation is due to the background. The case of depth electrodes is different. Due to the high quantity of spikes integrated in the background activity, which can account as much as 33 % of the record time, alert that full signal synchronization may be due to the occurrence of synchronized spikes instead of the background activity. In this case, it would be mandatory to carefully analyze the correlation values obtained from signals. As it is clear from the Figure 5, values of correlation around 0.4 (x-axis) for instance, for depth electrodes (dot-dashed line), gives a value approximately of 0.7 (right scale). This value implies that 70% of the correlation is due to the background activity and 30% due to the synchronized spikes. These values are rather high and must be analyzed in detail.

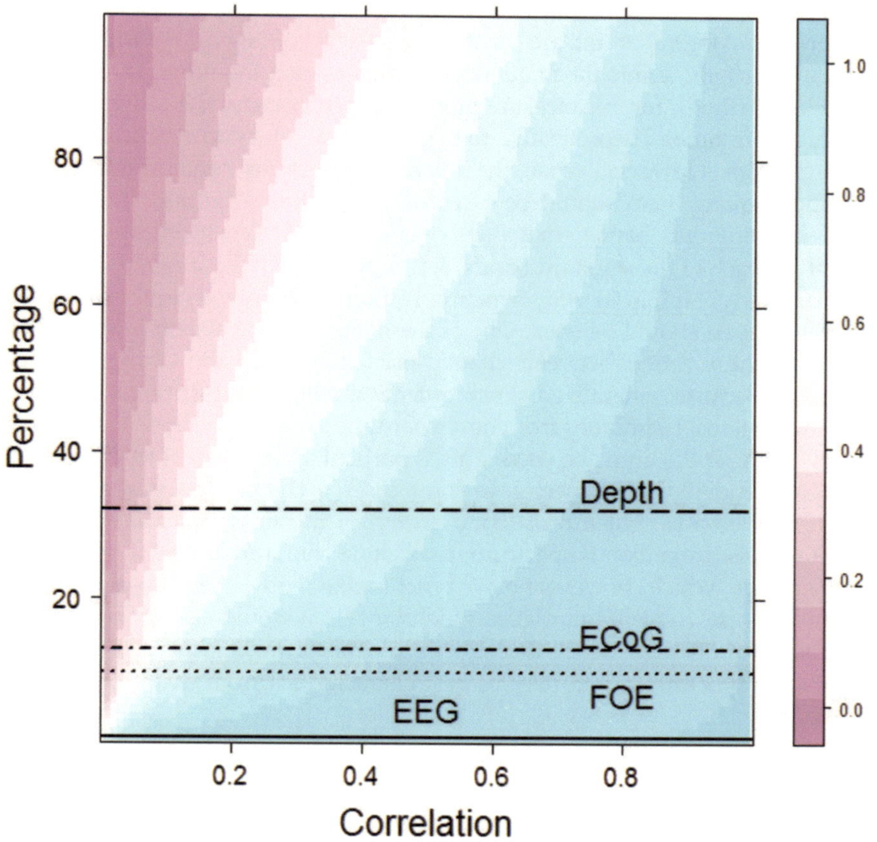

Fig. 5. Background interictal and IED synchronization

The above procedure is aimed to demonstrate that neurophysiological records from epileptic patients, which usually contain large quantities of IED, in the form of spikes or sharp waves, can be used safely for synchronization calculations because the presence of these IED hardly affect the results. For example, in the case of TLE, there exist reports that interictal synchronization is augmented in the hemisphere ipsilateral to epileptic seizures, calculated by using EEG records. From the above discussion, it must be discarded the influence of the greater quantities of IED in the ipsilateral side as a cause of the higher synchronization in the epileptic side. Furthermore, recent findings (Ortega et al., 2010) show the existence of lower values of synchronization in the ipsilateral side, as compared against the contralateral one, measured by FOE at the entorhinal cortex. In this case, the ipsilateral side contain higher quantities of IED, being the epileptic side, but with a decreased value of synchronization. This last fact seems soundly corroborate the minimal influence of IED over the background activity for the synchronization calculation purposes.

For the sake of completeness, we may generalize the above issue, in which we have compared IED synchronization with background synchronization, with the issue of broadband synchronization versus narrow-band synchronization. In general, cortical and scalp data, which are weighted space averages of many cortical rhythms (Nunez & Srinivasan, 2005) display broadband spectrum, from 0 to 80 Hz. This range is even greater, up to 500 Hz, when considering fast ripples in epileptic patients. However, many studies on synchronization, whether neurophysiological or not, are carried out on data coming from narrow-band systems, where typically a single dominant frequency prevails among the other. In these cases phase synchronization methods, for instance, are more sensitive, because the prevailing frequency is relatively stable in time. Theoretically, this is the case of synchronization between two nonlinear oscillators. However, when the interacting systems contain several frequencies, synchronization among them should be assessed with extreme caution. Interaction between frequencies in different bands may be overlooked if synchronization is calculated independently in each of the standard bands, δ, θ, α, β and γ. In these cases, standard methods of synchronization will fail in identify synchronization, and more general methods, or higher order spectral analysis should be used. One of them, for instance, which is applied in the case of assessing synchronization between chaotic oscillators, uses the concept of generalized synchronization. Because generalized synchronization relies in a functional relation between the underlying systems, techniques from nonlinear time series analysis, as it is the embedding methodology (Stam, 2005), must be used. This is particularly important in the case of systems with variable-dependent power spectra. In some cases, a chaotic system may shows a sharp frequency in one variable, but a broadband spectrum in the other variables (Ortega, 1995, Ortega, 1996). This broadband spectrum may hide the fundamental frequency of the underlying system, which however may synchronize with other systems through that frequency. In this case, a solely time series synchronization would also ignore this interaction. From the above comments, when dealing with neurophysiological data, especially from epileptic patients, synchronization may be best characterized from this broadband approach (Netoff & Schiff, 2002).

3.5 Seizures: Hypersynchronization, desynchronization, propagation or lag-synchronization?

Although seizures are the most prominent symptoms in the epileptic patient, few attempts to characterize seizures from a dynamical point of view have been done. It would be very valuable to know how synchronization evolves during the whole period of a typical seizure. The classical association of hypersynchronization with seizure events does not seems appropriate today, in the light of recent advances in this field (Schiff et al., 2005; Netoff & Schiff, 2002; Schindler et al., 2007). Augmented synchronization, as compared with pre-ictal stages, is only a small part of the seizure dynamic, where desynchronized activity dominates in the first part (Schindler et al., 2007), and in some cases at the end (Schiff et al., 2005) of the seizure. Bearing in mind that synchronization is usually measured by the zero-lag correlation coefficient, one may speculate that desynchronization is actually due to the spread of seizure activity. If seizures, starting at the particular focus, propagate by different routes and with different lags (Milton et al., 2007), one would expect that epileptogenic activity at different cortical points be zero-decorrelated, but lag-correlated instead. Although we have not enough data to support that desynchronization is caused by an "inhomogenous" propagation, we have some clues to support the hypothesis that interictal desynchronization, at least in TLE patients, may be the cause of desynchronized spread. Recently (Ortega et al., 2010), we have shown the existence of an imbalance in the mesial synchronization in TLE patients by using a cluster methodology (see sec. 2.4.4.). FOE records of interictal activity at the entorhinal cortex show the existence of lower levels of synchronization in the ipsilateral side of the epileptic seizures than the contralateral one. Figure 6 display the distance matrix, from a right TLE patient, among all the electrodes calculated from the correlation matrix (Equation (13)). It is apparent the existence of a great synchronization cluster among almost all the FOE electrodes, except three right electrodes (small circles), R4, R5 and R6. These electrodes are declusterized and desynchronized from the rest. If, at is believed (Kandel et al., 2000), the spread of seizures follows normal cortical circuits, it would be expected that seizure propagation will be desynchronized, at least at the onset stage, due to the synchronization inhomogeneity of the cortical substrate. Moreover, interictal synchronization inhomogeneities have been reported in the lateral cortex (Ortega et al., 2008) in TLE patients. Therefore, it is highly expected that during the seizure onset, the seizure spreading, both in the mesial and in the lateral side of the temporal lobe, display a desynchronized activity, at least through the intracortical propagation. As the seizure develops, and in the case of secondary generalization, ictal activity spread to sub-cortical structures which came back to cortex, perhaps reinforcing synchronization.

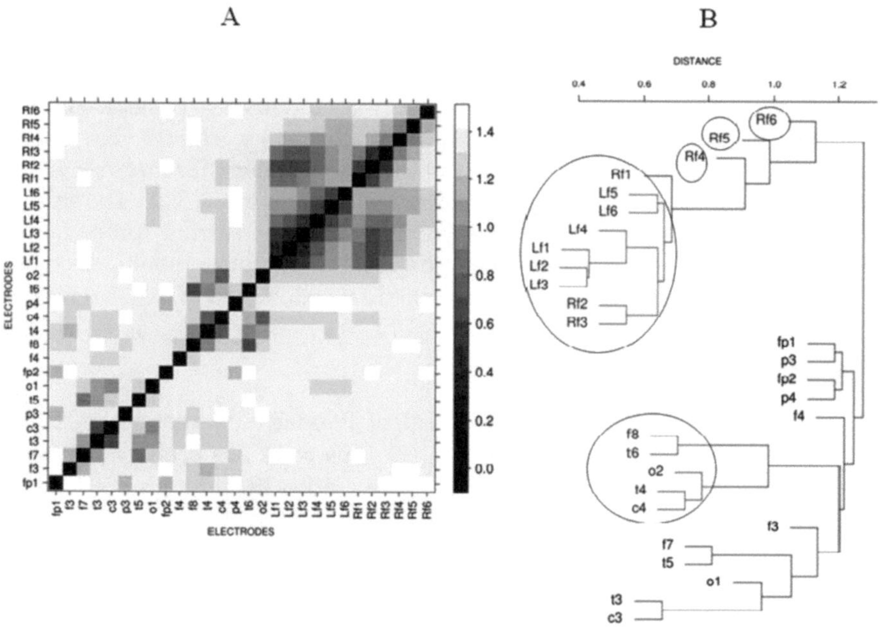

Fig. 6. Distance matrix (A) and hierarchical clusters (B) of scalp and FOE electrodes in TLE patients.

4. Conclusions

Recurrent seizures are the hallmark of epilepsy. They are usually termed as "hypersynchronized" events in the same fashion as the interictal epileptogenic discharges, which are the traditional EEG signatures of epilepsy. However, the term hypersynchronization seems to be misleading. In the case of seizures, a complex dynamics involving transitions from desynchronization to synchronization and perhaps distant and final lag synchronization seems to be the principal manifestation. In the case of interictal epileptogenic discharges, it seems more appropriate to justify its appearance due to the synchronizability of small cellular networks in the hippocampus and neocortex. In this last case perhaps, a correct term to be used would be full synchronization of all the participants in the underlying network.

With the support of chaos and network theories, the classical concept of synchronization has evolved substantially in the last decade. This fact has made possible recognize and differentiate new types of synchronizing mechanisms between two or more systems, even en the cases with very different corresponding signals. These new theoretical and methodological advances seem to be far from the epileptology realm yet. Although the number of new synchronization techniques applied in the analysis of neurophysiological data is growing, the step toward the introduction of these concepts in the daily language of clinical practice seems distant however. This is a disappointing situation, because the correct characterization of every synchronization process in epilepsy would boost new research, both in direction and deepness, certainly improving the fragmented knowledge we have of this disease today.

We believe that the recent advances in network theory and in particular, dynamics and synchronization in complex networks are crucial issue to be applied in the epilepsy field. Epilepsy is a neurological disorder and the brain is a network at its various scales. Seizure is intrinsically a synchronization/desynchronization manifestation, the correct characterization and dynamical properties of synchronization in the underlying cellular and cortical networks is the appropriate ground where this pathology should be considered. Characterizing the mesoscale synchronizability in the limbic system, for instance would be a step forward in the understanding of the appearance of seizures in the TLE. The study of the stability of synchronization, would further understand the seizure propagation and generalization. Synchronizability and synchronization stability should be therefore considered as new paradigms and future efforts should be addressed in that direction.

5. Acknowledgments

GJO is grateful to Rosario Ortiz de Urbina, head of Fundación Investigación Biomédica Hospital de la Princesa for her encouraging support. This work has been funded by grants from Fundación Mutua Madrileña, Instituto de Salud Carlos III, through PS09/02116 and PI10/00160 projects, and PIP N° 11420100100261 CONICET. GJO is member of CONICET, Argentina.

6. References

Arenas, A., Díaz-Guilera, A., Kurths, J. Moreno, Y. & Zhoug, C. (2008). Synchronization in complex networks. *Physics Reports*, Vol. 469, pp.93-153.

Arroyo, S., Lesser, R.P., Fisher R.S., Vining, E.P., Krauss, G.L., Bandeen-Roche, K., Hart, J., Gordon, B., Uematsu, S. & Webber, R. (1994). Clinical and electroencephalographic evidence for sites of origin of seizures with diffuse electrodecremental pattern. *Epilepsia*, Vol.35, pp.974-987.

Babb, T.L., Wilson, C.L. & Isokawaakesson, M. (1987). Firing patterns of human limbic neurons during stereoencephalography (SEEG) and clinical temporal-lobe seizures. *Electroencephal. Clin. Neurophysiology*, Vol.66, pp. 467–482.

Bettus, G., Wendling F., Guye M., Valton L., Regis J., Chauvel P. & Bartolomei F. (2008). Enhanced EEG functional connectivity in mesial temporal lobe epilepsy. *Epilepsy Research,* Vol.81, pp. 58-68.

Boccaletti, S., Latora, V., Moreno, Y., Chavez, M. & Hwanga, D.-U. (2006). Complex networks: Structure and dynamics. *Physics Reports,* Vol.424, pp.175–308.

Bourien, J., Bartolomei, F., Bellanger, J.J., Gavaret, M., Chauvel, P. & Wendling, F. (2005). A method to identify reproducible subsets of co-activated structures during interictal spikes. Application to intracerebral EEG in temporal lobe epilepsy . *Clinical Neurophysiology*, Vol.116, pp.443–455.

Chang, B.S., Schomer, D.L. & Niedermeyer, E. (2011). Epilepsy in Adults and the Elderly. In: *Niedermeyer's Electroencephalography: Basic Principles, Clinical Applications, and Related Fields, 6th Edition*. Schomer, D.L. & Lopes da Silva, F. Lippincott Williams & Wilkins.

Clemens, Z., Janszky, J., Szucs, A., Békésy, M., Clemens, B. & Halász, P. (2003). Interictal epileptic spiking during sleep and wakefulness in mesial temporal lobe epilepsy: a comparative study of scalp and foramen ovale electrodes. *Epilepsia*, Vol.44, No.2, pp.186–92.

Cover, M. & Thomas, J.A. (2006). *Elements of information theory, 2nd ed.* John Wiley & Sons, Inc.

Friston, K.J., Frith, C.D. & Frackowiak, R.S.J. (1993). Time-dependent changes in effective connectivity measured with PET. *Hum. Brain Mapp.* Vol.1, pp.69–80.

Gastaut, H., Roger, J., Ouahchi, S., Timsit, M. & Broughton, R. (1963). An electro-clinical study of generalized epileptic seizures of tonic expression. *Epilepsia,* Vol.4, pp.15-44.

Girvan, M., & Newman, M.E.J. (2002) Community structure in social and biological networks. *Proc. Natl. Acad. Sci. USA*, Vol.99, pp.7821-7826.

Granger, C.W.J. (1969). Investigating causal relations by econometric models and cross-spectral methods. *Econometrica*, Vol.37, pp.424–438.

Guevara, R., Velazquez, J.L., Nenadovic, V., Wennberg, R., Senjanovic, G. & Dominguez, L.G. (2005). Phase synchronization measurements using electroencephalographic recordings: what can we really say about neuronal synchrony?. *Neuroinformatics*, Vol.3, No.4, pp. 301–314.

Gyimesi, C., Pannek, H., Woermann, F.G., Elsharkawy, A.E., Tomka-Hoffmeister, M., Hortsmann, S., Aengenendt, J., Horvath, R.A., Schulz, R., Hoppe, M., Janszky, J. and Ebner, A. (2010). Absolute spike frequency and etiology predict the surgical outcome in epilepsy due to amygdala lesions. *Epilepsy Research* Vol. 92, pp.177-182.

Hilgetag, C. & Kaiser, M. (2004) Clustered organization of cortical connectivity, *Neuroinformatics*, Vol.2, pp.353-360.

Kandel, E.R., Schwartz, J.H. & Jessell, T.M. (2000) *Principles of Neural Science, 4th edition*, McGraw-Hill, New York

Krendl, R., Lurger, S. & Baumgartner, C. (2008) Absolute spike frequency predicts surgical outcome in TLE with unilateral hippocampal atrophy. *Neurology*, Vol.71, pp.413–8.

Kuramoto, Y. (1975). Self-entrainment of a population of coupled nonlinear oscillators, In: *International Symposium on Mathematical Problems in Theoretical Physics, in: Lecture Notes in Physics, vol. 39*. H. Araki, pp. 402-422, Springer, New York, NY, USA.

Lachaux, J.-P, Rodriguez, E., Martinerie, J. & Varela, F.J. (1999). Measuring Phase Synchrony in Brain Signals. *Human Brain Mapping*, Vol.8, pp.194–208.

Le Van Quyen, M., Navarro, V., Martinerie, J., Baulac, M., & Varela, F.J. (2003). Toward a neurodynamical understanding of ictogenesis. *Epilepsia*, Vol.44, No.S12, pp.30–43.

Lehnertz, K., Bialonski, S., Horstmann, M.T., Krug, D., Rothkegel, A., Staniek, M. & Wagner, T. (2009) Synchronization phenomena in human epileptic brain networks. *Journal of Neuroscience Methods*, Vol.183, pp. 42–48.

McCormick, D.A. & Contreras, D. (2001). On the cellular and network bases of epileptic seizures . *Annu. Rev. Physiol.* Vol. 63, pp.815–46

Milton, J.G., Chkhenkeli, S.A. & Towle, V.L. (2007). Brain Connectivity and the Spread of Epileptic Seizures. In: *Handbook of Brain Connectivity*. Jirsa, V.K. & McIntosh, A.R. pp. 477-503. Springer Berlin Heidelberg New York.

Morgan, R.J. & Soltesz, I. (2008). Nonrandom connectivity of the epileptic dentate gyrus predicts a major role for neuronal hubs in seizures, *PNAS* , Vol. 22, No.16, pp.6179-6184.

Mormann, F., Kreuz, T., Andrzejak, R.G., David, P., Lehnertz, K. & Elger, C.E. (2003) Epileptic seizures are preceded by a decrease in synchronization. *Epilepsy Research*, Vol.53, No.3, pp.173-85.

Mormann, F., Lehnertz, K., David, P. & Elger, C.E. (2000). Mean phase coherence as a measure for phase synchronization and its application to the EEG of epilepsy patients. *Physica D*, Vol.144, pp.358-369.

Netoff, T.I. & Schiff, S.J. (2002). Decreased neuronal synchronization during experimental seizures. *J. Neuroscience*, Vol.22, pp.7297-7307.

Noachtar, S. & Rémi, J. (2009). The role of EEG in epilepsy: A critical review. *Epilepsy & Behavior*, Vol.15, pp. 22–33

Nunez, P. L. & Srinivasan, R. (2005). *Electric Fields of the Brain: the Neurophysics of EEG 2nd.* Oxford Univ. Press, New York.

Ortega, G.J. (1995). A new method to detect hidden frequencies in chaotic time series. *Phys. Lett. A*, Vol.209, pp.351–355.

Ortega, G.J. (1996). Invariant measures as lagrangian variables: Their application to time series analysis. *Physical Review Letters. Lett. A*, Vol.77, No.2, pp.259-262.

Ortega, G.J., Herrera Peco, I., García de Sola, R. & Pastor, J. (2010) Impaired mesial synchronization in temporal lobe epilepsy. *Clinical Neurophysiology*, Vol.122, No.6, pp.1106-1116.

Ortega, G.J., Menendez de la Prida, L., García de Sola, R. & Pastor, J. (2008). Synchronization Clusters of Interictal Activity in the Lateral Temporal Cortex of Epileptic Patients: Intraoperative Electrocorticographic Analysis. *Epilepsia*, Vol.49, No.2, pp.269-280.

Pastor, J., Hernando, V., Domínguez-Gadea, L., De Llano, I., Meilán, M.L., Martínez-Chacón, J.L. & Sola, R.G. (2005) Impact of experience on improving the surgical outcome in temporal lobe epilepsy. *Revista de Neurología*, Vol.41, No.12, pp. 709-716.

Pastor, J., Menéndez de la Prida, L., Hernando, V. & Sola, R.G. (2006). Voltage sources in mesial temporal lobe epilepsy recorded with foramen ovale electrodes. *Clinical Neurophysiology*, Vol. 117, No.12, pp. 2604-2614.

Pastor, J., Sola, R.G., Hernando-Requejo, V., Navarrete, E.G. & Pulido, P. (2008). Morbidity associated with the use of foramen ovale electrodes. *Epilepsia*, Vol.49, No.3, pp.464-9.

Pastor, J., Wix, R., Meilán, M.L., Martinez-Chacon, J.L., de Dios, E., Dominguez-Gadea, L., Herrera-Peco, I. & Sola, R.G. (2010). Etomidate accurately localizes the epileptic area in patients with temporal lobe epilepsy. *Epilepsia*, Vol.51, No.4, pp. 602-609.

Penfield, W. & Jasper, H. (1954). *Epilepsy and the Functional Anatomy of the Human Brain.* Brown and Company Boston, MA.

Pereda, E., Quian Quiroga, R. & Bhattacharya, J. (2005). Nonlinear multivariate analysis of neurophysiological signals. *Progress in Neurobiology*, Vol.77, pp. 1–37.

Pikovsky, A., Rosenblum, M. & Kurths, J. (2001). *Synchronization: A Universal Concept in Nonlinear Sciences.* Cambridge Nonlinear Science Series.

Press, W.H., Teukolsky, S.A., Vetterling, W.T. & Flannery, B.P. (2007). *Numerical Recipes, The Art of Scientific Computing, 3rd Edition*, Cambridge University Press.

Quian Quiroga, R., Kraskov, A., Kreuz, T. & Grassberger, P. (2002) Performance of different synchronization measures in real data: A case study on electroencephalographic signals. *Physical Review E*, Vol.65, pp.041903.

Rosenblum, M.G., Pikovsky, A.S. & Kurths, J. (1996). Phase Synchronization of Chaotic Oscillators. *Physical Review Letters*, Vol.76, pp.1804-1807.

Schiff, S.J., Sauer, T., Kumar, R. & Weinstein, S.L. (2005). Neuronal spatiotemporal pattern discrimination: the dynamical evolution of seizures. *Neuroimage*, Vol.28, pp.1043–55.

Schindler, K., Leung, H., Elger, C.E. & Lehnertz, K. (2007) Assessing seizure dynamics by analyzing the correlation structure of multichannel intracranial EEG. *Brain*, Vol.130, pp.65–77.

Schreiber, T. (2000). Measuring information transfer. *Phys. Rev. Lett.* Vol.85, pp.461- 464.

Speckmann, E.-J., Elger, C.E. & Gorji, A. (2011). Neurophysiologic Basis of EEG and DC Potentials. In: *Niedermeyer's Electroencephalography: Basic Principles, Clinical Applications, and Related Fields, 6th Edition*. Schomer, D.L. & Lopes da Silva, Lippincott Williams & Wilkins.

Spencer, S.S. & Spencer, D.D. (1994) Entorhinal-Hippocampal Interactions in Medial Temporal Lobe Epilepsy. *Epilepsia*, Vol.35, No.4, pp721-727.

Stam, C.J. (2005). Nonlinear dynamical analysis of EEG and MEG: Review of an emerging field . *Clinical Neurophysiology*, Vol.116, pp.2266–2301.

Tao, J.X., Ray, A., Hawes-Ebersole, S. & Ebersole, J.S. (2005). Intracranial EEG substartes of scalp EEG interictal spikes. *Epilepsia*, Vol.46, pp.669–676.

Tatum, W.O., Husain, A.M. Benbadis, S.R. & Kaplan, P.W. (2008). *Handbook of EEG interpretation*. Demos Medical Publishing, LLC.

Timofeev, I. & Steriade, M. (2004). Neocortical seizures: initiation, development and cessation. *Neuroscience*, Vol.123, pp.299–336.

Townsend, T.N. & Ebersole, J.S. (2008). *Source localization of electroencephalography spikes,* In: Textbook of epilepsy surgey. Luders, H.O. Informa UK Ltd.

Traub, R.D., Bibbig, A., LeBeau, F.E.N. Buhl, E.H. & Whittington, M.A. (2004). Cellular mechanisms of neuronal population oscillations in the hippocampus in vitro. *Annual Review of Neuroscience,* Vol. 27, pp.247-278.

van Drongelen, W., Lee, H.G., Hereld, M., Chen, Z.Y., Elsen, F.P. & Stevens, R.L. (2005). Emergent epileptiform activity in neural networks with weak excitatory synapses. *IEEE Trans. Neural Sys. Rehab. Engn.*, Vol.13, pp.236–241.

Walczak, T.S., Jayakar, P. & Mizrahi, E.M. (2008). Interictal Electroencephalography. In: *Epilepsy: A Comprehensive Textbook, 2nd Edition.* Engel, J. & Pedley, T.A., Lippincott Williams & Wilkins.

Wendling, F., Bartolomei, J., Bellanger, J., Bourien, J. & Chauvel, P. (2003). Epileptic fast intracerebral EEG activity: evidence for spatial decorrelation at seizure onset. *Brain,* Vol.126, pp.1449-1459.

Winfree, A.T. (1967). Biological rhythms and the behavior of populations of coupled oscillators, *J. Theoret. Biol*ogy, Vol.16, pp.15-42.

Zhou, C., Zemanova, L., Zamora, G., Hilgetag, C.C. & Kurths, J. (2006). Hierarchical Organization Unveiled by Functional Connectivity in Complex Brain Networks. *Physical Review Letters,* Vol.97, pp.238103.

Zumsteg, G.H., Friedman, A., Wieser, H.G. & Wennberg R.A. (2006) Propagation of interictal discharges in temporal lobe epilepsy: correlation of spatiotemporal mapping with intracranial foramen ovale electrode recordings. *Clinical Neurophysiology,* Vol. 117, pp. 2615-2626.

7

Long-Term Monitoring: An Overview

B. Mesraoua[1], D. Deleu[1,2] and H. G. Wieser[3]
[1]*Hamad Hospital, Doha,*
[2]*Weill Cornell Medical College, Doha,*
[3]*University of Zurich,*
[1,2]*Qatar*
[3]*Switzerland*

1. Introduction

This chapter consists of two parts. In the first part, we will review the indications for long-term EEG monitoring in the diagnosis and management of epilepsy. This will include the differentiation of seizures and epilepsy, appropriate diagnostic and treatment options, the techniques and methods used in the EEG monitoring (ambulatory - hospital-based - noninvasive and invasive video/EEG telemetry), commercially available monitoring systems, and presurgical evaluation of candidates for epilepsy surgery. In the second part, we will focus on the continuous EEG monitoring in the Intensive care unit (ICU). In doing so, we will describe which ICU patients should undergo continuous EEG (cEEG) monitoring; how long such patients should be monitored? We will review the epidemiology of nonconvulsive status epilepticus (NCSE) and the pathological EEG patterns commonly occurring in the critically ill patient, including NCSE and periodic epileptiform discharges (PEDs). We will discuss the methods which enable real time monitoring of seizures, and the adequate treatment of these seizures in a timely fashion. We will also discuss the hypothesis that progressive brain damage is occurring with NCSE or PEDs and also whether ICU patients with "controversial EEG patterns" should receive aggressive treatment. Finally, we discuss the hypothesis that biomarkers, such as neuron- specific enolase (NSE), and magnetic resonance imaging (MRI) techniques (diffusion and perfusion MRI) as well as MR-Spectroscopy, allow to detect the EEG patterns which might result from brain damage or are associated with it.

2. Long-term EEG monitoring in the diagnosis and management of epilepsy

Long-term Monitoring (LTM) is useful and mandatory in the diagnosis and management of epilepsy in epilepsy surgery centers and Intensive care units (ICU). It is used for differentiating between epilepsy and paroxysmally occurring nonepileptic conditions, for assessing seizure type and frequency, for diagnosing and treatment of status epilepticus, and in noninvasive and invasive video/EEG investigations for epilepsy surgery. The Subcommittee of the International League against Epilepsy (ILAE) recommends use of hospital-based LTM in these conditions. The management of patients with head trauma, stroke, subarachnoid hemorrhage and other cerebrovascular disorders benefit from the continuous monitoring of the EEG.

Techniques/methods. From the techniques/methods one distinguishes: *ambulatory* versus *hospital-based*; and *noninvasive* versus *semi-invasive* and *invasive* video/EEG telemetry.

The ambulatory EEG monitoring uses a small portable EEG recorder which is worn on a waist belt. Recording can thereby be made during regular daytime activities and sleep, in the natural environment (see *Figure* 3). The EEG recorder records events that can be reviewed later on a special machine in an EEG laboratory. The patient should be asked to keep an account of daily activities, so that they can be related to the EEG recordings made at the time.

In comparison to an inpatient monitoring, ambulatory outpatient monitoring may be less informative because: (a) reduction of medication to provoke seizures may not be safe in an outpatient, (b) faulty electrode contacts cannot quickly be noticed and repaired, (c) the patient may move out of video surveillance, and (d) duration of ambulatory monitoring can be limited by technical constraints.

Video-telemetry uses a video camera linked to an EEG machine. The camera will visually record the behavior of the patient, in particular objective signs and symptoms of the seizure, and at the same time the EEG machine will record the brainwave pattern. Both the video and EEG are stored onto a computer that can be reviewed once the test is finished. The consultant will be able to see any episodes/seizures that the patient may have had, as well as any changes in the EEG at that time. The test is often carried out over a number of days in order to increase the chances of recording habitual and/or all seizure types. Long-term video-EEG monitoring is particularly important in the presurgical evaluation of potential candidates for surgical epilepsy therapy, and in the ICU.

Requirements for LTM in general are that it should be suitable for all patient ages, including neonates, infants, children, and adults. It should be suitable for hard-to-record populations as well, such as children with autism. Superior patient safety is mandatory. The user should be able to view all data in standard clinical EEG as well as customizable montages. The computer system should allow for data sharing and data archiving. The system should use ultra-compact amplifier, integrated software design, digital video integration, and world class customer support. Depending on local clinical and research needs the integration of specific software for research use, e.g., source estimation software, and recordings with 32- to 256-channel (dense array EEG) should be available.

Requirements for EEG analysis are online methods that can be potentially used for records of any duration, with any number of channels, any number of channel groupings (different topologies), and in a variety of situations (ICU, sleep, coma, etc.). The system should display samples of original EEG that represent the long-term EEG along with their temporal distribution. The actual EEG should be presented to the user in a familiar way, so that no new interpretive skills are required, i.e., the method can be employed by anyone familiar with EEG. An undisputed advantage is a simple graphical display allowing a non-EEG specialist to identify abnormal changes, i.e., the emergence of focal changes, bursts or sustained asymmetries, gradual or sudden changes, and cycling of EEG patterns. The quick identification of problems will allow such personnel to contact the EEG specialist for a detailed assessment.

Commercially available LTM System are equipped with special hardware , software, and training items, including an amplifier, acquisition computer, LCD display, net station software for EEG acquisition and physician review. In certain conditions a HydroCel

Geodesic Sensor Net (HCGSN) may be useful. HCGSN does not require any scalp preparation or abrasion and is suitable for all patient ages. Application times for the sponge-based HCGSN range between 5 minutes for 32 channels to 15 minutes for 256 channels.

A digital video system is crucial, as is warranty and support. Equally important are on-site installation and training visit by a support engineer providing reference materials and manuals. Regular training workshops of the staff are usually required and helpful.

A reviev of some commercially available Monitoring Systems is given in Table 1.

AirEEG WEE-1000	Nihon Kohden Corporation	The patient wears a transmitter in a pouch and an electrode junction box in a shoulder strap. The transmitter sends up to 64 channels of EEG data to one or more receivers that are connected to an electroencephalograph by LAN. Wireless is more comfortable.
Beehive® Horizon - Long-term Epilepsy Monitoring Neurotrac® Neuromonitoring Software	Grass Technologies	Windows-based digital EEG/Video instrument with complete data acquisition, recording and review capabilities that are specialized for epilepsy monitoring applications. Neurotrac Neuromonitoring software is designed for computing and displaying long-term trends of EEG features during continuous EEG monitoring in the ICU, or Seizure Monitoring units. The software module can be used with acquisition systems configured with any of Grass Technologies' amplifiers. The number of EEG channels recorded is dependent on the amplifiers used, the electrodes applied to the patient, and the selected montage.
Ceegraph VISION EEG Ceegraph VISION ICU Ceegraph VISION LTM	Natus Bio-logic Systems	This system incorporates the latest developments in networking, remote communications and information management. Ceegraph VISION is a Windows® XP Pro system that can be expanded to integrate all EEG needs: epilepsy monitoring, routine EEG, ambulatory recordings, and ICU monitoring. Ceegraph VISION ICU is designed to virtually eliminate the complexities of recording and interpreting continuous EEG data used to assess neurological status using sophisticated EEG analysis software.

Table 1. Commercially Available Monitoring Systems

2.1 Diagnostic yield of LTM in the differential diagnosis between epilepsy and paroxysmally occurring nonepileptic conditions

Events that may cause diagnostic confusion with epilepsy are (a) Non-epileptic pathophysiological events, such as autonomic disorders, cardiac arrhythmias, drug toxicity, metabolic disorders, migraine, orthostatic hypotension, sleep disorders, valvular heart disease, vasovagal syncope, vestibular disorder, and (b) Non-epileptic psychopathological and psychiatric events, such as anxiety, depression, panic attacks, psychogenic "seizures" and other somatoform disorders, and psychosis.

Approximately 20% of patients referred to comprehensive epilepsy programs because of medically "refractory" seizures do not have epilepsy. In the study of Benbadis et al **[1]**, 75 patients (30%) of 251 patients with video EEG monitoring had non-epileptic attacks: 69 without epilepsy and 6 patients combined with epileptic attacks. 61 patients from the 69 patients had so-called "psychogenic" and 8 patients had "other" attacks.

Yogarajah et al. **[2]** studied the value of the ILAE-recommended use of LTM. All admissions to the Sir William Gowers Unit at the National Society for Epilepsy (an epilepsy tertiary referral unit) in the years 2004 and 2005 were included. In this study LTM was primarily responsible for a change in the diagnosis in 133 (58%) and a refinement of diagnosis in 29 (13%) patients. The most common change was in distinguishing epilepsy from non-epileptic attack disorder (NEAD) in 73 (55%) and in distinguishing between focal and generalized epilepsy in 47 (35%). LTM was particularly helpful in differentiating frontal lobe seizures from generalized seizures and non-epileptic attacks.

Besides the diagnosis of a seizure disorder and the classification of seizure types, the evaluation of precipitating factors and the quantification of seizures is often very important. Surgical localization is another issue and briefly dealt below.

2.2 LTM in the context of presurgical evaluation of candidates for epilepsy surgery

LTM in specialized institutions offering epilepsy surgery is often divided into a *Phase 1* and *Phase 2*. Phase 1 is *noninvasive.* In this early stage of examination, three main questions have to be answered: (1) Is it epilepsy? (2) What type of epilepsy? And (3) Is it pharmacotherapy resistant? *Phase 2* consists of semi-invasive, and invasive techniques to pinpoint the area(s) where the seizures originate if pharmacotherapy-resistant epilepsy has been determined. Only when a surgically remediable seizure onset zone is determined,then curative epilepsy surgery can be considered.

Epilepsy surgery includes the categories *curative* (=causal =resective) and *palliative* epilepsy surgery. Curative means resection of the primary epileptogenic zone, i.e. resection of the actual and potential seizure onset zone. The goal is postoperative seizure freedom. Types of resections include more or less *standardized* resections as opposed to *tailored* resections. Standardized resections include interventions such as the "selective amygdalohippocampectomy" **[3-4]**, anterior temporal lobe resection, and functional hemispherectomy (see *Figure 1*).

Not infrequently the localization of the seizure onset zone in so-called "eloquent", i.e. indispensable brain area and/or the size of the seizure onset zone prevent its radical therapeutic excision. In such a case the "secondary pacemaker areas" (e.g. "ipsilateral discharge amplifiers") may be targeted. For example, in a left posterior neocortical temporal lobe epilepsies with rapid propagation of the discharge into the ipsilateral mesial temporal structures (-which act then as a secondary pacemaker -) a so-called "palliative hippocampectomy" has sometimes been performed with good results **[5]**. However, palliative interventions usually cannot provide postoperative seizure freedom, but may

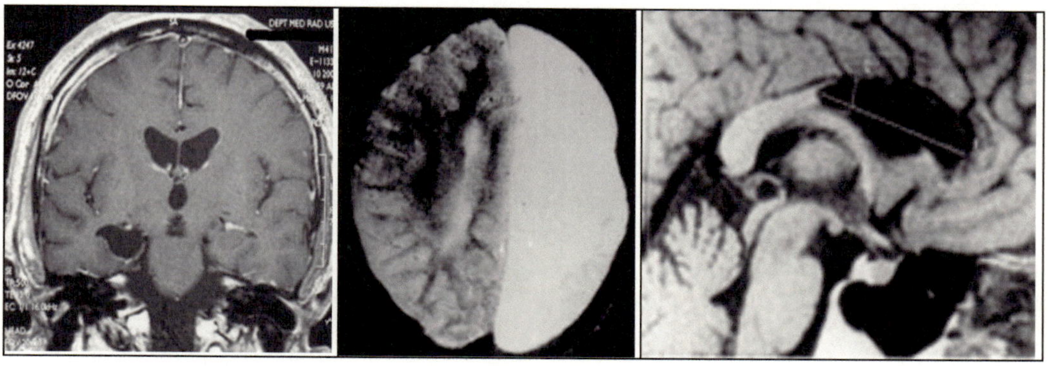

Fig. 1. Illustration of types of resective epilepsy surgery showing the MRI after selective amygdalohippocampectomy (*left*), after hemispherectomy (*middle portion*) and after anterior callosal section (*right*).

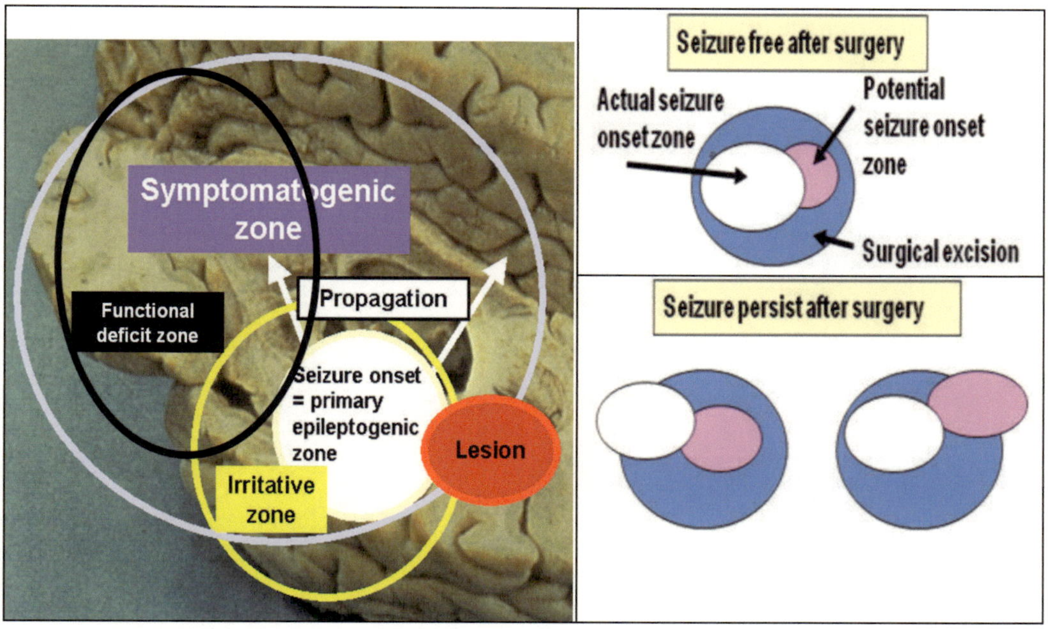

Fig. 2. Illustration of the conceptual zones of the "epileptic" focus. The *lesional zone* is characterized by the anatomical, i.e. morphological lesion; the *irritative zone* by the territory exhibiting "interictal spiking". The *functional deficit zone* is characterized by neurological and/or neuropsychological deficits and/or functional imaging (PET,SPECT) abnormalities; the *symptomatogenic zone*, by the "primictal", i.e. very early, ictal signs and symptoms. Depending whether the patient is rendered seizure-free or not after surgical excision, the seizure onset zone can be further divided into an *actual* and *potential seizure onset zone*.

ameliorate the patient`s seizure situation. Palliative procedures include callosal section (i.e., corpus callosum section, CCS; including Gamma Knife CCT), vagal nerve stimulation (VNS) and deep brain stimulation (DBS).

Curative epilepsy surgery relies on the concept of the "*epileptic focus*". At least for practical purposes, the "focus"- concept is still valuable, although the *network hypothesis* might be more realistic. Using stereo-electroencephalography (SEEG) Talairach and Bancaud have described a "zone lésionelle", a "zone irritative", and a "zone épileptogène" **[6]**. Wieser **[7-8]** and Lüders **[9]** enlarged the focus concept to the following zones (1) *lesional* (2) *primary epileptogenic,* (3) *secondary epileptogenic* (including the "mirror focus", and the "ipsilateral discharge amplifier"), (4) *irritative,* (5) *functional deficit,* and (6) *symptomatogenic.* Lüders et al. **[10]** then divided the epileptogenic zone into the (a) *actual,* and (b) *potential seizure onset* zones, depending whether the patient is rendered seizure-free or not after surgical excision. It is very important to note that the seizure onset zone is not identical with the lesion. Often it is in the vicinity of a lesion, but can be located in or involve remote areas [*Figure* 2].

Long-term video-EEG monitoring can be performed using conventional scalp electrodes, "true temporal electrodes", sphenoidal electrodes, and in the context of presurgical evaluation the so-called *semi-invasive* and *invasive techniques. Foramen ovale (FO) electrodes* are semi-invasive **[11]**. The so-called *invasive techniques* include *stereo-electroencephalography* **[12]** and *subdural strip-* and *grid electrodes.* Not infrequently non-invasive, semi-invasive and invasive techniques are combined with the goal to optimize the diagnostic yield in an individual patient. **[8-9,13-14]**.

Figures 3 to 6 illustrate some aspects of LTM in the context of presurgical evaluation of candidates for epilepsy surgery. *Figures 7 to 9* some research aspect, for example the question

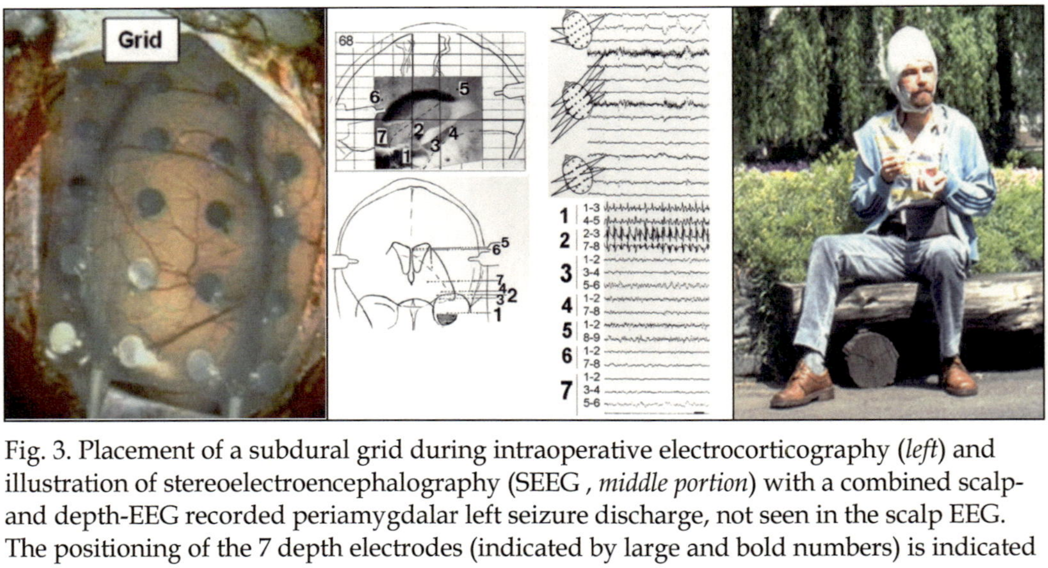

Fig. 3. Placement of a subdural grid during intraoperative electrocorticography (*left*) and illustration of stereoelectroencephalography (SEEG , *middle portion*) with a combined scalp- and depth-EEG recorded periamygdalar left seizure discharge, not seen in the scalp EEG. The positioning of the 7 depth electrodes (indicated by large and bold numbers) is indicated in the brain map. Each depth electrode has 10 contacts numbered 1 to 10 from inside to outside (small numbers). *At the right* ambulatory SEEG monitoring is illustrated (University Hospital Zurich 1984).

of spike activation and seizure precipitation associated with different sleep stages (*Figure 8*) and seizure prediction (*Figure 9*). For the latter the pre-ictal spike frequency, the surface negative DC-shift, the inter- and intrahemispheric coupling strength , and some indicators derived from the "chaos theory" (e.g. Lyapunov-exponents and correlation dimension) have been examined. *Figure 9* is an example that nonlinear time-series analysis with neuronal complexity loss in interictal EEG recorded with foramen ovale electrodes can predict a seizure. However, in the illustrated case the calculation was done off-line **[15-16]**.

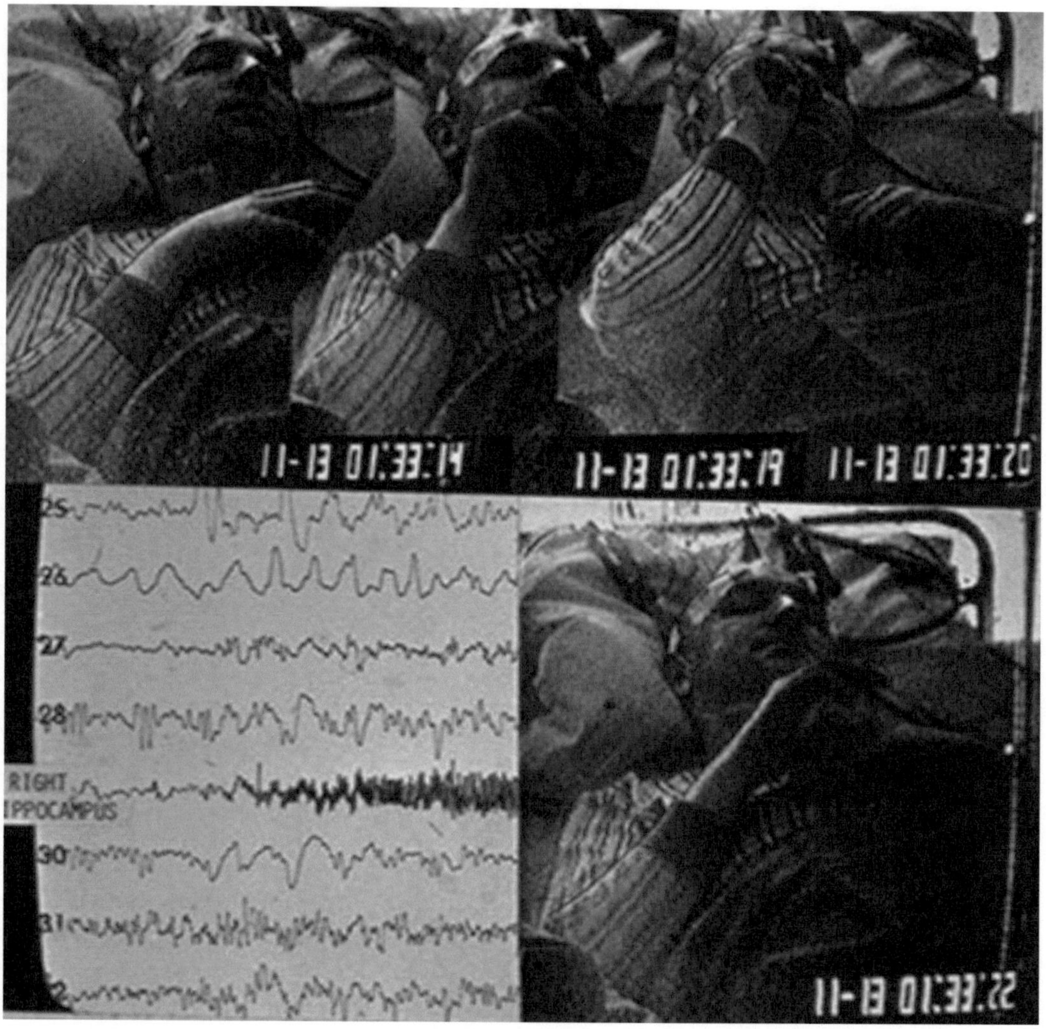

Fig. 4. llustration of invasive video-SEEG monitoring in a patient with pharmaco-resistant temporal lobe epilepsy. Depicted is an example of electroclinical seizure analysis: During the right hippocampal high-frequency "tonic" seizure discharge (of 8 seconds, channel 29; the discharge starts at 01:33:14, see *left top* video-picture) the drowsy patient experienced an olfactory-gustatory aura with ipsilateral nose wiping (modified from Wieser 1983 **[7]**).

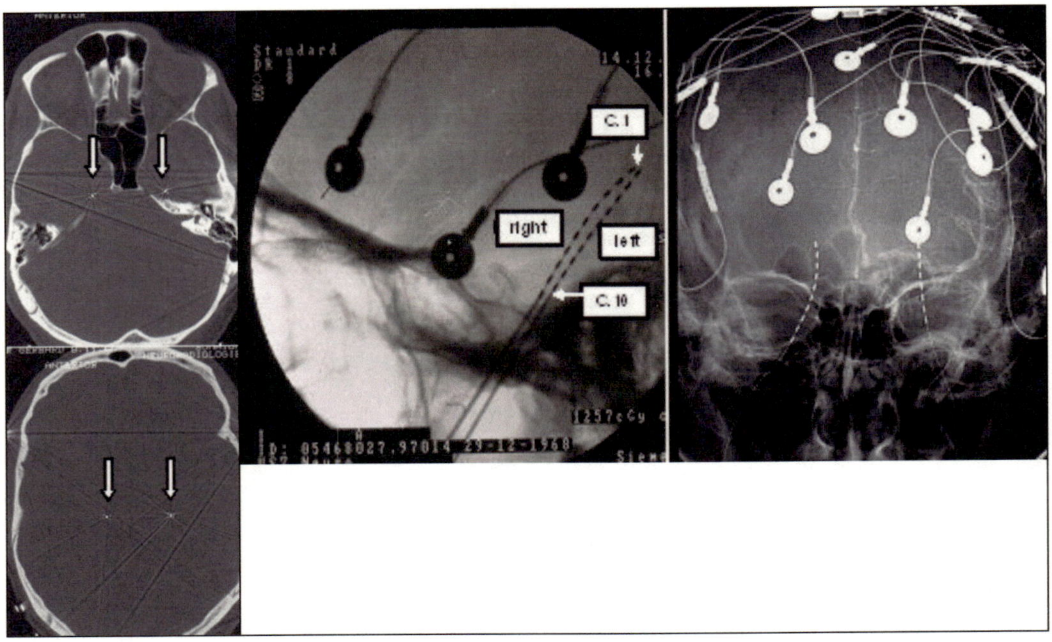

Fig. 5. CT (*left*) and x-rays with lateral (*middle portion*) and ap view (*right*) showing the position of the foramen ovale (FO) electrodes with 10 contacts. Contact 1 is in the ambient cistern and contact 10 close to the foramen ovale.

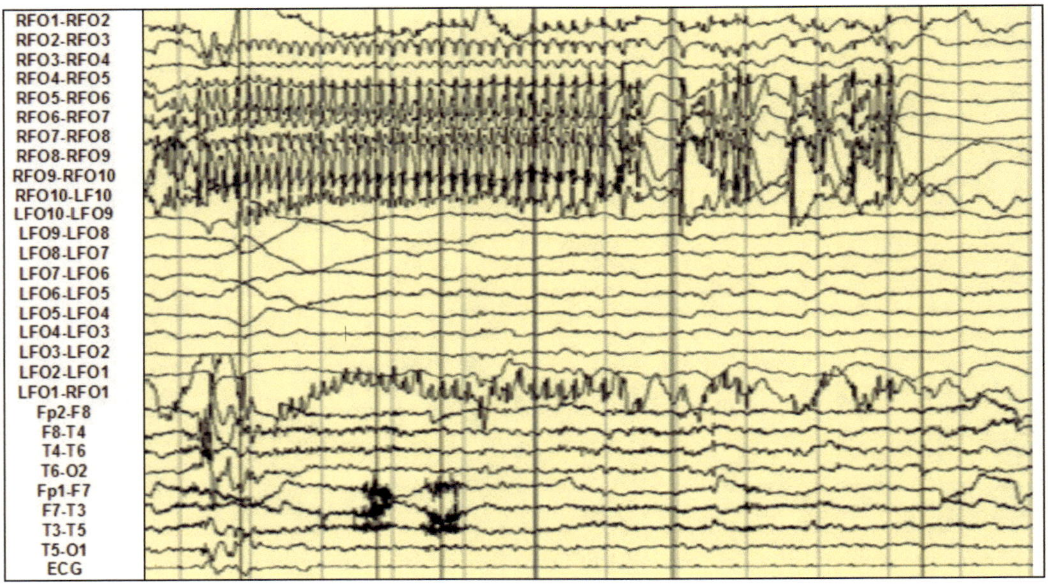

Fig. 6. Illustration of a foramen ovale (FO) recorded right mesiotemporal seizure. Note that the conventional scalp EEG does not reliably pick up this seizure discharge. A total of 12 seconds are shown towards the end of this seizure. Recording from the 10-contact FO electrodes is in a closed chain starting with contact 1 of the right FO electrode (RFO1). R, right; L left.

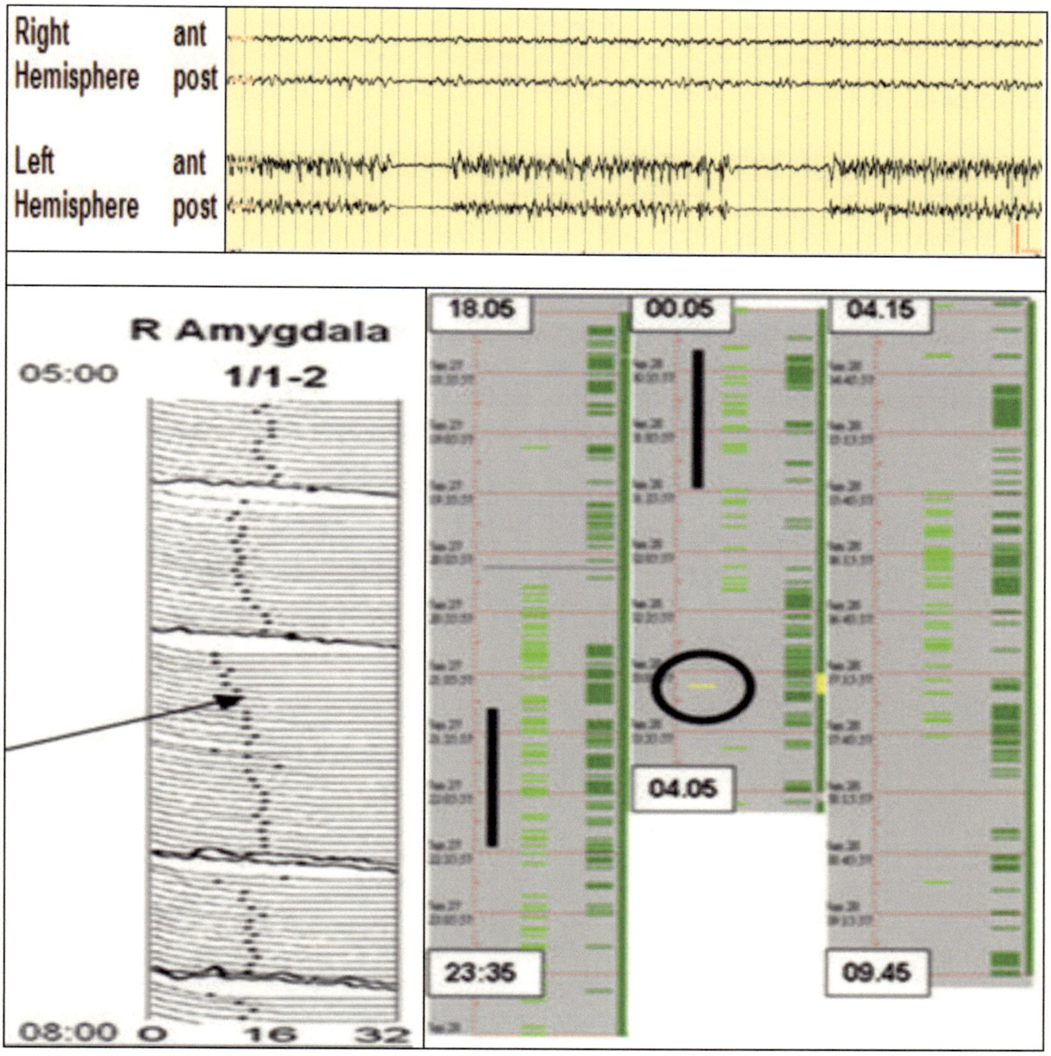

Fig. 7. *Top*: Example of LTM during an intermittent left hemispheric status epilepticus, showing the actual EEG with seizure discharges of 1.5 - 4 min and intervals of 5 -14 minutes. *Bottom left*: Graphical display of compressed spectral array over 3 hours (05.00 to 08.00 a.m) showing 4 seizure dicharges in the depth EEG recording from right amygdala. The arrow indicates the spectral edge frequency (SEF) marker set at 97%. Measurement time is 120 msec, power range is -20 to -60d. Frequency is from 0 to 32 Hz. *Bottom right:* Example of the summary display of a surface EEG-LTM (18:05 to 09:45) using the automatic seizure detection Grass-Telefactor TWin 2.60 software. The patient had numerous subclinical seizure discharges (indicated by green bars) and one clinical seizure (indicated by the yellow bar; the circle denotes alarm button press). Quantification of seizure discharges during the black vertical bars revealed more than 20 discharges/hour.

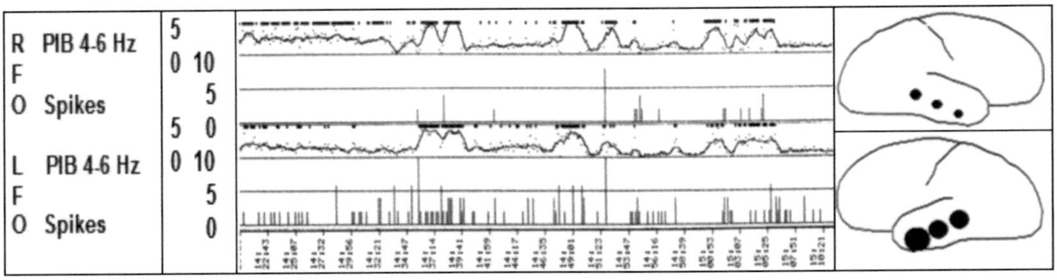

Fig. 8. Chrono-spectrogramm (illustrated are 45 minutes), depicting the vigilance-related activation of spikes using an automatic spike detection algorithm and power in band (BIP) 4-6 Hz (indicating superficial SWS-sleep). Illustrated are the results of two foramen ovale channels, one right (RFO) and one left (LFO). At the *right of the figure* a display of quantified FO electrode recorded spikes is shown. There is a marked left anterior mesiotemporal preponderance of spiking. Spiking (measured in units from 0 to 10) is activated during superficial slow-wave-sleep with increase of theta waves (PIB 4-6 Hz is measured in units 0 to 5).

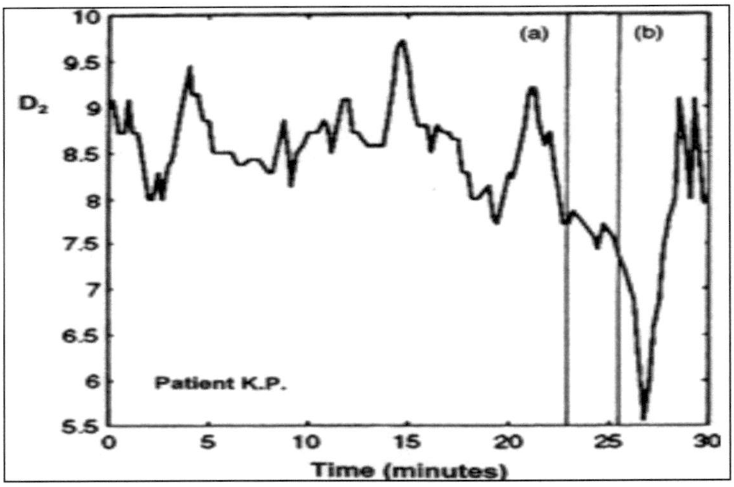

Fig. 9. Seizure prediction: Plot of the correlation dimension D2 over 30 minutes indicating that this measure predicted a seizure about 3 minutes before onset of the foramen ovale recorded seizure discharge. Beginning of the seizure is at the vertical line (b). The D2 transgresses a critical treshold at the vertical line (a). (See also **[15 and 16]**).

2.3 LTM in the diagnosis and treatment of status epilepticus

Continuous EEG (cEEG) monitoring of critically ill patients is frequently utilized to detect non-convulsive seizures (NCS) and non-convulsive status epilepticus (NCSE). Abend et al. **[17]** studied the current practice of cEEG in critically ill patients. The authors conducted an international survey of neurologists focused on cEEG utilization and NCS management: 83% of 330 physicians who completed the survey used cEEG at least once per month and 86% managed NCS at least five times per year. The use of cEEG in patients with altered mental status was common (69%), with higher use if the patient had a prior convulsion (89%) or abnormal eye movements (85%). Most respondents would continue cEEG for 24 h.

If NCS or NCSE was identified, the most common anticonvulsants administered were phenytoin/fosphenytoin, lorazepam, or levetiracetam, with slightly more use of levetiracetam for NCS than NCSE.

3. Continuous EEG monitoring in the Intensive Care Unit

In the ICUs, cEEG monitoring reveals the presence of electrographic seizures in a significant percentage of critically ill patients and in patients with depressed levels of consciousness **[18-29]**. cEEG monitoring is primarily conducted to detect epileptic seizures. It helps assessing the long-term EEG trends and the level of sedation; both essential for treatment and prognosis. Another argument to perform cEEG monitoring in the ICU is cost-containment: In a retrospective study of *refractory status epilepticus* (RSE) following NCSE (RSE defined as seizures lasting more than 60 minutes and failure of two AEDs), Mayer et al. **[30]** concluded that it is imperative to diagnose and treat NCS/NCSE as early as possible since the prognosis worsens with increasing duration of seizures which will prolong these patients stay in the ICU and consequently increase the cost on health care ressources.

3.1 ICU patients that should undergo cEEG monitoring

NCSE is usually diagnosed when facing a patient with an abnormal mental status and diminished responsiveness, a "compatible" EEG, and frequently a clinical response to antiepileptic therapy. Kaplan **[27]** asks for both, an alteration in the level of consciousness for 30 to 60 minutes and some seizure activity in the EEG to diagnose NCSE. The EEG remains the most important tool to diagnose NCSE. NCSE can be found in different patient populations, for example in psychiatric and mentally retarded patients on psychotropic medications, and in patients with a history of epilepsy after a tonic-clonic seizure. NCSE/NCSs are common in critically ill patients **[19,21]**. Therefore, we believe that ICU patients with impaired consciousness, especially those patients with unclear etiology, should undergo cEEG monitoring. Clinical conditions where it is appropriate to look for NCSE, are: (a) unexplained diminished level of consciousness (with or without limb and facial twitching and/or eye deviation) in the ICU; (b) confusional states or altered mental status in patients in the emergency room **[31]**; (c) patients with persisting lethargy or behavioral changes after a tonic-clonic seizure **[32]**, (d) psychomotor retardation or behavioral changes beyong baseline in a psychiatric or mentally retarded patient;

- Lethargy and confusion attributed to a postictal state
- Ictal confusion mistaken for metabolic encephalopathy
- Unresponsiveness and catalepsy presumed to be psychogenic
- Obtundation thought to be due to alcohol or drug intoxication
- Hallucinations and agitation mistaken for psychosis or delirium
- Lethargy presumed secondary to hypoglycemia
- Mutism attributed to aphasia
- Laughing and crying ascribed to emotional lability

Table 2. Clinical examples in which the diagnosis of NCSE was missed or delayed Reproduced from Kaplan PW. Behavioral manifestations of nonconvulsive status epilepticus. Epilepsy and Behavior 2002; 3:122-139

(e) comatose patients with metabolic / infectious or toxic problems in the ICU **[23]**; (f) ICU patients in coma with known epilepsy, aged <18 years, with a history of convulsive seizures **[22]**; (g) patients with alteration of level of consciousness after Benzodiazepine withdrawal **[33]**; (f) patients with drug-induced confusional states such as lithium **[34]**, tiagabine **[35]** or ifosfamide **[36]**. *Table 2* lists clinical examples in which the diagnosis of NCSE was missed or delayed.

3.2 Epidemiology of NCSE

There have been no large-scale population-based studies regarding selectively NCSE. Dunnes et al. [37] estimate that NCSE constitutes 20% of all status epilepticus in adults. Recent populations-based studies have reported a higher percentage of NCSE among all status epilepticus (30% to 60%) [38-42]. This wide range is explained by the difference in NCSE definition, classification and probably the differences in population characteristics. Knake et al. [40] reported the highest proportion of NCSE in the adult population. Both the large number of patients over 65 years, and the history of stroke and epilepsy might explain this unusually elevated percentage of NCSE in this study. Shorvon [43] reports an annual incidence of NCSE ranging from 5 to 11 per 100,000. Shorvon`s figures are likely underestimates because of the challenges in diagnosing NCSE and also the need for continuous EEG monitoring.

3.3 Duration of monitoring in the ICU

There is still no general rule on how long cEEG monitoring should be performed to efficiently detect seizures. Claassen et al. **[19]** looked for NCSE in comatose patients: in 13% of these patients, the seizures were recorded during the third day. Unless seizures are recorded during the first few hours, or there is strong suspicion for seizures occurence based on the clinical manifestations and the EEG data or both, we usually record for 24 hours; one example are the *periodic discharges* (PD). Claassen et al. **[19]** have shown that the occurence of PD during EEG monitoring is a risk factor for seizures and, in such cases , recording might go beyond 48 hours. Once the seizures are detected and treatment started, EEG monitoring should run uninterrupted from the time continuous IV AEDs is started, until the patient is successfully treated. Frequency of further EEG monitoring depends on the patient's condition; less often EEG recording may be sufficient if the seizures are stopped, while persistent or recurrent seizures require more intensive monitoring.

3.4 Pathological EEG patterns commonly occurring in the critically ill patient

A large spectrum of ictal EEG abnormalities may be seen in the ICU, particularly with NCSE **[45-46]**: *Periodic short-interval diffuse discharges* (PSIDD) are usually seen in anoxia, metabolic encephalopathy, Creutzfeldt Jakob disease, and toxic encephalopathy (lithium and baclofen). *Periodic long-interval diffuse discharges* (PLIDD), also called Radermecker complexes, are seen in patients suffering from Subacute Sclerosing Panencephalitis (SSPE) and in patients with Phencyclidine ("angel dust") or ketamine intoxication **[47-51]**. PLIDD-like patterns are also seen in anoxia and barbiturate intoxication.

The following *benign electroencephalographic variants* and *patterns of uncertain significance* may appear during cEEG monitoring but have to be separated from the EEG findings described above **[47,52]**. These patterns include: *"Rhythmic temporal Theta bursts of drowsiness [psychomotor variant]", "Midline theta rhythms [formerly Ciganek-rhythm]", "Subclinical rhythmic electrographic discharge in adults [SREDA]", "Wicket spikes", "Small sharp spikes [SSS- previously called Benign epileptiform transients of sleep [BETS]", "14 and 6 Hz positive bursts"* and *"6 Hz spike and wave [also called Phantom spike and wave]"* .

Generalized periodic burst suppression, generalized periodic slow-wave complexes (GPSC), generalized repetitive sharp transients , and generalized periodic triphasic waves are seen in patients with anoxic/metabolic encephalopathies. GPSC may also occur in subacute sclerosing panencephalitis and in other encephalitis. In comatose patients after cardiorespiratory arrest, burst suppression usually indicates a poor outcome.

Other EEG patterns that are often seen in the ICU are classified according to their predominant EEG frequencies, their morphology or distribution **[52]**. The following patterns involve mainly the alpha frequencies: Alpha squeak, retained alpha **[53]**, alpha delta sleep **[54]**, unilateral decrease in reactivity of alpha activity and extreme spindles. Beta frequencies include the fast alpha variant, posterior temporal fast activity in children, fast spiky spindle variant, central fast activity, and diffuse paroxysmal or continuous fast activity.The theta frequencies comprise slow alpha variant, frontal arousal rhythm, rhythmic temporal theta activity of drowsiness, middle theta rhythms, and focal parietal theta activity.The delta frequencies include the transient rhythmic slowing occurring after eye closure and more continuous posterior rhythmic slowing . Claassen et al. **[19]** summarized some EEG patterns seen in the critically ill ICU patient **[Table 3]**

Electrographic seizures	Rhythmic discharge or spike and wave pattern with definite evolution in frequency, location or morphology, lasting at least 10s; evolution in amplitude alone does not qualify
Periodic epileptiform discharges (PEDs)	Repetitive sharp waves, spikes, or sharply contoured waves at regular or nearly regular intervals and without clear evolution in frequency or location (includes, PLEDs, GPEDs, BiPLEDs, triphasic waves)
Periodic lateralized epileptiform discharges (PLEDs)	Consistently lateralized PEDs
Generalized PEDs (GPEDs)	Bilateral and synchronous PEDs with no consistent lateralization
Bilateral PLEDs (BiPLEDs)	PLEDs occurring bilaterally, but independently and asynchronously
Triphasic waves	Generalized periodic sharp waves or sharply contoured delta waves with triphasic morphology at 1-3 Hz with/without anterior-posterior or posterior-anterior lag
Frontal intermittent rhythmic delta activity (FIRDA)	Moderate to high voltage monorhythmic and sinusoidal 1-rhythmic delta 3 Hz activity seen bilaterally maximal in anterior leads, no evolution

Table 3. Definitions of EEG patterns in ICU seizures given by Claassen [19]

3.5 Controversial EEG patterns

NCSE is accompanied with EEG patterns that are often seen with clinical seizures; however many EEG patterns are called "controversial", because the question whether they are ictal in nature or not, is still debated.

3.6 Periodic patterns

These include periodic epileptiform discharges (PEDs) which can be generalized (GPEDs) or lateralized (periodic lateralized epileptiform discharges, PLEDs); or bilateral independent periodic epileptiform discharges (BIPLEDs).

PLEDs are lateralized complexes made up of sharp waves or spikes possibly followed by a slow wave and recurring every one to two seconds **[55]**. PLEDs are often associated with clinical seizures and also with destructive hemispherical lesions **[56]**. A great proportion of ICU patients showing these patterns present clinical/electrographic seizures during EEG monitoring **[57]**. Treiman et al. **[58]** reported a sequence of five EEG patterns of ictal discharges observed during an untreated generalized convulsive status epilepticus: discrete electrographic seizures, waxing and waning, continuous with flat periods, and PLEDs on a relatively flat background. According to Garzon et al. **[59]** these patterns should be considered as ictal when they evolve from lower to faster rates(>2 Hz), change their frequencies, morphology or field (PLEDs-plus). PLEDs are shown in *Figure 10.*

BIPLEDs are less common than PLEDs; the complexes are asynchronous; their morphology, amplitude, frequency and location generally differ from one hemisphere to the other. They

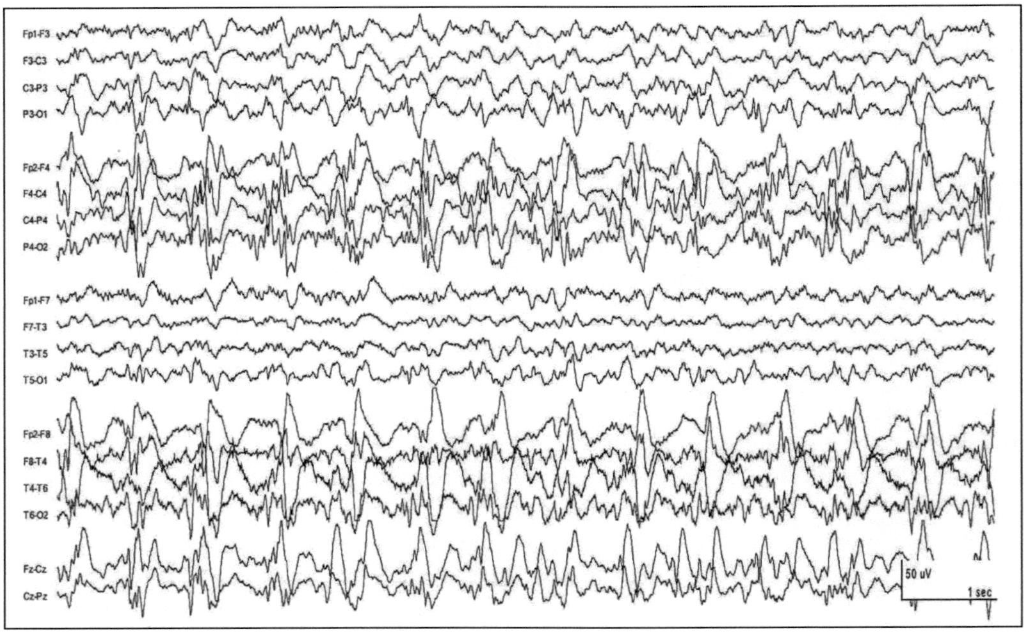

Fig. 10. Periodic lateralized epileptiform discharges (PLEDs) in a 64-year old woman with a right frontal astrocytoma. PLEDs display 1,5 Hz with intervening fast frequencies (PLEDs-plus). PLEDs-plus is extremely likely to evolve into electrographic seizures; whether this pattern represents nonconvulsive status epilepticus is debated.

are found in hypoxic encephalopathy, encephalitis/meningitis and chronic seizure disorders **[60-62]**. BIPLEDs are shown in *Figure 11.*

GPEDs *[Figure 12]* occur in anoxic-ischemic coma, Creutzfeld-Jacob disease, metabolic encephalopathies, CNS infections such as SSPE, HSV, and in final stages of status epilepticus. GPEDs may represent NCSE **[63]** or an epileptic encephalopathy **[64-65]**.

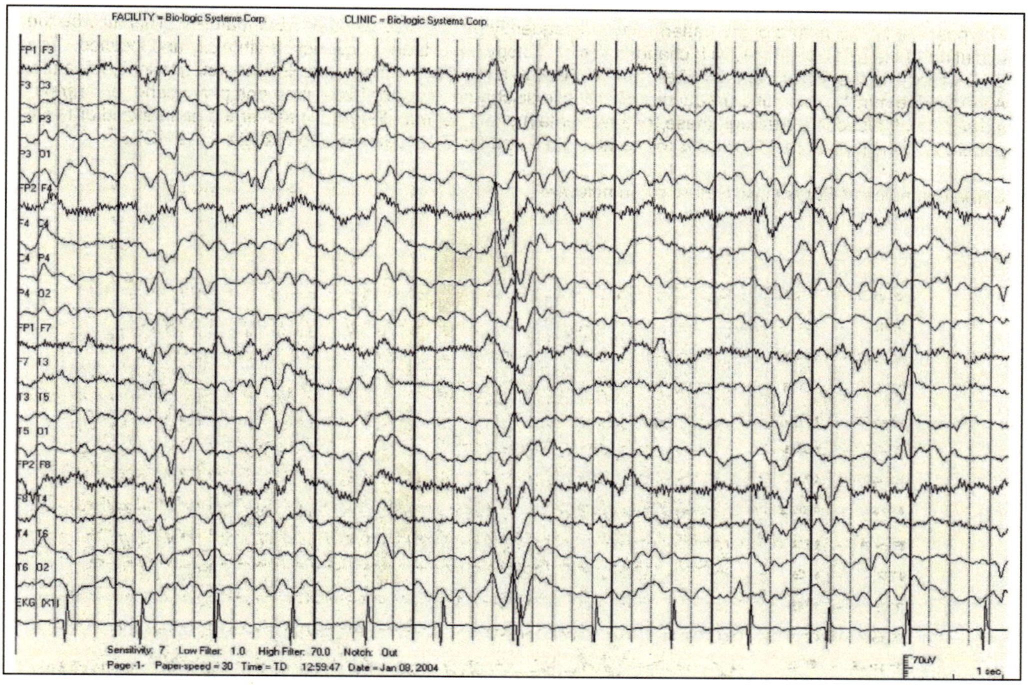

Fig. 11. This EEG shows bilateral independent peudoperiodic lateralized epileptiform discharges in an obtunded patient with left hemisphere complexes every several seconds, independent from right-sided discharges.

4. Triphasic waves (TWs)

In NCSE/NCSs, Kaplan **[66]** describes the TWs as follows: Surface-negative, blunted triphasic complexes with (a) low amplitude, blunted, negative first phase (often wide-based); (b) dominant, steep positive second phase, and (c) slow rising third slow-wave component. No polyspike component. Complex duration: 400-600 msec; amplitude: 100 to 300 uV on referential montage; smaller on bipolar. Frequency: 1.0-2.5 Hz (typically 1.8 Hz); persistence: wax and wane, but more than 10% of a standard recording (20 min). Evolution/reactivity: decrease in sleep, drowsiness or after BZPs; increase and reappear with arousal or noxious stimulation. May exhibit phase-lag best seen on referential montage. TWs may present an epileptiform morphology with spike-like ongoing first phases, and spike-slow-wave complexes. In such cases, it is not easy to differentiate NCSE from metabolic encephalopathy, particularly when TWs run at a frequency exceeding 1 per second. Hence there exist the terms triphasic-like waves and nonepileptic true TWs **[63].** Indeed, in NCSE, TWS and generalized spike-wave discharges look similar. In metabolic encephalopathies, in Creutzfeld-Jacob disease, and in the course of status epilepticus, bisynchronous periodic sharp complexes with a triphasic configuration are frequently found **[58,67-69].** The generalized periodic patterns, associated with lithium, baclofen, ifosfamide, metrizamide, tiagabine, and levodopa, might represent NCSE **[70-75].** TWs are also reported in Alzheimer disease **[76].** TWs are shown in *Figure 13 and* " Epileptiform triphasic waves" in **Figure 14**.

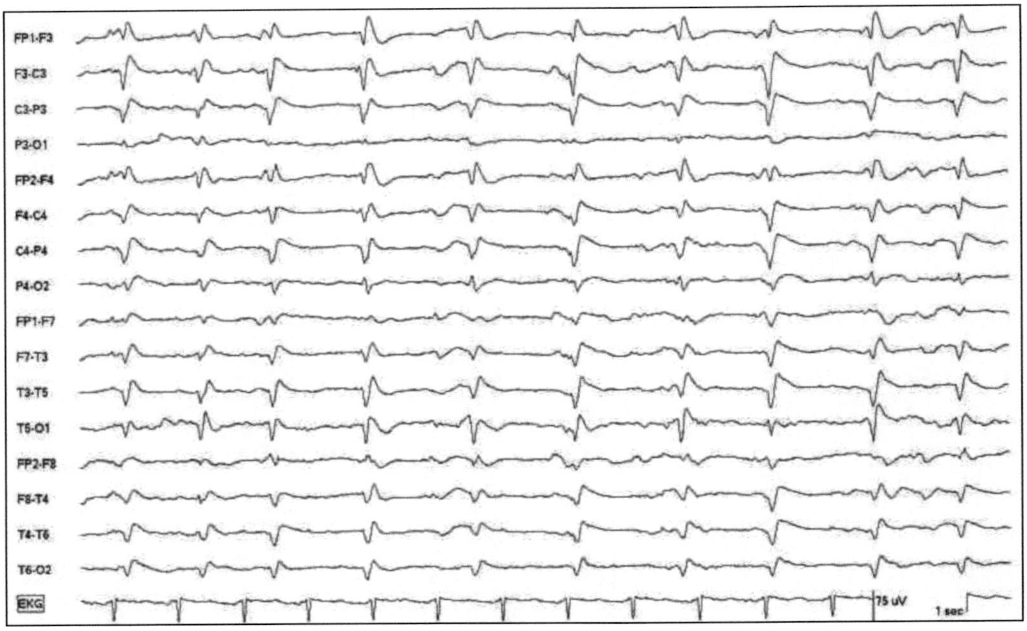

Fig. 12. Generalized periodic epileptiform discharges (GPEDs) in a 45-year-old-man

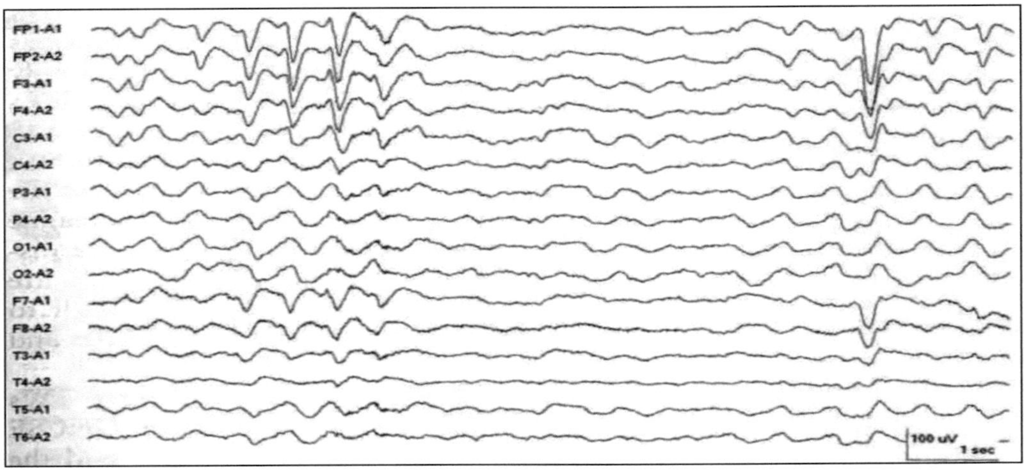

Fig. 13. Triphasic waves in a 51-year-old man with hepatic failure.

SIRPIDs. *Stimuli-induced rhythmic periodic, or ictal discharges* [SIRPIDs] were first described by Hirsch et al **[77]**. These are periodic discharges, focal or generalized, frequently seen immediately after stimulation or arousal of a comatose or stuporous patient. SIRPIDs occur with acute brain disorders (brain injury) and severe metabolic disorders. Is not clear whether SIRPIDs represent reflex seizures, abnormal arousal patterns, or a combination of both. It has been speculated that faster discharges (more than 3Hz) may be ictal, whereas the slower ones are less likely to be ictal. Few patients have clinical correlates to their SIRPIDs. SIRPIDs are shown in *Figure 15.*

4.1 Nonconvulsive status epilepticus in the critically ill ICU patient

There exists a large literature dealing with classification and electroclinical accompaniments of nonconvulsive status epilepticus (NCSE). NCSE is a heterogenous condition. Its diagnosis is difficult on the basis of clinical semiology alone. An impaired mental status, a reduced responsiveness, a supportive EEG and a response to antiepileptic medications are important criteria to diagnose NCSE. A continuous EEG ictal activity and impaired consciousness for one hour are necessary according to Tomson et al **[78];** some form of EEG seizure activity and an impaired consciousness for 30 to 60 minutes according to Kaplan **[27]**. Because NCSE is not accompanied by generalized jerky movements, cEEG monitoring is essential for detecting nonconvulsive seizures (NCSs)/ NCSE particularly in the ICU **[21-29,79]**. EEG characteristics of NCSs/NCSE are heterogeneous, with a highly variable morphology; *Table 4* describes the EEG patterns occuring in NCSE **[25]**.

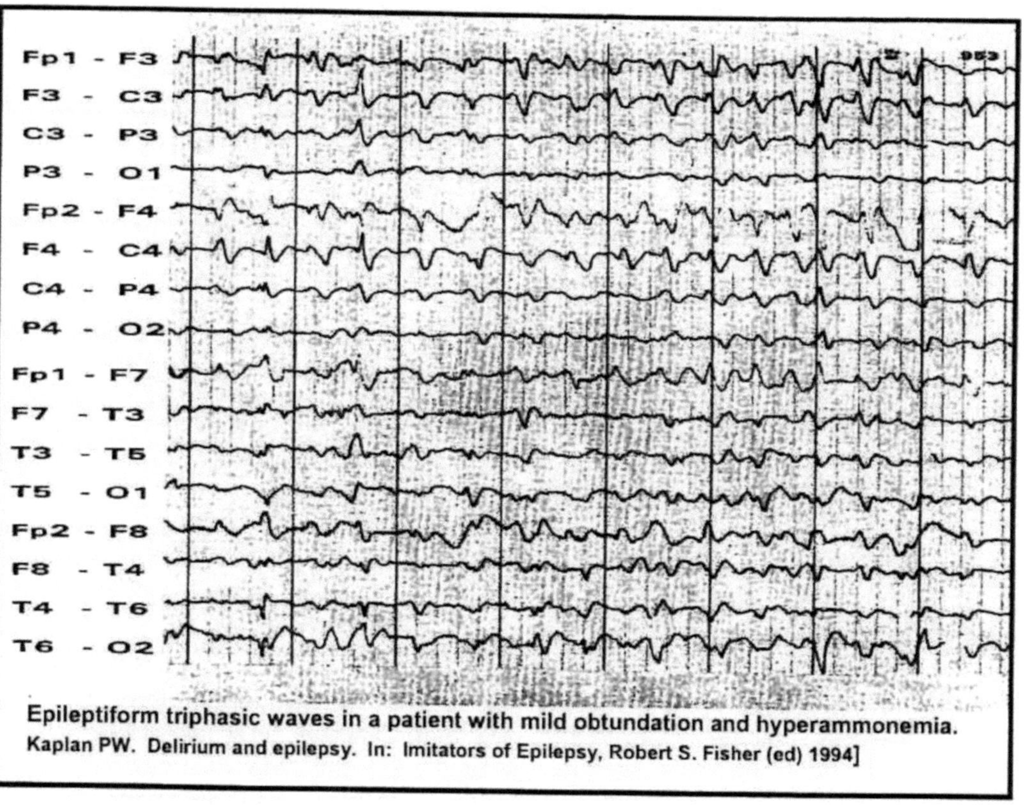

Epileptiform triphasic waves in a patient with mild obtundation and hyperammonemia.
Kaplan PW. Delirium and epilepsy. In: Imitators of Epilepsy, Robert S. Fisher (ed) 1994]

Fig. 14. Epileptiform triphasic waves

Of particular interest is the distinction between NCSE (as described above) and an epileptic encephalopathy **[80]**. In this later condition, the impaired level of consciousness is related to the underlying brain pathology causing the seizures rather than the seizures themselves **[22,79,81-83]**.

Although nonconvulsive seizures and NCSE are common in the critically ill patient, many neurologists may be challenged by the frequent "absence" of specific clinical signs, and the difficulty in interpreting the EEG when confronted to such patients. A trial of rapid acting benzodiazepine might be helpful.

Not infrequently NCSE follows a convulsive status epilepticus **[32]**. In the study of DeLorenzo et al **[32]**, using cEEG recording after the control of convulsive status epilepticus, in 164 patients, the authors found that 48% showed persistent electrographic seizures, with NCSE in 14% of them. The authors concluded that the use of cEEG monitoring in these patients was essential in detecting NCSE. *Figure 16* shows an example of NCSE.

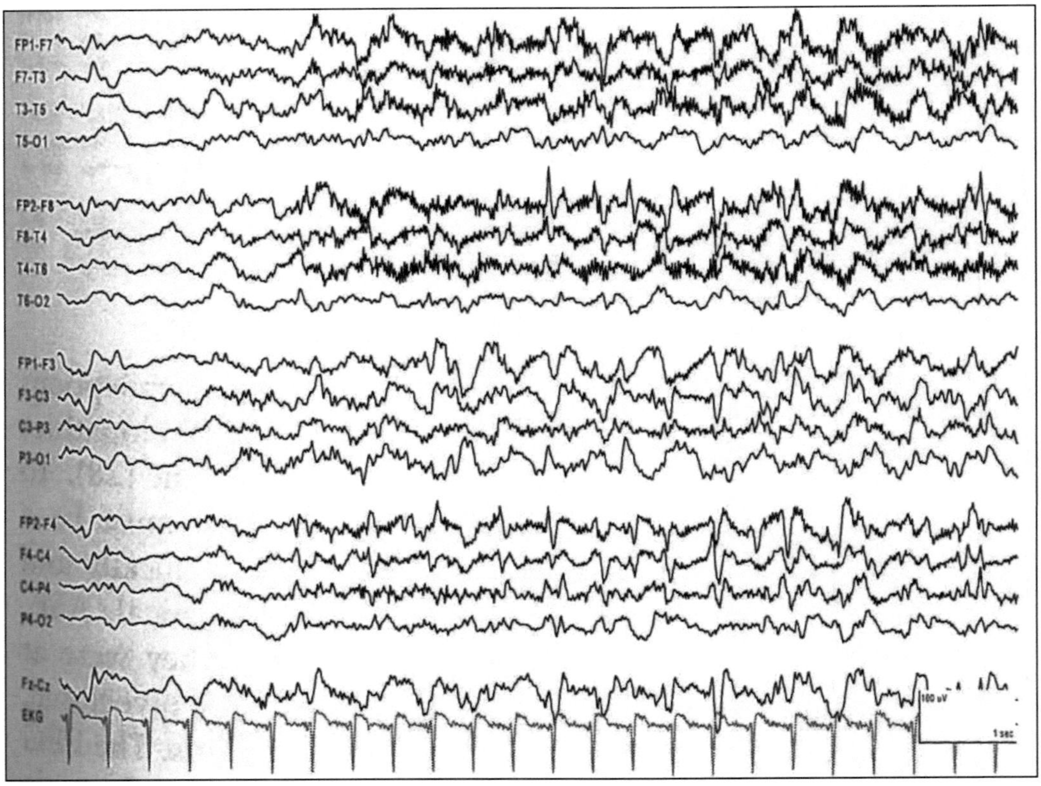

Fig. 15. Stimulus-induced rhythmic periodic or ictal discharges (SIRPIDs). Repetitive right-hemispheric spikes at 2 Hz were provoked by sternal rub in a patient with subarachnoid hemorrhage.There is simultaneous semi-rhythmic delta activity over the left hemisphere. Electromyographic artifacts indicate a probable change in the patient's state of alertness.

In patients without a known epileptic encephalopathy

1. Repetitive generalized or focal spikes, poly-spike, sharp waves, spike–and-wave or Spike-and-slow wave complexes at > 2.5/second.
2. Above with discharges < 2.5/second but with EEG and clinical improvement after rapid onset anti-epileptic drugs, typically benzodazepines (BZPs). [Testing for patient responsiveness and improvement in EEG! EEG: Increases in EEG reactivity and appearance of EEG background activity].
3. Above, with discharges < 2.5/second with focal ictal phenomena (e.g. facial twitching, nystagmus, limb myoclonus).
4. Rhythmic waves (theta-delta) at >0.5 Hz with (a) incrementing onset (increase in voltage with increase or decrease in frequency), (b) evolution in pattern (increase or decrease in frequency (>1Hz), or location [changes in discharge voltage (amplitude) or morphology are not sufficient], or (c) decrementing termination (voltage or frequency), (d) post-PEDs background slowing or attenuation. (a), (b), (c) or (d) may be acutely induced by IV BZPs.

In patients with known epileptic encephalopathy

1. Frequent or continuous generalized spike-wave discharges which show an increase in profusion or frequency when compared to baseline EEG with observable change in clinical state
2. Regression (improvement) of clinical or EEG features with BZPs.

Caveat: these are working definitions, exceptions may occur. This classification is not a rote endorsement of the use of parenteral BZPs, which carry clear risk of hypotension and hypopnea. (Classification modified from Kaplan 2005; Young et al. 1996; Hirsh et al. 2005). From Kaplan PW. EEG criteria for nonconvulsive status epilepticus. Epilepsia. 2007; 48 Suppl 8:39-41.

Table 4. EEG patterns suggestive of Nonconvulsive status epilepticus

4.2 Do NCSE or some EEG patterns (PEDs) damage the brain ?

PLEDs are often seen with destructive hemispherical lesions and associated with clinical seizures. Baroque & Purdy assessed 32 patients who had undergone computed tomography (CT) and/or magnetic resonance imaging (MRI). PLEDs and BIPLEDs correlated with postmortem data that showed consistent localization of lesions in the gray matter **[56]**. Investigators using serial EEG, SPECT, and FDG-positron emission tomography (PET) findings argue that PEDs following SE are often ictal **[84]**. Others classify PEDs as postictal or simply as markers (an epiphenomenon) of severe injury or encephalopathy. In several studies PLEDS were associated with refractory SE **[49]**. Jaitly et al **[85]** reported that ictal discharges which follow GCSE may be harmful. Indeed these discharges are associated with an increase in morbidity and mortality **[85]**. The mortality in NCSE is similar to that of GCSE. Seizures duration and delay in diagnosis were the factors which correlated the most with mortality rate (33%) **[86]**. Claassen et al. **[87]** reported even a higher mortality rate (75%) when NCSE was the first clinical presentation. NCSE could cause long term sequelae from neuronal injury , e.g., impaired cognition or memory loss might result from cortical or hippocampal injury **[88]**.

As mentioned above, NCSE following generalized convulsive status epilepticus is often a refractory status epilepticus and carries a high mortality **[32]**. Therefore, it is imperative to diagnose and treat patients with this condition as early as possible because the prognosis worsens with increasing duration of seizure activity, inducing then more complications and ending with high morbidity and mortality.

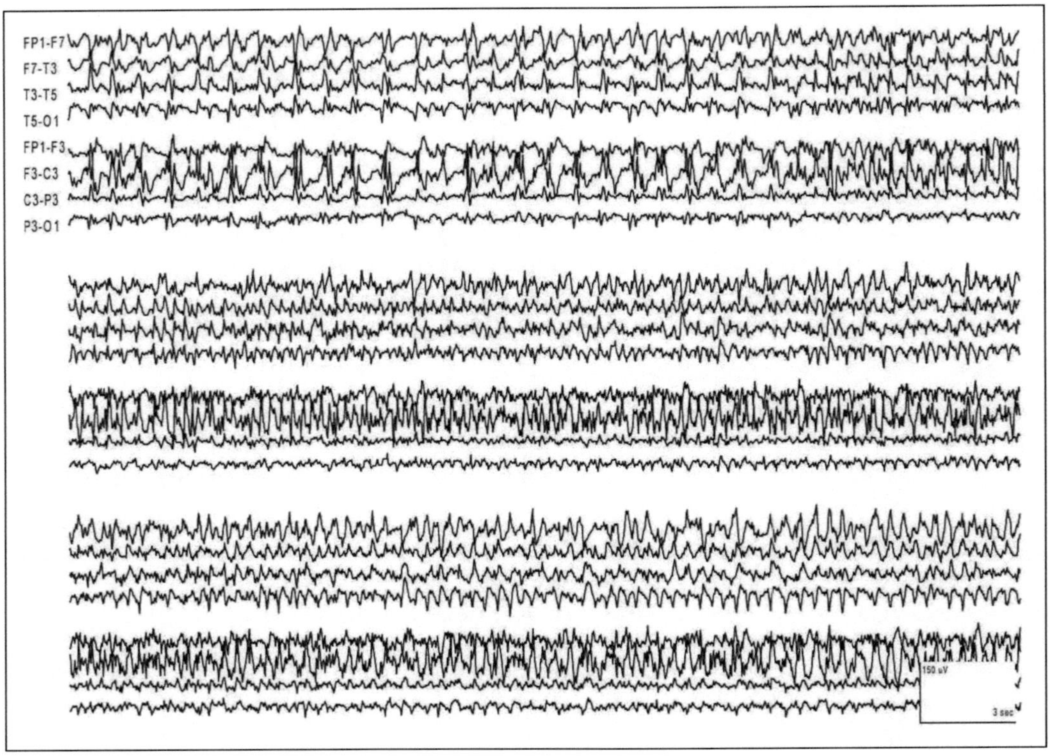

Fig. 16. Nonconvulsive status epilepticus (NCSE) evolving from left-hemisphere periodic lateralized epileptiform discharges (PLEDs) in a 66-year-old man with a left frontal intracerebral hemorrhage. Initially the EEG shows periodic discharges at 1,5 per second, which are invariant for several minutes, then gradually increase in frequency and merge to electrographic theta activity over the entire left hemisphere.This pattern repeated every 5 to 10 minutes for several hours.

There is a strong evidence that the marker of neuronal injury, *neuron-specific enolase* (NSE) is elevated during NCSE **[89-91].** In these studies the level of NSE correlated with the duration of SE favorizing an early treatment of NCSE.

Autopsies of patients with NCSE have shown focal neuronal loss. Three patients, who died 11 to 27 days after the onset of focal motor status epilepticus, but without pre-existing epilepsy, showed neuronal loss and gliosis in the hippocampus, amygdala and periamygdalar area, dorsomedial thalamus, Purkinje cells, pirifrom and entorhinal cortices **[92]**. The neuronal injury might be responsible for the seizures and NCSE recurrence **[93].**

In accordance with autopsy studies, nonspecific CT and MRI findings have been reported in NCSE. For example, MRI has shown a transient increased signal in the hippocampus after NCSE **[94].** In NCSE, FDG-PET changes are not very consistent, but one reported case showed a focal hypermetabolism which corresponded to the localization of EEG seizure activity **[95]**. The use of SPECT during the seizures ("ictal SPECT") is easier and helpful in localizing the epileptic focus **[96]**.

In humans, the underlying pathophysiology of the various subtypes of NCSE has not been investigated in detail **[97-98]**. The correlation with underlying pathology, EEG patterns, type of treatment and prognosis remains to be studied. Valid data regarding which EEG patterns cause ongoing neuronal damage are still lacking. Therefore we do not know which patient suffering from NCSE should be treated aggressively. Subtle cases of NCSE may resolve spontaneously. Patients with NCSE with minimal obtundation have a morbidity and mortality of 7%, whereas those with deep lethargy or coma have a mortality rate of 39%. **[86]**. Whereas Treiman **[99]** explicitly recommends that a patient actively having seizures or being comatose and who exhibits any of the periodic epileptiform discharge patterns on EEG should be treated aggressively to stop all clinical and electrical seizure activity to prevent further neurological morbidity and mortality, Kaplan **[100]** believes that aggressive ICU management of NCSE in the elderly critically ill patient may be harmful rather than beneficial .

4.3 Future study: To develop a novel cEEG monitoring system able to monitor seizures in real time, and adequately treat these seizures in a timely fashion

Available seizure-detection software algorithms are geared towards patients with normal EEG backgrounds and high-frequency rhythmic seizure discharges. Unfortunately these computer algorithms are not very successful to detect NCSs/NCSE in ICU patients. This because (a) these kind of seizures have a tendency to wax and wane and do evolve more slowly when compared to the fast activity seen in convulsive seizures, and (b) these seizures are harder to separate from the abnormal background activity related to the impaired vigilance and the associated metabolic disturbances often found in ICU patients **[45]**.

In an ongoing study (Mesraoua et al; see also **[101]**) we are planning to develop a novel cEEG monitoring system to be used in the ICU. This system should be able to detect pathological EEG patterns commonly occurring in the critically ill patient (NCSE and PEDs, TWs). This system should also be able to quantify the frequency and duration of seizures and will make possible an adequate antiepileptic drug (AED) treatment, in a timely fashion, using the recommended antiepileptic drugs in nonconvulsive seizures/NCSE **[102-108]**. Using the available neuroradiologic tools (MRIs, MRS) and the CNS biomarkers (Neuron-specific Enolase , NSE), we will examine the possibility that brain damage is occuring with NCSE or PEDs. This way we hope to answer the most important question, i.e., whether NCSE and PEDs should be treated aggressively. This study is a combined effort between neuroscientists/neurologists and signal processing specialists in order to develop a software for the *cEEG*. The technique will use *wavelet transforms (WTs) and wavelet packet energy ratio (WPER)* which have been proven to be successful tools for epileptic seizure detection, particularly in temporal lobe epilepsy **[109-100]**. The proposed seizure detection algorithm will perform the following:

1. Subdivide the EEG signal into segments for individual interpretation,
2. Extract features which characterize the individual segments based on EEG behavior in the ICU,
3. Identify the onset of seizures, specifically NCS and NCSE seizures, using the extracted features,

4. Identify the offset of a seizure, and
5. Provide summaries of the seizure activity over the monitored time frame.

This 3-year research study has just started and no conclusion can be drawn at the present time. Hopefully, cEEG monitoring with the developed algorithm will be able to identify NCSs/NCSE in critically ill patients and thus will help to improve treatment protocols to prevent morbidity.

5. References

[1] Benbadis SR. Outcome of prolonged video-EEG monitoring at a typical referral epilepsy center. Epilepsia 2004; 45:1150-3.
[2] Yogarajah M, Powell HW, Heaney D, Smith SJ, Duncan JS, Sisodiya SM. Long term monitoring in refractory epilepsy: the Gowers Unit experience. J Neurol Neurosurg Psychiatry. 2009;80:305-10
[3] Wieser HG, Yasargil MG. Selective amygdalohippocampectomy as a surgical treatment of mesiobasal limbic epilepsy. Surg Neurol 1982:17:445-57
[4] Wieser HG. Selective amygdalohippocampectomy: Indications, investigative technique and results. Advanc Techn Stand Neurosurg. Springer: Vienna 1986; Vol.13:39-133.
[5] Wieser HG. Selective amygdalohippocampectomy has major advantages In: Epilepsy Surgery. Principles and Controversies. Miller JW, Silbergeld DL (eds.) Taylor & Francis Group: New York 2006:465-78.
[6] Kahane P, Landré E, Minotti L, Francione S, Ryvlin P. The Bancaud and Talairach view on the epileptogenic zone: a working hypothesis. Epileptic Disord. 2006; Suppl 2:S16-26.
[7] Wieser HG. Electroclinical Features of the Psychomotor Seizure. A Stereo-electroencephalographic Study of Ictal Symptoms and Chronotopographical Seizure Patterns Including Clinical Effects of Intracerebral Stimulation. G. Fischer-Butterworths: Stuttgart-London 1983 (242 pp).
[8] Wieser HG, Elger CE. Presurgical Evaluation of Epileptics. Basics, Techniques, Implications. Springer: Berlin, Heidelberg, New York 1987 (415 pp)
[9] Lüders HO, Comair YG (eds.) Epilepsy Surgery, 2nd ed. Lippincott Williams & Wilkins, 2001
[10] Lüders HO, Najm I, Nair D, Widdess-Walsh P, Bingman W. The epileptogenic zone: general principles. Epileptic Disord. 2006;Suppl 2:S1-9.
[11] Wieser HG, Elger CE, Stodieck SRG. The 'Foramen Ovale Electrode': A new recording method for the preoperative evaluation of patients suffering from mesio-basal temporal lobe epilepsy. Electroenceph Clin Neurophysiol 1985;16:314-22.
[12] Wieser HG. Stereoelectroencephalography. In: Presurgical Evaluation of Epileptics. Wieser HG, Elger CE (eds.) Springer: Berlin-Heidelberg-New York 1987:192-204.
[13] Engel J jun (ed) Surgical Treatment of the Epilepsies. 2nd ed. Lippincott Williams & Wilkins, 1993
[14] Wieser HG. Stereoelectroencephalography and Foramen ovale electrode recording. In: Electroencephalography: Basic Principles, Clinical Applications, and Related Fields. Niedermeyer E, Lopes da Silva F (eds.) Fourth Edition. Williams & Wilkins: Baltimore 1999:725-40.

[15] Moser HR, Meier PF, Wieser HG, Weber B. Pre-ictal changes and EEG analyses within the framework of Lyapunov theory. *In:* Chaos in Brain? Lehnertz K., Arnhold J, Grassberger P, Elger CE (eds.) World Scientific: Singapore: 2000:96-111.
[16] Weber B, Lehnertz K, Elger CE, Wieser HG. Neuronal complexity loss in interictal EEG recorded with foramen ovale electrodes predicts side of primary epileptogenic area in temporal lobe epilepsy: a replication study. Epilepsia. 1998;39:922-7
[17] Abend NS, Dlugos DJ, Hahn CD, Hirsch LJ, Herman ST. Use of EEG monitoring and management of non-convulsive seizures in critically ill patients: a survey of neurologists. Neurocrit Care. 2010;12:382-9.
[18] Pandian JD, Cascino GD, So EL, Manno E, Fulgham JR. Digital videoelectroencephalography monitoring in the neurological-neurosurgical intensive care unit: clinical features and outcome. Arch Neurol 2004; 61:1090-4.
[19] Claassen J, Mayer S, Kowalski G, et al. Detection of electrographic seizures with continuous EEG monitoring in critically ill patients. Neurology 2004;62 :1743-8.
[20] Scheuer M. Continuous EEG monitoring in the intensive care unit. Epilepsia 2002; 43 (Suppl 1):114-27.
[21] Hirsch LJ. Continuous EEG monitoring in the intensive care unit: an overview. J Clin Neurophysiol 2004; 21:332-340
[22] Privitera M, Hoffman M, Moore JL, Jester D. EEG detection of non-tonic-clonic status epilepticus in patients with altered consciousness. Epilepsy Res 1994; 18:155-66
[23] Towne AR, Waterhouse EJ, Morton LD et al. Prevalence of nonconvulsive status epilepticus in comatose patients. Neurology 2000; 54:340-5.
[24] Brown A, Garnett L. Persistent non-convulsive status epilepticus after control of convulsive status epilepticus. Epilepsia 1998; 39:833-40.
[25] Kaplan PW. EEG criteria for nonconvulsive status epilepticus. Epilepsia. 2007;48 Suppl 8:39-41.
[26] Vespa PM, Nuwer MR, Nenov V, Ronne-Engstrom H, Hovda DA, Bergsneider M.et al. Increased incidence and impact of nonconvulsive and convulsive seizures after traumatic brain injury as detected by continuous electroencephalographic monitoring.J.Neurosurg. 1999:91:750-60.
[27] Kaplan PW. Nonconvulsive status epilepticus. Seminars in neurology 1996; 16:33-40.
[28] Bottaro FJ, Martinez OA, Pardal MM, Bruetman JE, Reisin RC. Nonconvulsive status epilepticus in the elderly: a case-control study. Epilepsia 2007; 48:966-72.
[29] Jordan KG. Emergency EEG and continuous EEG monitoring in acute ischemic stroke. J Clin Neurophysiol 2004; 21:341-52
[30] Mayer SA, Claassen J, Lokin J, Mendelsohn F, Dennis LJ, Fitzsimmons BF. Refractory status epilepticus: frequency, risk factors, and impact on outcome. Arch Neurol 2002; 59:205-10.
[31] Kaplan PW. Nonconvulsive status epilepticus in the emergency room. Epilepsia 1996;37:643-50
[32] DeLorenzo RJ, Waterhouse EJ, Towne AR, Boggs JG, Ko D, DeLorenzo GA, et al. Persistent nonconvulsive status epilepticus after the control of convulsive status. Epilepsia 1998; 39:833-840.
[33] Thomas P, Lebrun C, Chatel M. De novo absence status epilepticus as a benzodiazepine withdrawal syndrome. Epilepsia 2005, 34, 2:355-8

[34] Roccatagliata L, Audenino D, Primavera A, Cocito L. Nonconvulsive status epilepticus from accidental lithium ingestion. Am J Emerg Med. 2002;20(6):570-2.
[35] Jette N, Cappell J, Van Passel L, Akman CI. Tiagabine-induced nonconvulsive status in an adolescent without epilepsy. Epilepsy Res. 2007; 73: 1-52.
[36] Primavera A, Audenino D, Cocito L. Ifosfamide encephalopathy and nonconvulsive status epilepticus. Can J Neurol Sci 2002; 29(2): 180-3
[37] Dunnes JW, Summers QA, Stewart-Wynne EG. Non-convulsive status epilepticus : a prospective study in an adult general hospital. Q J Med 1987;62:117-26.
[38] Delorenzo RJ, Pellock JM, Towne AR, Boggs JG. Epidemiology of status epilepticus . J Clin Neurophysiol 1995;12:316-25.
[39] DeLorenzo RJ, Hauser WA,Towne AR,Boggs JG,Pellock JM, Penberthy L, Garnett L,Fortner CA,Ko D. A prospective, population-based epidemiologic study of status epilepticus in Richmond ,Virginia. Neurology 1996;46:1029-35.
[40] Knake S, Rosenow F, Vescovi M, Oertel WH, Mueller HH, Wirbatz A, Katsarou N, Hamer HM. Incidence of status epilepticus in adults in Germany:a prospective, population-based study.Epilepsia 2001;42:714-18.
[41] Coeytaux A, Jallon P, Galobardes B, Morabia A. Incidence of status epilepticus in French-speaking Switzerland (EPISTAR). Neurology 2000;55:693-7.
[42] Vignatelli L, Rinaldi R, Galeoti M, de CP, D'Allessandro R. Epidemiology of status epilepticus in a rural area of northern Italy: a 2-year population-based study. Eur J Neurol 2005;12:897-902.
[43] Shorvon S.The definition, classification and frequency of NCSE. Epileptic Disord 2005;7:255-8
[44] Claassen J, Jette N, Chum F et al: Seizures and periodic discharges after intracerebral hemorrhage. Neurology 2007; 69:1356-65
[45] Drislane FW. Presentation, evaluation, and treatment of nonconvulsive status epilepticus. Epilepsy Behav. 2000; 1:301-14.
[46] Kaplan PW.The clinical features, diagnosis, and prognosis of nonconvulsive status epilepticus. Neurologist. 2005; 11(6):348-61. Review.
[47] Zumsteg D, Hungerbühler HJ, Wieser HG. Atlas of Adult Electroencephalography. Hippocampus-Verlag, Bad Honnef (Germany), 2004
[48] Brenner RP, Schaul N. Periodic EEG patterns: classification, clinical correlation, and pathophysiology. J Clin Neurophysiol. 1990;7:249-67.
[49] Yemisci M, Gurer G, Saygi S, Ciger A. Generalised periodic epileptiform discharges: clinical features, neuroradiological evaluation and prognosis in 37 adult patients. Seizure 2003; 12:465-72.
[50] Fernandez Torre JL, Arce F, Martinez-Martinez M, Gonzalez-Rato J, Infante J,Calleja J. Necrotizing leukoencephalopathy associated with nonconvulsive status epilepticus and periodic short-interval diffuse discharges: a clinicopathological study. Clin EEG Neurosci. 2006; 37:50-3
[51] Radermecker J. Leucoencéphalite subaigu sclérosante avec lésions des ganglions rachidiens et des nerfs. Rev. neurol. 1949; 81:1009-17.
[52] Westmoreland BF, Klass DW. Unusual EEG patterns. J Clin Neurophysiol 1990;7:209-28

[53] Sauseng P, Klimesch W, Doppelmayr M, Pecherstorfer T, Freunberger R, Hanslmayr S. EEG alpha synchronization and functional coupling during top-down processing in a working memory task. Hum Brain Mapp 2005; 26: 148-55.

[54] Van Hoof E, De Becker P, Lapp C, Cluydts R, De Meirleir K.Defining the occurrence and influence of alpha-delta sleep in chronic fatigue syndrome. Am J Med Sci 2007; 333: 78-84

[55] Chatrian GE, Shaw CM, Leffman H. The significance of periodic lateralized epileptiform discharges in EEG: an electrographic, clinical and pathological study.Electroencephalogr Clin Neurophysiol 1964;17:177-93.

[56] Baroque HG, Purdy P. Lesion localization in periodic lateralized epileptiform discharges: grey or white matter. Epilepsia,1995:36:58-62.

[57] Snodgrass SM, Tsuburaya K, Ajmone Marsan C. Clinical significance of periodic lateralized epileptic discharges: relationship with status epilepticus.J Clin Neurophysiol 1989;6:159-72.

[58] Treiman DM, Walton NY, Kendrick C. A progressive sequence of elctroencephalographic changes during generalized convulsive status epilepticus. Epilepsy Res 1990;5:49-60.

[59] Garzon E, Fernandes RM, Sakamoto AC. Serial EEG during human status epilepticus : evidence for PLEDs as an ictal pattern.Neurology 2001;57:1175-83.

[60] De la Paz, Brenner RP.Bilateral independant periodic lateralized discharges. Clinical significance. Arch Neurol 1981;38:713-5

[61] Walker M, Cros H, Smith S. Nonconvulsive status epilepticus: *Epilepsy Research Foundation Workshop Reports.*Epileptic Disord 2005;7:253-93.

[62] Treiman DM, Meyers PD, Walton NY, Collins JF, Colling C, Rowan AJ, Hand-forth A, Faught E, Calabrese VP, Uthman BM, Ramsay RE, Mandani MB. A comparison of four treatments for generalized convulsive status epilepticus. *Veterans Affairs Status Epilepticus Cooperative study Group.*N Eng J Med 1998;339:792-8.

[63] Hussain AM, Mebust KA, Radtke RA. Generalized periodic epileptiform discharges: etiologies, relationship to status epilepticus, and prognosis. J Clin Neurophysiol 1999:16:51-8.

[64] Krumholz A. Epidemiology and evidence for morbidity of nonconvulsive status epilepticus.Clin Neurophysiol 1999;16:314-2.

[65] Niedermeyer E, Ribeiro M. Considerations of nonconvulsive status epilepticus.Clin Electroencephalogr 2000;31:192-5.

[66] Kaplan PW. EEG criteria for nonconvulsive status epilepticus. Epilepsia. 2007; 48 Suppl 8:39-41.

[67] Brenner RP, Schaul N. Periodic EEG patterns: classification, clinical correlation, and pathophysiology. J Clin Neurophysiol 1990;7:249-67.

[68] Yemisci M, Gurer G, Saygi S, Giger A. Generalized periodic epileptiform discharges: clinical features,neuroradiological evaluation and prognosis in 37 adults patients. Seizures 2003;12:465-72

[69] Wieser HG, Schindler K, Zumsteg D. EEG in Creutzfeldt-Jacob disease.Clin Neurophysiol 2006;117:935-51

[70] Hormes JT, Bennaroch EE, Rodriguez M, Klass DW. Periodic sharp waves in baclofen-induced encephalopathy. Arch Neurol 1988;45:814-5

[71] Neufeld MY.Periodic triphasic waves in levodopa-induced encephalopathy. Neurology 1992;42:444-6.
[72] Kaplan PW, Birbeck G. Lithium-induced confusional states: nonconvulsive status epilepticus or triphasic encephalopathy? Epilepsia 2006;47:2071-4.
[73] Primavera A, Audenino D, Cocito L. Ifosfamide encephalopathy and nonconvulsive status epilepticus. Can J Neurol Sci 2002; 29: 180-3
[74] Drake ME, Erwin CW. Triphasic EEG discharges in metrizamide encephalopathy. J Neurol Neurosurg Psychiatry 1984;47:324-5.
[75] Jette N, Cappell J, Van Passel L, Akman CI. Tiagabine-induced nonconvulsive status epilepticus in an adolescent without epilepsy. Epilepsy Res. 2007; 73:1-52.
[76] Sundaram MB, Blume WT. Triphasic waves: clinical correlates and morphology. Can J Neurol Sci 1987;14:136-40
[77] Hirsch LJ, Claassen J, Mayer SA, Emerson RG. Stimulus-induced rhythmic, periodic, or ictal discharges (SIRPIDs): a common EEG phenomenon in the critically ill. Epilepsia 2004; 45:109-23
[78] Tomson T, Lindbom U, Nilsson BY. Nonconvulsive status in adults : thirty-two consecutive patients from a general hospital population.Epilepsia 1992;33:829-35.
[79] Litt B, Wityk RJ, Hertz SH, Mullen PD, Weis H, Ryan DD, Henry TR. Nonconvulsive status epilepticus in the critically ill elderly. Epilepsia.1998; 39:1194-202
[80] Kaplan PW. Assessing the outcomes in patients with nonconvulsive status epilepticus: nonconvulsive status epilepticus is underdiagnosed, potentally overtreated, and confounded by comorbidity. J Clin Neurophysiol 1999;16:341-52
[81] Krumholtz A, Sung G, Fisher RS, Barry E, Bergey GK, Grattan LM. Complex partial status epilepticus accompanied by serious morbidity and mortality. Neurology 1995; 45:1499-504.
[82] So El, Ruggles Kh, Ahmann PA, Trudeau SK, Weatherford KJ, et al. Clinical significance and outcome of subclinical status epilepticus in adults. J Epilepsy 1995;8:11-5.
[83] Lowenstein DH, Aminoff MJ. Clinical and EEG features of status epilepticus in comatose patients. Neurology 1992;42:100-4
[84] Handforth A, Cheng JT, Mandelkem MA, Treiman DM. Markedly increased mesiotemporal lobe metabolism in a case with PLEDS: further evidence that PLEDs are manifestations of partial status epilepticus. Epilepsia 1994; 35:876-81.
[85] Jaitly R, Sgro JA, Towne AR, Ko D, DeLorenzo RJ. Prognostic value of EEG monitoring after status epilepticus: a prospective adult study. J Clin Neurophysiol. 1997;14:326-34
[86] Young GB, Jordan KG, Doig GS. An assessment of nonconvulsive seizures in the intensive care unit using continuous EEG monitoring: an investigation of variables associated with mortality. Neurology 1996;47:83-9.
[87] Claassen J, Hirsch LJ, Emerson RG, Bates JE, Thompson TB, Mayer SA. Continuous EEG monitoring and midazolam infusion for refractory nonconvulsive status epilepticus.Neurology. 2001;57:1036-42
[88] Shneker BF, Fountain NB. Assessment of acute morbidity and mortality in nonconvulsive status epilepticus. Neurology. 2003;61:1066-73.
[89] DeGiorgio CM, Heck CN, Rabinowicz AL, Gott PS, Smith T, Correale J. Serum neuron-specific enolase in the major subtypes of status epilepticus. Neurology 1999; 52:746-9.

[90] DeGiorgio CM, Correale J, Gott PS, Ginsburg DL, Bracht KA, Smith T. et al. Serum neuron-specific enolase in human status epilepticus. Neurology 1995;45:1134-7.
[91] DeGiorgio CM, Gott PS, Rabinowicz AL, Heck CN, Smith T, Correale J. Neuronspecific enolase, a marker of acute neuronal injury, is increased in complex partial status epilepticus. Epilepsia 1996;37:606-9
[92] Fujikawa DG, Itabashi HH, Wu A, Shinmei SS. Status epilepticus-induced neuronal loss in humans without systemic complications or epilepsy.Epilepsia. 2000;41:981-91
[93] Tomson T, Lindbom U, Nilsson BY. Nonconvulsive status epilepticus in adults: thirty-two consecutive patients from a general hospital population.Epilepsia. 1992;33:829-35.
[94] Bauer J, Stefan H, Huk WJ, Feistel H, Hilz MJ, Brinkmann HG, Druschky KF, Neundörfer B. CT, MRI and SPECT neuroimaging in status epilepticus with simple partial and complex partial seizures: case report. J Neurol. 1989;236:296-9.
[95] Maeda Y, Oguni H, Saitou Y, Mutoh A, Imai K, Osawa M, Fukuyama Y, Hori T, Yamane F, Kubo O, Ishii K, Ishiwata K. Rasmussen syndrome: multifocal spread of inflammation suggested from MRI and PET findings. Epilepsia. 2003; 44 :1118-21
[96] Kutluay E, Beattie J, Passaro EA, Edwards JC, Minecan D, Milling C, Selwa L, Beydoun A. Diagnostic and localizing value of ictal SPECT in patients with nonconvulsive status epilepticus. Epilepsy Behav. 2005;6:212-7
[97] Wieser HG. Aura continua In: Gilman S, editor. MedLink-Neurobase. San Diego: Arbor Publishing, 3rd 2001 ed. [updated Oct 29,2006]
[98] Wieser HG. Dyscognitive focal (psychomotor, complex partial) status epilepticus.In: Gilman S, editor. MedLink-Neurobase. San Diego: Arbor Publishing,3rd 2001 ed.[updated Oct 29,2006]
[99] Treiman DM. Generalized convulsive status epilepticus in the adult. Epilepsia 1993; 34:S2-S11
[100] Kaplan PW. No, some types of nonconvulsive status epilepticus cause little permanent neurologic sequelae (or: "the cure may be worse than the disease"). Neurophysiol Clin. 2000; 30:377-82.
[101] Mesraoua B, Wieser HG. Non-convulsive seizures and non-convulsive status epilepticus monitoring in the intensive care unit. A real need for the Gulf Cooperation Council countries. Neurosciences (Riyadh). 2009;14:323-37
[102] Hirsch LJ, Claassen J. The current state of treatment of status epilepticus. Curr Neurol Neurosci Rep 2002; 2:345-56.
[103] Alldredge BK, Gelb AM, Isaacs SM, Corry MD, Allen F, Ulrich S, Gottwald MD, O'Neil N, Neuhaus JM, Segal MR, Lowenstein DH. A comparison of lorazepam, diazepam, and placebo for the treatment of out-of-hospital status epilepticus. N Engl J Med. 2001;345:631-7.
[104] Leppik IE, Derivan AT, Homan RW, Walker J, Ramsay RE, Patrick B. Doubleblind study of lorazepam and diazepam in status epilepticus. JAMA 1983; 249:1452-4.
[105] Rossetti AO, Bromfield EB. Levetiracetam in the treatment of status epilepticus in adults: a study of 13 episodes. Eur Neurol.2005; 54:34-8.
[106] Atefy R, Tettenborn B. Nonconvulsive status epilepticus on treatment with levetiracetam. Epilepsy Behav. 2005; 6:613-6.
[107] Patel NC, Landan IR, Levin J, Szaflarski J, Wilner AN. The use of levetiracetam in refractory status epilepticus. Seizure. 2006; 15:137-41.

[108] Bensalem MK, Fakhoury TA. Topiramate and status epilepticus: report of three cases. Epilepsy Behav.2003; 4:757-60.
[109] Tafreshi R, Dumont G, Gorss D, Ries CR, Puil E, MacLeod BA. Seizure detection by a novel wavelet packet method. Conf Proc IEEE Eng Med Biol Soc 2006; 76: 6141-4.
[110] Khan YU, Gotman J. Wavelet based automatic seizure detection in intracerebral electroencephalogram. Clin Neurophysiol 2003; 114: 898-908.

Part 3

Psychological Aspects

8

Neuropsychological Evaluation in Epilepsy Surgery – A Cross-Cultural Perspective

Ahmed M. Hassan
Department of Neurosciences, King Faisal Specialist Hospital & Research Center, Riyadh Kingdom of Saudi Arabia

1. Introduction

Epilepsy is a chronic neurological disorder that affects around 50 million people globally (Boer, et al., 2008; Meyer, et al., 2010). It is probably the most universal of all medical disorders, occurring in both men and women, and affecting all ages, races, social classes, and nations (Reynolds, 2000). In its report on epilepsy, the Eastern Mediterranean Regional Office (2010) of the World Health Organization (WHO) estimates that 85% of the 50 million cases of epilepsy worldwide are living in the 'developing countries', of which an estimated 4.7 million cases live in the WHO Eastern Mediterranean region that comprises 22 'developing countries', which extend from Morocco in the west to Iran in the east, and include Saudi Arabia the focus of this chapter.

The term 'developing country', however, warrants some verification in order to correct a misleading connotation associated with the countries that the term indiscriminately labels, leading to a false implication of homogeneity between these countries. The term is generally used loosely to describe a nation with a low level of material well-being. Since, at least in reality, no single definition of the term *developing country* is recognized internationally, the level of development may vary widely within the so-called developing countries. In fact, according to the United Nation Statistics Division, *"There is no established convention for the designation of 'developed' and 'developing' countries or areas in the United Nations system"*. These terms are *"intended for statistical convenience and do not necessarily express a judgment about the stage reached by a particular country or area in the development process"* (United Nation Statistics Division, 2008).

On the other hand, the developing countries have been strongly associated in the literature (Scott, et al., 2001; Meinardi, et al., 2001; Coleman, et al., 2002; Sridharan, 2002) with a high proportion of 'seizure treatment gap', a term defined as "the difference between the number of people with active epilepsy and the number of those whose seizures are being appropriately treated in a given population at a given point of time, presented as a percentage. The definition includes diagnostic and therapeutic deficits" (Meinardi, et al., 2001). The definition is said to propose a useful parameter to compare access to and quality of care for epilepsy patients across populations (Kale, 2002). The frequent link in the literature between high proportions of seizure treatment gap and the 'developing countries' has formed a strong general impression that the estimated high proportions of epilepsy treatment gap in 'developing countries' is largely a function of the low-income and resource-poor economies of these countries (Meyer, et al., 2010; Mbuba, 2008). No doubt the adverse effects of a poor economy in some of such countries can affect the availability and quality of health care systems. Nevertheless, the case to be discussed in this chapter is that, other than an economic factor, there are diverse sociocultural realities and beliefs, as well as some indigenous customs and traditions, which prevail in some of such 'developing countries' with the consequent effects of significantly contributing towards the creation of a 'seizure treatment gap'.

Saudi Arabia is a vast country occupying most of the Arabian Peninsula, with a population of nearly 27 million who are entirely Muslims. All Saudi citizens speak Arabic. Nearly all the Saudi citizens are Arab, although there has been considerable ethnic mixture, especially in Al-Hejaz region, because of centuries of immigration connected with performing the rituals of Hajj, the fifth pillar of Islam. This rich oil-producing country is listed in the World Bank classification of world economies as a 'high income country', with a gross national income (GNI) of US$ 24,020 per capita (World Bank, 2009). The country generally provides high standards in health care, and the medical technologies are continually upgraded. In addition, the governmental hospitals provide free medical treatment to all citizens. The country has a rate of 9.4 physicians and 21.0 nurses/midwives per 10,000 individuals (WHO, 2009). Accordingly, although it is often cited among the developing countries, Saudi Arabia can hardly be considered as a low-income or a resource-poor economy.

There is a scarcity in the published studies related to epidemiological and sociocultural dimensions of epilepsy in Saudi Arabia. A similar condition applies to studies on the role of neuropsychology in epilepsy surgery. Apart from Escandall's (2001) narrative account about neuropsychological evaluation in Saudi Arabia, no study so far has addressed the unique sociocultural features of the Saudi society in relation to the practice of neuropsychology in general, and the neuropsychological evaluation in epilepsy surgery in particular. However, epilepsy is a common neurological disorder in Saudi Arabia, as is the case with regard to other neurological disorders, due to the high rate of consanguinity (Jan, 2005; Abduljabbar, 1998). The published rate of prevalence of epilepsy in Saudi Arabia is 6.54 per 1000 population (Al Rajeh et al., 2001), which may help in approximating an estimate for seizure treatment gap in the country. It is argued that the country, based on its solid economic stand, is capable in principle of providing its epilepsy patients with adequate treatment including epilepsy surgery. The challenge is whether the cultural and social aspects related to epilepsy will facilitate or hinder optimal delivery of such care. One of the aims in this chapter, however, is to discuss the social and cultural variables that interact with the role of neuropsychological evaluation in epilepsy surgery being a viable management strategy for intractable focal epilepsies.

It is estimated that up to a third of all individuals with epilepsy are refractory to drug therapy, and surgery is widely accepted as an effective therapy for selected individuals with drug resistant epilepsy (Spencer and Huh, 2008; Helmstaedter, 2004). Despite major challenges, including a limited number of epilepsy surgery centers that are placed in one or two cities hence requiring patients to travel long distances to seek the service, the surgical treatment of epilepsy in Saudi Arabia has witnessed some notable advances in the last 10 years. Epilepsy surgery has scored considerable successes in the ultimate control of seizures in focal epilepsies resistant to drug interventions. This chapter is built around a retrospective research work concluded this year in the comprehensive epilepsy surgery center of a tertiary medical institution that receives referrals of epilepsy patients from regional medical institutions in the country. Surgical resection procedures performed in the center for temporal lobe epilepsies represented two thirds of all epilepsy surgical operations, with the frontal lobe epilepsies ranking second in order of frequency. These figures are not different from those reported in other centers worldwide (e.g., Helmstaedter, 2004). The compiled data for this chapter came from the Adult Neurology section in the medical institution, and the discussion, therefore, is pertinent to patients of 16 years of age or older.

This chapter also deals with the role of neuropsychological evaluation as one of the primary presurgical evaluation modalities in epilepsy surgery programs. In doing so, the social and cultural aspects of epilepsy that may influence the practice of neuropsychological evaluation of this category of patients are reviewed. The ultimate cause of this work is to elucidate the need for developing an indigenous set of neuropsychological philosophy, material, and practice that are both evidence-based and reflective of social and cultural characteristics of the Saudi people. There is recognition that some important concepts and assumptions embedded in the neuropsychological testing material are derived from studies and experiences in the 'developed countries' and extrapolated to developing world practice. Such 'imports' are often proving to be inappropriate. One important factor behind this situation is that indigenous neuropsychological products of the developing countries are sparse, and what do exist may be sometimes of doubtful authenticity or originality. The account reported here in this respect is derived in most parts from hospital-based observations and clinical encounters, which are inevitably incomplete, and occasionally impressionistic. This and similar accounts may eventually pave the way to comprehensive data based on systematic reviews of sociocultural variables influencing epilepsy sufferers in the country.

2. The social and cultural aspects of epilepsy in Saudi Arabia

There is generally limited data related to the developing countries on the social, psychological, and cultural aspects of epilepsy (Shorvon and Farmer, 1988; Boer, et al., 2008). In addition, data comes primarily from general surveys, which are less reliable than censuses or large-scale demographic and health surveys. There is a more obvious lack in the literature for data on the epidemiological and sociocultural aspects of epilepsy in Saudi Arabia. For example, obtaining systematic data on such aspects of epilepsy is particularly problematic due to the hard-core conservatism of most of the population of the country.

Saudi Arabia is governed according to the Islamic Sharia law whereby the Islamic orthodoxy and conservatism generally dominate the social and cultural mechanisms of the Saudi society. Traditional tribal principles and customs, however, remain significant in influencing a wide range of behavioral and attitudinal manifestations in the society. Accordingly, religious beliefs, social customs and traditions have both positive and negative impacts on epilepsy in this country. On the positive side, it is rare in Saudi Arabia that sufferers of chronic diseases or disorders find themselves abandoned by their families, regardless of the severity or burden of the ailments on either the patients or the families. Families, regardless of their financial capabilities, usually support their affected members. It is quite frequent to discover, while interviewing an adult patient with intractable epilepsy since childhood (who, because of early onset epilepsy, has no education or a job) that he is being supported by family members into marriage and in raising his children over the years. This deeply rooted familial solidarity and generosity are frequently attributed to Islamic teachings and are reinforced by the traditional norms. This trend, however, is aided by the fact that necessities such as medical treatment and antiepileptic drugs are provided in governmental hospitals free of charge to all Saudi patients. Colleagues and superiors at a workplace also usually have understanding towards the needs of their co-workers who have epilepsy. In a clinic, many people with epilepsy on full-time jobs reported that their colleagues usually helped them by willingly covering their duties whenever they had their seizures at work. The bosses are also usually lenient with individuals with chronic diseases such as epilepsy, permitting them to keep flexible or shorter working hours. These accounts were the norm with the majority of epilepsy patients encountered in the clinical practice. It is not clear, however, if this widely reported positive support enjoyed by epilepsy patients at the workplace in Saudi Arabia is a function of a feeling of pity towards patients with epilepsy, or it is a psychosocial practice supported by a religious teaching. It is probably both.

Ironically, epilepsy in Saudi Arabia, on the other hand, is socially and culturally a devalued condition, similarly in other societies (Boer, 2008). The social disadvantage associated with epilepsy has its roots in the belief system of the society. Outside the family, a Saudi epilepsy patient and the members of his family may fear what people might think of the patient if a seizure occurs in the public. Such fear leads to noticeable efforts by the patient and his family to hide the disorder from the outsiders. Many of the epilepsy patients encountered in the clinic gave one of two reasons for why they quit school: either due to repeated school failures caused by cognitive impairment, or because of recurring seizures at school, leading to either dismissal by the school or withdrawal from school by his family to hide the fact of epilepsy in the family. After all, epilepsy is a socially sensitive and negative matter in this traditional society that over-values conformity to public norms, beliefs, and values. Related to stigmatization is the fact that epilepsy is often blamed solely on heredity. Attribution to hereditary processes is a common belief among many individuals in the Saudi society. Such an attribution often creates a social disadvantage for the patient with epilepsy, and probably for his or her healthy siblings as well, with regard to marriage prospects. In other words, the stigma occasionally extends beyond the individual to family members. It is interesting that some Saudi researchers (Jan, 2005; Abduljabbar, M. 1998) attributed the prevalence of epilepsy in Saudi Arabia to the high rate of consanguinity among Saudi people.

According to Islamic teachings, every disease of body or mind, including epilepsy, is attributed to the will of Allah (God) and to His wisdom. Probably very few Saudi individuals will dispute this belief. However, there is another widely prevailing belief that, for example in the case of epilepsy, the will of God dictates that a particular person be inflicted with a particular ailment, and the wisdom behind that only Allah knows. It is common, therefore, that some patients with epilepsy seek faith healing as the first line of intervention. Faith healers usually recite verses of the Holy Quran upon the patient who may be subjected to many sessions of faith healing over several days, weeks, or months. This trend is found among individuals from different socioeconomic strata in Saudi Arabia. Seeking faith healing to treat epilepsy in many developing countries is caused by the non-availability or non-affordability of medical treatments (e. g., Baskind and Birbeck, 2005). Seeking faith healing for treatment of epilepsy in Saudi Arabia, where medical treatment is available free of charge, is based on an indigenous belief system, rather than on economic or availability factors. However, the precise impact of preference for faith healing on the seizure treatment gap in Saudi Arabia is unknown. Nevertheless, giving faith healing priority over medical treatment of epilepsy is believed to considerably delay the initiation of medical treatment for those patients. In the study reported in this chapter, nearly half of the patients in our sample reached the center for an advanced treatment of epilepsy after 15.4 years, in average, since the onset of their seizure disorder. Although the time of referral to the center may not be an accurate indicator in this respect, it gives an estimate about the possible role of sociocultural factors in influencing aspects of epilepsy treatment.

3. Neuropsychological evaluation in epilepsy surgery

Neuropsychology and epilepsy has enjoyed a special relationship with mutual benefit to both (Loring, 1997). Neuropsychology has been involved with preoperative and postoperative evaluations of patients undergoing focal resection of a variety of brain regions to control their epilepsy (Novelly, 1992). Such an evaluation remains the best method of identifying and quantifying the nature and degree of cognitive malfunctioning that arises from epilepsy (Helmstaedter, 2004; Jones-Gotman, et al., 2000), and it provides information that other assessment modalities, such as electroencephalography and magnetic resonance imaging, do not provide (Jones-Gotman, et al., 2000). Evaluating cognitive abilities is an important part of the comprehensive presurgical clinical workup in epilepsy surgery programs. The primary function of neuropsychology in epilepsy surgery, then, is the evaluation of epilepsy-related, or lesion-related, impairments before surgery in the direction of functionally defining possible focus or foci of the epileptogenic zones in the brain. The second, and equally important, function is to employ neuropsychology as a tool for assessing a surgical treatment outcome.

Epilepsy causes a set of cognitive impairments in people with epilepsy. Memory impairment of various severities is an important cognitive problem in patients with epilepsy, especially in temporal lobe epilepsy. Epilepsy affects learning, for example, by reducing alertness and by interfering with short-term information storage and abstraction. Frequent and uncontrolled seizures impair learning of new information, disrupt consolidation of memory, and affect language functions. Cognitive impairments can also be a side effect of the various antiepileptic drugs used to control seizures. Such drugs are associated with learning difficulties, behavior changes, and memory deficits (Meador, 2006; Motamedi and Meador, 2004; Ortinski and Meador, 2004; Brunbech and Sabers, 2002; Aldenkamp, 2001; Kwan and Brodie, 2001; Devinsky, 1995). People with intractable seizures endure the additional burdens of social discrimination, stigmatization, social embarrassment when seizing or falling in public. As with any chronic condition, epilepsy can be linked to demoralization and a negative perspective on life and to disturbances of affect and mood (Baker, 2002).

The above combination of potential symptoms is what a clinical neuropsychological assessment may encounter during a presurgical evaluation of patients with intractable epilepsy considered for surgical intervention. As such, a neuropsychological testing for a pre-operative evaluation in epilepsy surgery consists, by necessity, of a battery of tests, not just one test. A typical neuropsychological assessment, thus, consists of a comprehensive evaluation of cognitive functioning that includes intelligence, executive skills, memory, attention, visuospatial abilities, language, and motor skills. Different epilepsy centers use different tests. Apart from a neuropsychological battery for epilepsy proposed by Dodrill more than 30 years ago (Dodrill, 1978), there is not an established battery of tests that is designed and calibrated for the sole purpose of presurgical neuropsychological evaluation of patients with epilepsy. However, the question of which tests are used in epilepsy presurgical evaluation, and according to what protocol these tests are administered, is discussed below while reviewing three major neuropsychological assessment approaches. These approaches, which developed over the long history of the discipline, have informed the practice of clinical neuropsychology.

The *behavioral neurological approach* is European in origin, and it is best reflected in the Luria-Nebraska neuropsychological battery (Luria, 1973). It relates distinct functional abnormalities (signs) to neuropsychological entities (conditions; syndromes). A sign is a dichotomous state, being either present or absent, and there is no continuum of performance between the 'normal' and the 'pathological'. The patient either passes or fails the test. This approach requires a skilled clinician, depending heavily on identification of valid and easily assessable signs as indicators of abnormal function.

The *psychometric approach* is North American in origin. A psychometric example for this approach is the Halstead-Reitan neuropsychological test battery (HRB; Reitan and Wolfson, 1986). The HRB involved Reitan's observations that the relationship between tests differed in patients with different kinds of pathology. Therefore, in a pattern analysis method, the neuropsychologist is concerned with the interpretation of test performances as an index of neuropsychological functioning. This approach, accordingly, uses a number of batteries that survey a broad range of psychological functions. Then, it draws conclusions from the detailed pattern of results, employing psychometric methods. It is a thorough and systematic approach. However, this approach takes significant clinical time, and it is inflexible in its application to individual patients (Horton, 2008). One obvious limitation of such a method is exposed in unique cases for which there are no significant patterns of performance established. This limitation imposes serious problems when the battery of tests is employed in multicultural settings to assess individuals who are not represented in the normative sample on which characteristics the tests have been developed.

Approaches that employ a fixed set of tests have been criticized on the ground that neuropsychology is not a "one size fits all". That is, the use of a standardized fixed battery approach has the potential disadvantage that a particular set of tests may be inappropriate for particular patients (Horton, 2008) for a variety of reasons. On the other hand, several researchers in the field of cross-cultural neuropsychology (e.g., Artiola, et al., 1997) have been outspoken critics of those who would simply take existing measures and attempt to transform them into a foreign language measure without adequate understanding of the second language or the prevailing culture in which the test will be used. Artiola, et al. (1997) emphasized that the relevant literature afford numerous examples of how well intended practitioners have erroneously assumed that tests that work for a typical North American cohort should be adequate for use with samples or individuals that differ in important ways.

As the practice of neuropsychology became more entrenched in the clinical settings, another major division in neuropsychology occurred as a function of the kind of battery or method employed by clinical neuropsychologists. Russell (1997) described the fundamental distinction between current schools of thought as adhering either to 'pattern analysis' or 'hypothesis testing' methods. Hypothesis testing assessments are driven by a series of individual hypotheses to be tested with regard to neuropsychological functioning in an individual patient. The *individual-centered normative approach* is originated in the United Kingdom, and it adheres to the 'hypothesis testing' school. This approach employs specific tests, combined flexibly in the investigation of a hypothesis about the patient's difficulties. It is a scientific detective work rather than medical examination (as in the behavioral neurological approach) or formal assessment (as in the psychometric approach). This approach utilizes a general initial screening procedure (e.g., the Wechsler Adult Intelligence Scale), in addition to the current clinical information about the patient, to construct hypotheses about present deficits, which are consequently tested using individual test procedures or single-case empirical investigations. This approach requires unusually high knowledge and expertise on the part of the clinician. On the other hand, the hypotheses generated are largely a function of the individual neuropsychologist's knowledge base. This, unfortunately, can lead to a considerable lack of uniformity in the assessment product. That is, if the individual administering the tests does not ask a question, it probably goes unanswered. Obviously, this approach requires from the neuropsychologist an adequate understanding of the language and the culture of the individuals being examined.

From a technical perspective, neuropsychology has undergone evolutionary changes since the initial involvement in epilepsy surgery at the Montreal Neurological Institute. Neuropsychological findings were used to help in identifying suitable candidates for surgery, deciding whether Wada testing was necessary, and to determine surgery outcome in terms of postoperative cognitive changes. Despite of the development of better technologies, such as structural imaging capability, sophisticated electroencephalography including video monitoring, and refinement of Wada test, neuropsychology has continued to provide quantitative measurements of cognitive abilities and, thus, to contribute to the evaluation, management, and surgery outcome issues of epilepsy surgery patients (Spencer and Huh, 2008; Tellez-Zenteno, et al., 2007; Loring, 1997).

4. Cross-cultural issues in neuropsychological evaluations

The ability to localize impairment on the basis of neuropsychological performance is dependent upon the sensitivity and specificity of the tests employed for that purpose (Jones-Gotman, et al., 2000). In addition to appropriate psychometric properties, such tests need to be void of cultural biases in order to allow an accurate categorization of a patient's deficits. The WMS-III and the WAIS-III were adapted and validated in Saudi Arabia using primarily clinical samples (Escandall, 2002). Apart from the shortcoming of lacking validation with normal subjects (Escandall, 2002), the use of these scales in a cultural setting such as Saudi Arabia, whose population is significantly different from the normative population upon which these scales were originally developed, imposes important problems. These tests have been developed and validated with primarily non-Muslim white Western normative cohorts and, therefore, there always is the fear that some differences in test performance may not be directly attributable to differences in brain functioning but rather a reflection of test biases (Gasquoine, 1999). In fact, within the standardization samples of the WAIS-III and the WMS-III, African American and Hispanic subjects scored lower than Caucasian subjects, and that the African American subjects were more likely than other groups to be misclassified as impaired using a cut-off score of 1 SD below the mean (Heaton, et al., 2003). Cultural differences on the Wechsler scales have also been demonstrated in patient groups (e.g., Boone, et al., 2007). Most of these studies have been conducted with cultural groups such as African American and Hispanic subjects, which are prominent in North America and have been educated in the West with proficient English language skills, yet the cultural differences have impact on their neuropsychological test performance. Obviously, cultural differences entail much more factors than just a different language. Several definitions have been given to the term 'culture'. Handwerker (2002) states that 'culture is an arrangement of cognition, emotion, and behavior', while Padilla (1999) defines culture as similar thoughts, feelings, behaviors including, but not limited to, traditions, customs, and ways of life. Defined as the way of living of a human group, culture includes also values, attitudes, and beliefs (Harris, 1983). In addition, the physical elements characteristic of a particular culture (such as, in Saudi Arabia, traditional weapons such as swords, houses including desert tents, animals such as camels, characteristic male and female dress, traditional hobbies and games played by children or adults) are cultural elements and are also included in the definition of culture. Some researchers (e.g., Ardila, et al., 2000) include education, literacy, and schooling as element of culture, since both culture and formal education have significant effects on cognition (Cole, 1997).

There have been numerous attempts over the years to construct psychological testing measures that would be "culture free" (see, e.g., Anastasi, 1988). There has been evidence that culture-bias cannot be controlled only by eliminating verbal items from a test (Anastasi, 1988; Irvine and Berry, 1988). In fact, also the nonverbal tests may be culturally biased (Rosselli and Ardila, 2003). It has been observed in our neuropsychological practice in Saudi Arabia that a task involving drawing or copying figures of various level of complexity on paper using a pencil (as in the Visual Reproduction subtest of the WMS-III; the Complex Figure Test) can represent an enormous task for, for example, a rural illiterate healthy lady in her forties or fifties who has never touched a pencil in her life. Such a task for normal individuals in the Western societies is effortless. Such a task surely requires more than just some motor coordination. Non-verbal tests often require specific strategies and cognitive styles characteristic of middle-class Western cultures (Cohen, 1969). The notion that the performance of some nonverbal tasks, such as drawing a map or copying figures, in neuropsychological assessment are universal skills of most normal adults (Lezak, 1995), is contrasted with the argument that drawing a map or copying figures represent abilities that are absent in many cultures (Ardila and Moreno, 2001; Berry et al., 1992; Irvine and Berry, 1988). In addition, such abilities are argued to be highly school-dependent skills (Ardila, et al., 1989).

Given the fact that language is an ultimate cultural element, the translation process of a neuropsychological test from one language to another may raise a host of issues, including copyright, literal versus cognitive and emotional equivalence, and norms. Often the items are translated literally without their culturally correct equivalents, thus limiting their validity (Salazar, et al., 2007). Even when a test is properly translated, it means a little without an appropriate normative data. Related to this point is a situation whereby the neuropsychological evaluation is administered through a translator due to a language barrier between the neuropsychologist, or the examiner, and the individual evaluated. It poses ethical issues as well as testing validity concerns. There are, unfortunately a few neuropsychological tests in Saudi Arabia that were originated in the West, then literally translated into Arabic language in other countries, apparently without author permissions, and eventually found their way to be used in the clinic with Saudi patients. Such practices have caused, among many other problems, an obstacle in the face of consolidating the efforts towards constructing indigenous Saudi neuropsychological tests.

There are important differences between the Saudi culture and the North American culture within which these tests have been developed and validated. For example, the rate of literacy (defined in terms of ability to read and write among ages from 15 and over) in Saudi Arabia for the total adult population is 85.5% (WHO, 2008), compared to 99% in the United States (2004). However, the concept of 'functional' illiteracy determines that any person with less than five years of schooling is considered functionally illiterate, or unable to engage in social activities in which literacy is assumed (UNESCO General Conference, 1978). By either definition of literacy, illiteracy rate is expected to be much higher among patients with intractable seizure disorder in Saudi Arabia, due to various reasons preventing them from seeking or continuing education, such as cognitive difficulties, stigma, and other psychosocial factors. Individuals with minimal or no formal schooling are believed to be disadvantaged with regard to neuropsychological test performance, especially when such tests are validated for a culturally different normative population. Several studies (e.g., Ardila, et al., 1989; Ardila, et al., 1992) have shown strong association between educational level and performance on various neuropsychological measures. The effect of schooling on neuropsychological test performance has been reported for various types of abilities including memory, language, problem soling, constructional abilities, motor skills, and calculation abilities (Ardila, et al., 1989; Rosselli, et al., 1990; Lecours, et al., 1988). Ostrosky and his colleagues (1998), in a study to analyze the effects of education on neuropsychological test performance across different age ranges, they compared 64 totally illiterate normal subjects with two barely schooled control groups (1-2 and 3-4 years of schooling). They found a robust education effect on most of the tests, including constructional abilities (figure copying), language (comprehension), phonological verbal fluency, and conceptual functions (similarities, calculation abilities, and sequences). They concluded by calling for caution with the education variables in neuropsychological assessment in order to avoid the risk of finding brain pathology where there are only educational differences.

The Family Picture subtest of the WMS-III has deservedly received frequent criticism in the literature for failing to assess its intended non-verbal memory construct, since the test content to be memorized can be verbally mediated (Djordjevic and Jones-Gotman, 2004). This subtest is comprised of pictorial representations of four different scenes containing members of a family including their dog. The four pictorial representations themselves contain items and acts that can be shown here that they are culturally biased. For example, the first scene (the Picnic) contains Frisbee play, which is a flying disc play that about 90% of Americans have played it at one time or another in their life (Malafronte, 1998). This recreational play, which is essentially an outdoor play, is by no means familiar among the bulk of Saudi population, whether as a physical item or as a game. It is not seen played in this country by Saudi nationals, probably due to environmental reasons since the weather is extremely hot during most of the year. Therefore, most of the patients who completed the Family Picture subtest gave different, occasionally funny, responses when prompted at the recall phase. One of the frequent responses given by some patients was that the grandfather was throwing a piece of meat to the dog! The few patients who did recognize a Frisbee were in fact those who had seen or even played the game during their frequent travels abroad. A Frisbee game is probably a familiar practice in the Western countries and, being featured as a stimulus item in a memory test, may represent in such countries a culturally familiar scene that may aid both memory consolidation and subsequent recall. Further, the third scene (the Yard) in the same subtest involves a family member giving the dog a bath. Dogs in Saudi Arabia, at least among most of the mainstream social strata, do not enjoy the same preference or intimacy that these creatures receive in Western societies. For example, it is very unlikely that an average Saudi individual considers a dog as a member of the family. The Islamic teachings prevent harming animals, dogs included; in the same time, these teachings discourage the Muslims from keeping dogs in the houses where they live, because they can violate the purity of the living quarters, where the household live and pray, by their urine, saliva, etc. Accordingly, dogs are rarely found in the houses of Saudi individuals, let alone to give a dog a bath in public. It was noticed, during testing for this subtest, that the common practice among the vast majority of patients examined was to ignore mentioning the dog altogether, leaving the door open for guessing whether it was a recall failure or just disregarding the dog as a culturally dictated response. In general, the unsuitability of the use of pictorial representations, such as in Family Picture and Faces, has been reported in the literature (Miller, 1973). For example, the problem confronted by a neuropsychologist in Saudi Arabia when administering the Faces subtest has a religious background. Since the faces used in this subtest are real life pictures, several patients refused or were very reluctant to look at the faces that were for adult females, emphasizing that their religion prohibited looking at foreign women (a foreign woman is a woman who does not belong to one of the categories of women a muslim is prohibited by the Islamic Sharia law from mating, including sisters, aunts, mothers-in-law, etc.), unless it was an accidental glance and only once. It would have been probably acceptable for such patients if all the faces used in this subtest were for men and/or children.

5. Neuropsychological evaluation: A Saudi experience

The practice of neuropsychology within the context of the Comprehensive Epilepsy Surgery Program (CESP) has been informed by the individual-centered approach. This approach is deemed flexible enough to allow the introduction of new individual tests that are locally developed and standardized for clinical use. It is considered also efficient in focusing on targeted areas of dysfunction. It can be finely tuned to investigate the exact parameters of abnormal performance. As stated earlier, temporal lobe epilepsies formed a majority of epilepsy conditions that the center received and treated. Therefore, a version of the Wechsler Memory Scale (Wechsler, 1945), which is the most common test battery used to evaluate learning and memory in individuals with epilepsy (e.g., Walker, et al., 2010; Loring and Bauer, 2010), had been used extensively in the presurgical neuropsychological evaluations. The main components of the evaluation protocol adopted in this use are outlined below.

The *clinical interview* plays a central role in the individual-centered approach towards obtaining demographic data, accounting for a possible impact of behavioral and emotional malfunctioning on a patient's cognitive performance, and capturing a longitudinal personal history that, along with the *general cognitive screening procedure,* facilitate the categorization of patients' specific and/or hypothesized cognitive deficits and impairments. The clinical interview took the shape of roughly two portions:

a. An initial part, which was dedicated mainly for understanding a patient's status, collecting demographic data, and explaining the purposes of current and subsequent clinical encounters. The demographic data collected include age, education, geographic origin, handedness, age of seizure onset, family history of epilepsy, frequency of seizures, etc. A patient's relative was usually present, and was occasionally allowed to attend this initial part of the interview depending on the state of the patient.
b. A subsequent part, which was aimed at obtaining and verifying the cognitive complaints of the patient, and any psychological problems related to mood, interpersonal difficulties, etc. This second part of the interview, and the subsequent testing sessions, involved the patient alone.

The general screening adopted in this study involved administering both the Wechsler Adult Intelligence Scale - Third Edition (WAIS-III; Wechsler, 1997a) and the Wechsler Memory Scale - Third Edition (WMS-III; Wechsler, 1997b) on all patients referred for presurgical neuropsychological evaluation. The exception was the group of patients who failed to complete the evaluation due to significant cognitive impairment.

Memory, being a temporal lobe function, is routinely evaluated in patients with epilepsy undergoing presurgical evaluations. An international survey of 82 epilepsy surgery centers found that 84% of centers routinely administered all or part of the WMS or the WMS-R in their pre-operative evaluations of epilepsy (Jones-Gotman, et al., 1993). On the other hand, some degree of intellectual decline is expected in patients with recurrent seizures for relatively long durations of their life. Therefore, an assessment of intellectual capacity in such patients usually forms an essential part of their neuropsychological profiling and hypothesis testing.

In this study, various neuropsychological tests were used to verify the findings that were obtained using the WAIS-III and the WMS-III. They include Wisconsin Card Sorting Test-64, Rey-Osterrieth Complex Figure Test, Verbal Fluency Test, Beck Depression Inventory, and Epstein-Fenz Anxiety Scale. However, the data presented here were limited to the subtests from the WMS-III and the full-scale IQ (FSIQ) derived from the WAIS-III, being the tests common to all patients in the sample.

The pre-surgical investigations involved the administration of intracarotid amobarbital procedure (ICAP), or Wada test, to determine language dominance in patients suspected of having atypical representation of language. Most of the patients, however, underwent an ICAP to assess memory in conditions with bilateral temporal lobe structural anomalies and/or neuropsychological test performance showing depressed verbal and nonverbal memory test scores. The standard Wada test protocol in the center entailed bilateral testing, whereby the suspected hemisphere was injected first. The protocol required the patient to be admitted the day before the date of the procedure. The one-night inpatient stay allowed adequate time for educating the patient on the proposed invasive procedure, and for obtaining an informed consent from him or her. Quite frequently, the informed consent had to be processed with the relative accompanying the patient or with both. The process of consenting for an invasive procedure such as the Wada Test is usually a family, not an individual, decision. Almost all patients admitted to the Epilepsy Monitoring Unit (EMU) are accompanied by a close relative who stays with the patient throughout the duration of admission. The practice of admitting patients for one night for Wada testing was informed by past experience that reinforced the importance of adequate preparation of patients and key relatives ahead of the procedure.

The patients included in this study were individuals with the diagnosis of a seizure disorder resistant to a drug therapy. Patients were referred from general hospitals in the country to the CESP for advanced management of their seizure disorder, where they were considered for epilepsy surgery intervention based on their clinical semiology and results of their presurgical investigations. These investigations were performed by different services in the hospital (i.e., Neurophysiology Section, Radiology Department, and neuropsychology service in Psychiatry Section). The inclusion criteria for this study were: (a) epilepsy was resistant to drug treatment; (b) the patient was an adult (16 years old or older); and (c) the patient had valid results in the four presurgery evaluation studies, namely: magnetic resonance imaging (MRI), positron emission tomography (PET), neuropsychological evaluation (NP), and ictal electroencephalography (iEEG).

From over 600 entries in the database, there were 330 adult patients with completed neuropsychological evaluations. Some of the excluded entries involved cases with the final diagnosis of pseudoseizure or uncompleted presurgical evaluations. Few patients failed to show up at their follow-up appointments for completing their neuropsychological evaluation. However, the final number of patients with valid sets of NP data and MRI, PET, and iEEG findings at the time of analyzing the data was 133 patients.

Patients were admitted to the EMU of the CESP from a waiting list. A CESP coordinator handled the referral of patients in the list to relevant services, and arranged hospital admissions of patients for EEG video monitoring. The patients received neuroimaging and neuropsychological evaluations, and other necessary investigative procedures, after their video monitoring was concluded. Neuropsychological evaluations were blinded from other presurgical evaluation results. A neuropsychologist administered and interpreted the neuropsychological evaluations of all patients in the sample. A research assistant then entered the neuropsychological data in a purpose-built computer database program, named "The Neuropsychology Database", which contained the data for all patients referred for neuropsychological evaluation from the CESP. The patients included in this report were those encountered in the period from July 2006 to July 2009 (during which one neuropsychologist performed all the evaluations) and satisfied the inclusion criteria stated above.

In summary, the outlined social, cultural, and ethnic characteristics of the Saudi population were expected to produce important implications on the neuropsychological test performance of the Saudi patient sample. This prediction was based on the fact that the test material involved were originally designed and standardized for use in a potentially different cultural context and with individuals having different cultural characteristics. Despite of the previous standardization work on these tests, there were still some salient aspects of cultural bias, as described above, that were expected to influence the ability of these tests in efficiently characterizing patient performances. The general purpose of the study, then, was to elucidate these implications by using the tests with a group of Saudi epilepsy patients in a presurgical neuropsychological evaluation, and then to examine their scores on these tests against the findings obtained for the same group of patients using other presurgical evaluation studies (i.e., MRI, PET, and iEEG). The specific objectives for this study were:

1. To examine the amount of agreement between the neuropsychological evaluation and PET, MRI, and iEEG findings iEEG in identifying the epileptogenic brain areas in presurgical evaluation of patients with intractable epilepsies.
2. To examine whether the scores obtained using the WMS-III were capable of differentiating the three patient groups with three categories of epilepsy: Temporal lobe epilepsy, frontal lobe epilepsy, and other epilepsies.
3. To examine whether the scores obtained using the WMS-III were capable of differentiating a subgroup of TLE patients with right and left temporal lobe foci.

The sample of 133 patients included 97 (72.93%) males and 36 (27.07%) females. The mean age of the sample was 27.5 (SD = 7.6) years and the mean educational level was 9.9 (SD = 4.4) years. The mean age at the onset of seizures was 10.7 (SD = 8.4) years, while the mean duration between seizure onset and time of evaluation was 15.4 (SD = 8.3) years. The mean WAIS-III FSIQ was 71.9 (SD = 12.6). A demand to interview and test a patient alone sometimes met with objections from patients' relatives, especially those of female patients, who frequently demanded to be present during both the interview and testing sessions. Other patients began the interview with feverish objection to being referred to a 'psychic' clinic. Almost always, an on-the-spot counseling and explanation were successful in defusing the dismay of such patients' relatives.

The 133 patients were classified, according to the brain lobe of their seizure origin, into three groups: temporal lobe epilepsy (TLE), frontal lobe epilepsy (FLE), and epilepsy originating in a brain area other than the temporal or frontal lobes (Other). The characteristics of the patients in each of the three groups are presented in Table 1. A comparison of groups according to demographic variables revealed no significant differences in group-composition for age, years of education, age at seizure onset, duration since seizure onset, FSIQ, gender, or hand dominance.

Group	N	Age (years) M (SD)	Education (years) M (SD)	FSIQ (WAIS) M (SD)	OnsetAge (years) M (SD)	Epilepsy Duration M (SD)	Gender		Handed.	
							M	F	R	L
TLE	90	28.21(7.81)	9.76(4.38)	72.49(12.75)	10.50(8.42)	16.22(8.17)	62	28	75	14*
FLE	12	27.08(8.54)	10.58(4.71)	67.82(11.79)	12.18(10.05)	17.75(8.28)	08	04	10	02
OTHER	31	25.58(6.36)	9.94(4.44)	71.81(12.78)	10.72(7.76)	12.29(8.18)	27	04	24	02
TOTAL	133	27.50(4.39)	9.87(4.39)	71.90(12.65)	10.71(8.36)	15.44(8.32)	97	36	109	23

TLE = Temporal Lobe Epilepsy – FLE = Frontal Lobe Epilepsy – Other: Epilepsies originating in brain lobes other than temporal or frontal (include: occipital & parietal lobe epilepsies, generalized epilepsies, and unclassified epilepsies with secondary generalized semiologies) – FSIQ = Full-scale Intelligence Quotient – WAIS = Wechsler Adult Intelligence Scale – Third Edition.
* = One male patient in this cell was not included because he was ambidextrous.

Table 1. Characteristics of Patient Groups

A caution is necessary when handling information given by patients about their age, education, and occupation in Saudi Arabia (see Escandall, 2002). An approximation of the patient's age is frequently given, which is often inconsistent with other chronological items of personal history, such as years of education or age of seizure onset, given by the same patients. Knowing this cultural factor, a clinician needs to skillfully and politely probe the best age approximation, especially with patients in their 40s or elder and/or of minimal formal education. With respect to educational level, it was noticed that an educational level given by a patient might be misleading. Sometimes, the figure given as the years of schooling might represent the number of years spent in grade one after which the patient was dismissed from the school or he quit. Such incidents were particularly often with patients from rural communities. One striking example was that of a patient who, during an interview, was found to have a primary school education, only to discover during testing that he was illiterate. In brief, education and occupation often did not correlate with the actual intellectual abilities assessed in the clinic. Such realities emphasize the importance of a thorough and careful clinical interviewing of such patients by a practitioner who is both familiar with the unique sociocultural variables of the population and proficient in the language of the people. The patients were classified, following the definition of the UNESCO (1978) for a functional illiterate (= any one with less than 5 years of formal schooling), into literates and illiterates. There were 114 (85.71%) literate patients and 19 (14.29%) illiterate patients. The chi-square test results (chi-square = 0.5141; DF = 2; p = 0.7733) indicated that there was no significant group difference in literacy. The group distribution on the literacy variable is presented in Table 2. The overall percentage of literate patients in the sample (85.71%) is similar to that for the adult population in Saudi Arabia (85.5%).

Education	TLE	FLE	Other	Total
Illiterates	14 (10.53%)	01 (0.75%)	04 (03.01%)	019 (14.29%)
Literates	76 (57.14%)	11 (8.27%)	27 (20.30%)	114 (85.71%)
Total	90 (67.67%)	12 (0.02%)	31 (23.31%)	133 (100.00%)

Table 2. Illiteracy Level in Epilepsy Groups

Regarding handedness, 109 (81.95%) patients claimed to be right-handed, whereby 23 (17.29%) declared themselves left-handed, and just one patient (0.75%) regarded himself as ambidextrous. Based on subsequent Wada test procedures, 124 (93.23%) patients were left hemisphere dominant for language functions, and 8 (6.01%) patients were right-hemisphere dominant. A single patient (0.75%) was proven by a Wada test procedure to have language functions in both hemispheres of the brain. These findings show that the number of patients in the sample who claimed to be left handed (N = 23), an attribute that was believed to be culturally and religiously unfavorable, was in fact higher than the number of patients with Wada-proven right hemisphere language dominance. The self-reported handedness compared with Wada-proven hemispheric language dominance of the sample is presented in Table 3.

	Hemisphere Dominance			
Handedness	Right	Left	Both	Total
Right-handed	2	106	1	109
Left-handed	6	17	0	23
Ambidextrous	0	1	0	1
Total	8	124	1	133

Table 3. Self-reported Handedness versus Wada Proven Hemispheric Language Dominance of the Sample

The types of seizure encountered in each of the three patient groups are presented in Table 4. The chi-square test performed (chi-square = 75.8544; DF = 26; p < 0.0001) showed significant difference between the three groups in terms of seizure types. In fact, 77.78% of the seizures in the TLE group were complex partial, which represents 52.63% of all complex partial seizures accounted for in this patient sample.

The temporal lobe epilepsies represented two thirds (67.67%) of epilepsy syndromes of the patients encountered in this Saudi sample, followed by frontal lobe epilepsies that mounted to 9.02% of the total number of patients. Other epilepsies represented 23.31% of all cases. The classification figures of epilepsies encountered in this center are not claimed to be a representation of the prevalence rates of epilepsy types in Saudi Arabia. In fact, there has been a bias in favor of temporal lobe epilepsies among the total number of referrals to this center. This bias is probably related in someway to the fact that temporal lobe epilepsy is globally the most common epilepsy syndrome (Weibe, et al., 2001), and that its resection procedures make up about two thirds of all operations (Helmstaedter, 2004).

Seizure Type	TLE N (%)	FLE N (%)	Other N (%)	Total
Complex Partial	70 (52.63%)	05 (3.76%)	14 (10.53%)	89 (66.92)
Simple Partial	05 (3.76%)	02 (1.50%)	05 (3.76%)	12 (09.02)
Tonic-Clonic	13 (9.77%)	05 (3.76%)	10 (7.52%)	28 (21.05)
Other	02 (1.50)	0	02 (1.50%)	04 (03.01)
Total	90 (67.67%)	12 (9.02%)	31 (23.31%)	133 (100.00)

Table 4. Classification of Seizure Types in the Patient Groups

Hereditary was reviewed as a possible risk factor among this sample of epilepsy patients. There were 45 (33.83%) patients who had at least one first-degree or second-degree relative with epilepsy, while 88 (66.17%) patients had no family history of the disorder. Among the patients with positive family history of epilepsy (N=45), 31 (68.89%) of them were temporal lobe epilepsy patients, compared to 14 (31.11%) patients with frontal lobe or other extra-temporal epilepsies.

5.1 Relative agreement between presurgical evaluation modalities

The amount of agreement between the four evaluation modalities (i.e., NP, PET, MRI, and iEEG) in identifying the lobe impaired by epilepsy in the patient sample was examined by computing kappa statistics. The results presented in Figure 1 show the relative amount of agreement between each two methods of evaluation on each brain lobe in the left and the right hemispheres. The figure indicates a moderate amount of agreement achieved between NP and both PET and MRI modalities (kappa values are around the 0.5) in identifying left temporal lobe involvement. The agreement amount between NP and iEEG in identifying frontal lobe involvement is even smaller (kappa = 0.43). However, other amounts of agreement between the evaluation modalities are of low magnitude (kappa < .40). The high amount of agreement pertinent to the left occipital lobe area is probably misleading since a very small number of patients had an occipital lobe seizure focus.

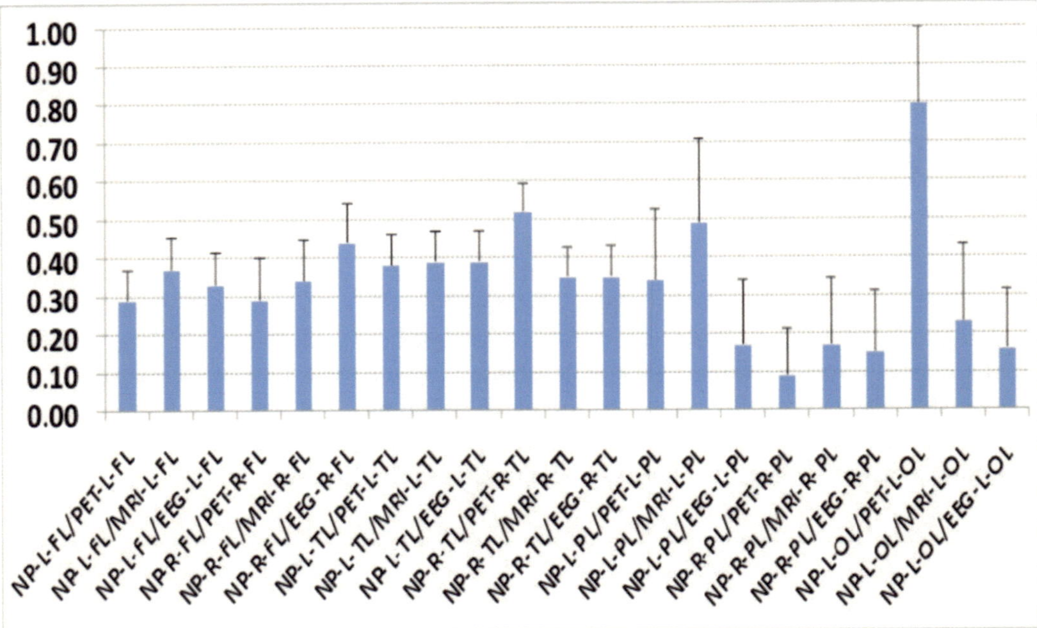

Legend: NP = Neuropsychology - PET = Positron Emission Tomography - MRI = Magnetic Resonance Imaging - iEEG = ictal Electroencephalography.
L-FL = Left Frontal Lobe - R-FL = Right Frontal Lobe - L-TL = Left Temporal Lobe -
R-TL = Right Temporal Lobe - L-PL = Left Parietal Lobe - R-PL = Right Parietal Lobe -
L-OL = Left Occipital Lobe - R-OL = Right Occipital Lobe.

Fig. 1. Amount of Agreement between the Four Raters (NP, PET, MRI, and iEEG) across the Four Lobes of the Brain Bilaterally.

The accurate neuropsychological localization of dysfunction is a function of several factors. Sensitivity and specificity of tests used (Jones-Gotman et al., 2000) is one of such factors. The already outlined cultural biases in the tests used in this study can partially explain the relative accuracy of results presented here. However, other factors not accounted for in the study, such as the differential impact of prolonged use of antiepileptic medications, or practically difficult to control, such as the factual level of literacy, may contribute to the varying neuropsychological performance of the patients in the sample and, hence, influence the localization accuracy.

5.2 Group differences across WMS-III indexes, subtests, and discrepancy scores

ANOVA procedures were computed to analyze group differences in the WMS-III Primary Indexes and Subtest raw scores (refer to Table 5 for the labels of these indexes and subtests). There were no significant differences between the three groups on any indexes or subtest scores. These results are not quite surprising for two reasons. First, the ability of the WMS-III to distinguish Saudi patients with temporal and non-temporal lobe epilepsies based on their memory scores has been disputed on cross-cultural considerations. Second, each of the three study groups contained patients with left and right hemisphere epileptic foci and, therefore, group differences on the primary indexes and subtest scores may be affected by 'within group' differences. This latter assumption was examined in the next section below after the TLE group was divided into right (RTL) and left (LTL) temporal lobe subgroups. Discrepancy scores were calculated by subtracting the Visual Memory Index from the Auditory Memory Index, and compared between the groups for both immediate and delayed indexes. An ANOVA procedure was then computed to analyze group difference in these Indexes. The result indicated no significant group difference on these discrepancy scores. Again, these results are attributed to reasons similar to those stated above.

WMS-III score	LTL			RTL		
	N	M	SD	n	M	SD
Indexes						
Auditory Immediate Index	44	66.20	(10.843)	34	93.00	(13.423)
Visual Immediate Index	44	71.73	(10.210)	34	78.12	(11.969)
Immediate Memory Index	44	62.05	(10.716)	34	82.97	(14.540)
Auditory Delayed Index	44	62.86	(13.861)	34	92.56	(15.673)
Visual Delayed Index	44	72.68	(12.799)	34	78.35	(12.982)
General Memory Index	44	63.82	(12.848)	34	85.47	(15.839)
Auditory Recognition Delayed Index	44	75.23	(12.802)	34	95.44	(16.939)
Working Memory Index	44	72.55	(14.534)	34	81.38	(13.253)
Auditory Immediate - Visual Immediate	44	-5.5227	11.5929	34	14.8824	8.72740
Auditory Delayed - Visual Delayed	44	-9.8182	13.7457	34	14.2059	12.3160
Subtests						
Logical Memory I	44	17.95	(8.482)	34	33.59	(11.206)
Logical Memory II	44	06.66	(6.961)	34	18.71	(10.406)
Faces I	44	32.34	(4.554)	34	33.56	(04.800)
Faces II	44	33.73	(5.173)	34	34.38	(03.742)
Verbal Paired Associates I	44	07.48	(5.263)	34	18.18	(06.264)
Verbal Paired Associates II	44	01.73	(1.796)	34	05.65	(02.073)
Family Pictures I	44	23.41	(8.809)	34	29.88	(11.635)
Family Pictures II	44	23.18	(9.473)	34	29.65	(12.185)
Letter-Number Sequencing	44	05.18	(2.863)	34	07.53	(02.915)
Spatial Span	44	11.02	(4.185)	34	12.50	(03.703)

WMS-III = Wechsler Memory Scale - Third Edition - LTL = left temporal lobe - RTL = right temporal lobe.

Table 5. WMS-III Mean Scores for the LTL and RTL Subgroup

5.3 LTL and RTL subgroup differences across WMS-III indexes, subtest, and discrepancy scores

The temporal lobe epilepsy (TLE) group was analyzed in terms of left and right lateralization of memory deficits. That is, the WMS-III indexes and subtest scores of the patients in the TLE group were analyzed after they were subdivided into left temporal lobe (LTL) and right temporal lobe (RTL) subgroups. The means and standard deviations for the WMS-III Primary Indexes and Subtest scores are provided in Table 5 (For auditory-visual discrepancy scores shown in Table 5, the net difference scores were in the positive direction for the RTL group indicating that Visual Index scores were lower than Auditory Index scores, whereas the opposite was the case for the LTL group). The independent t-tests of the Primary Index scores indicated that the LTL and RTL subgroups differed significantly from one another on all index scores except the Visual Delayed Index. The independent t-test results for the LTL and RTL comparisons on the WMS-III Subtest scores indicated that the LTL and RTL subgroups differed significantly from one another in 8 of the 10 Subtest scores. Faces subtest, a visually encoded test, failed to differentiate the two groups at both the immediate and delayed stages of assessment. Table 6 shows these independent t-test results, which reflect some sensitivity to lateralization in patients with right and left temporal lobe epilepsies. Discrepancy scores, calculated by subtracting the visual memory index from the auditory memory index, were compared between the LTL and RTL subgroups for both immediate and delayed indexes. Analyses of these discrepancies using independent t-tests revealed significant differences between the TLT and RTL subgroups on both the Immediate [t (75.962) = -8.868, p < .0001] and the Delayed scores [t (74.284) = -8.119, p < .0001]. Similar results were obtained in earlier studies (e.g., Wilde, et al., 2001).

WMS-III Primary Indexes	t	df	**Sig. (2-tailed)**
Auditory Immediate Index	- 9.491	62.476	0.000
Visual Immediate Index	- 2.491	64.818	0.015
Immediate Memory Index	- 7.043	58.593	0.000
Auditory Delayed Index	- 8.722	66.344	0.000
Visual Delayed Index	- 1.925	70.614	0.058
General Memory Index	- 6.490	62.660	0.000
Auditory Recognition Index	- 6.005	76	0.000
Working Memory Index	- 2.799	73.873	0.007
WMS-III Subtest Scores	t	df	**Sig. (2-tailed)**
Logical Memory I	- 6.772	59.707	0.000
Logical Memory II	- 6.115	76	0.000
Faces I	- 1.136	69.185	0.260
Faces II	- 0.649	75.713	0.519
Verbal Paired Associates I	- 8.011	64.165	0.000
Verbal Paired Associates II	- 8.772	65.489	0.000
Family Pictures I	- 2.701	59.715	0.009
Family Pictures II	- 2.554	60.837	0.013
Letter-Number Sequencing	- 3.554	70.480	0.001
Spatial Span	- 2.799	73.873	0.007

Table 6. LTL and RTL differences on WMS-III Primary Indexes and Subtest Scores

Finally, to examine the predictive power of the auditory-visual delayed index discrepancy, a univariate logistic regression was used to study the effects of the discrepancy on the probability of having temporal lobe epilepsy on the diagnostic groups, and consequently to plot the Receiver Operating Characteristics (ROC). The level of significance was set at p < 0.05. The resulting area for the auditory-visual delayed index difference score is 0.53, which is close to chance level. The area under the curve (AUC) that describes the ROC is presented in Figure 2. This result emphasizes the suspected issue of sensitivity and specificity of the neuropsychological evaluation tools used.

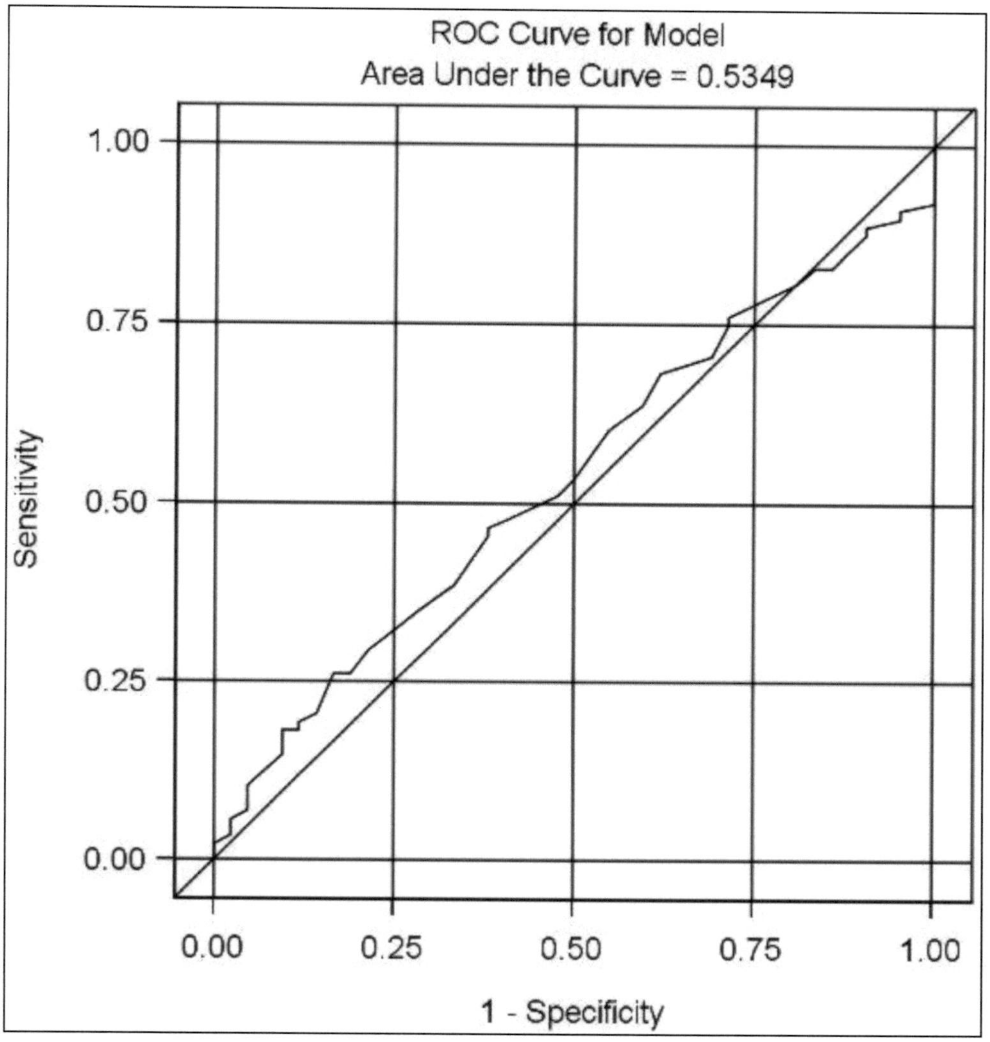

Fig. 2. The AUC for the Auditory-Visual Delayed Index Difference Score

6. Discussion

There are enormous similarities shared by the various groups in the human kingdom, yet cultural diversity is a dynamic reality. In a field like clinical neuropsychology, a practitioner cannot afford not to attend to cultural diversities. The practice of neuropsychology is affected by cultural differences. For a neuropsychologist in a 'developing country' where most of the substances used on daily practice are extrapolated from a different culture, the professional challenges are limitless. The purpose of this chapter was not to provide an in-depth analysis of the psychometric properties of neuropsychological testing material. It was rather to review some of the social and cultural characteristics in Saudi Arabia that are believed to dictate the relative relevance of much of the neuropsychological material that have been imported and used to evaluate Saudi patients. The review focused on the practice of neuropsychology and its role of evaluating epilepsy patients in an epilepsy surgery program. As the majority of epilepsy syndromes encountered in the presurgical workup were temporal lobe epilepsy, the discussion was devoted in part to neuropsychological issues relevant to temporal lobe epilepsies. The data presented above were aimed at elucidating the need for developing indigenous neuropsychological traditions and material not only for patient evaluation, but also for rehabilitation, teaching and training.

The results presented above may speak for themselves. They generally indicate the relatively limited ability of the WMS-III in identifying the brain areas that were implicated with dysfunction. The performance of TLE, FLE, and other seizures' groups on the WMS-III indexes and subtest scores was insensitive to localization of dysfunction. Nor was the auditory-visual discrepancy for both delayed and immediate scores. On the other hand, the WMS-III auditory-visual discrepancy scores were efficiently sensitive to laterality among patients with either left or right temporal lobe dysfunction. Hence, despite the several aspects of cultural bias identified in the WMS-III subtests, these discrepancy scores demonstrated ability to distinguish patients with left and right temporal lobe dysfunction associated with a unilateral seizure onset. In fact, apart from the failure to document "between group" differences, the WMS-III primary indexes and subtest scores were able to detect material-specific memory impairment as demonstrated by auditory-visual difference scores of temporal lobe patients. It is not clear, however, if the failure to demonstrate differences in the performance of patients with different epileptic foci is associated with the psychometric characteristics of the WMS-III subtests. There were studies that reported similar insensitivity of the WMS-III index and subtest scores to laterality (Wilde, et al., 2001). Comparing the results reported here with results published in similar previous studies is hampered by the fact that the sample of this study differed in a number of important aspects from the WMS-III normative population.

These results reinforced the importance of using the WMS-III scores in combination with other neuropsychological measures to improve the validity of evaluations. One of the inadequacies of the WMS-III is that, some of the subtests included in the scale may not allow accurate lateralization of dysfunction because they can be coded both verbally and visually (Jones-Gotman at al., 2000). In addition, the WMS-III protocol for verbal passages task, for example, requires recall after a single presentation of the material to be remembered. Jones-Gotman and colleagues (2000) argued that when material was presented to patients with lesions in either the left or the right temporal lobe only once, deficient recall might result for a variety of reasons including lapses of attention and misunderstanding. However, the demonstrated ability of the scale to relatively accurately differentiate between the LTL and RTL subgroups as reported above, may suggest that factors other than the psychometric properties of the scale are at work in influencing the performance of patients in this sample. On the other hand, the amount of agreement between the neuropsychological data and other evaluation modalities was at best 'fair' (Haut, et al., 2002) and limited to temporal lobe areas. Studies that compare and contrast the efficacy of different evaluation modalities in documenting dysfunction in epilepsy patients at the presurgical workup are quite rare (e.g., Akanuma, et al., 2003; Haut, et al., 2002). Such comparison studies can provide a gold standard against which neuropsychological evaluation tools can be calibrated.

To conclude, this chapter has emphasized the importance of developing indigenous neuropsychological tests and norms relevant to the cultural characteristics of individuals to whom the tests are administered. The reported study was meant to provide an empirical platform on which the concepts argued above were demonstrated. The WMS-III remains a useful tool for assessing memory, even if its ability to classify patients is limited.

7. Acknowledgement

The statistical analysis of data reported in this chapter was performed with assistance from Abdelmoneim Eldali, MSc, from the Biostatistics, Epidemiology and Scientific Computing Department of King Faisal Specialist Hospital & Research Center, Riyadh. The analyses were accomplished using SAS version 9.2 (2008) and SPSS for Windows, Release 17.0.1 (2008).

8. References

Abduljabbar, M., A., A. Ogunniyi, Abdulkarem, D., Abdulrahman, T., Al-Bunyan, M., and Al-Rajeh, S. (1998). Epilepsy classification and factors associated with control in Saudi Arabia patients. Seizure, 7:501-504.

Akanuma, N., Alarcon, G., Lum, F., Kissani, N., Koutroumanidis, M., Adachi, N., Binnie, C. D., Polkey, C. E., and Morris, R. G. (2003). Lateralizing Value of Neuropsychological Protocols for Presurgical Assessment of Temporal Lobe Epilepsy. Epilepsia, Volume 44, Issue 3, pp. 408-418.

Al Rajeh, S., Awada, A., Bademosi, O., and Ogunniyi, A. (2001). The prevalence of epilepsy and other seizure disorders in an Arab population: a community-based study. Seizure, 10:410-414.

Aldenkamp, A. P. (2001). Effects of Antiepileptic Drugs on Cognition. Epilepsia, Volume 42, Issue Supplement s1, pp. 46-49.

Anastasi, A. (1988). Psychological Testing. New York: Macmillan.

Ardila, A., Rosselli, M. and Rosas, P. (1989). Neuropsychological assessment in illiterates: Visuospatial and memory abilities. Brain and Cognition, 11, pp. 147-166.

Ardila, A., Rosselli, M. and Ostrosky, F. (1992). Sociocultural factors in neuropsychological assessment. In: A. E. Puente and R. J. McCaffrey (Eds.), Handbook of neuropsychological assessment: A biopsychosocial perspective (pp. 181-192). New York: Academic Press.

Ardila, A., Ostrosky, F. and Mendoza, V. (2000). Learning to read is much more than learning to read: A neuropsychologically based learning to read method. Journal of the International Neuropsychological Society, 6, pp. 789-801.

Ardila, A. and Moreno, S. (2001). Neuropsychological evaluation in Aruaco Indians: An exploratory study. Journal of the International Neuropsychological Society, 7, pp. 510-515.

Artiola, L. F. and Mullaney, H. (1997). Neuropsychology with Spanish speakers: Language use and proficiency issues for test development. Journal of Clinical & Experimental Neuropsychology, Vol. 19 (Issue 4), pp. 615-622.

Baker, G. A. (2002). The psychosocial burden of epilepsy. Epilepsia, 43(Suppl. 6): 26-30.

Berry, J. W., Poortinga, Y. P., and Segall, M. G. H. (1992). Cross-cultural psychology. Cambridge: Cambridge University Press.

Boer, H. M., Mula, M., and Sander, J. W. (2008). The global burden and stigma of epilepsy. Epilepsy & Behavior 12, 540-546.

Boone, K. B., Victor, T. L., Wen, J., Razani, J., and Ponton, M. (2007). The association between neuropsychological scores and ethnicity, language, and acculturation variables in a large patient population. Archives of Clinical Neuropsychology, 22, pp. 355-365.

Brunbech, L. and Sabers, A. (2002). Effects of Antiepileptic Drugs on Cognitive Function in Individuals with Epilepsy: A Comparative Review of Newer Versus Older Agents. Drugs, Volume 62, Number 4, pp. 593-604.

Cohen, R. A. (1969). Conceptual styles, culture conflict, and nonverbal tests. American Anthropologist, 71, pp. 828-856.

Cole, M. (1997). Cultural mechanisms of cognitive development. In: E. Amsel and K. A. Renninger (Eds.), Change and development: Issues of theory, method, and application. The Jean Piaget symposium series (pp. 250-263). Mahwah, NJ: Erlbaum.

Devinsky, O. (1995). Cognitive and Behavioral Effects of Antiepileptic Drugs. Epilepsia, Volume 36, Issue Supplement s2, pp. S46-S65.

Djordjevic, J. and Jones-Gotman, M. (2004). Psychological Testing in Presurgical Evaluation of Epilepsy. In: S. Shorvon, E. Perucca, D. Fish and E. Dodson (Eds.), The Treatment of Epilepsy, Second Edition (pp. 699-715). Malden, Massachusetts: Blackwell Publishing.

Dodrill, C. B. (1978). A neuropsychological battery for epilepsy, Epilepsia, Volume 19, Issue 6, pp. 611-623.

Escandall, V. A. (2002). Cross-cultural neuropsychology in Saudi Arabia. In: E. R. Ferraro (Ed.), Minority and Cross-Cultural Aspects of Neuropsychological Assessment. Lisse: Swets & Zeitlinger Publishers.

Gasquoine, P. G. (1999). Variables moderating cultural and ethnic differences in neuropsychological assessment: The case of Hispanic Americans. The Clinical Neuropsychologist, 13, pp. 376-383.

Handwerker, W. P. (2002). Quick ethnography. Walnut Creek, CA: Altamira.

Haut, S. R., Berg, A. T., Shinnar, S., Cohen, H. W., Bazil, C. W., Sperling, M. R., Langfitt, J. T., Pacia, S. V., Walczak, T. S., and Spencer, S. S. (2002). Interrater Reliability among Epilepsy Centers: Multicenter Study of Epilepsy Surgery. Epilepsia, Volume 43, Issue 11, pp. 1396-1401.

Heaton, R. K., Taylor, M. J., and Manly, J. (2003). Demographic effects and use of demographically corrected norms with the WAIS-III and WMS-III. In: D. S. Tulsky, D. H. Saklofske, G. J. Chelune, R. K. Heaton, R. J. Ivnik, R. Bornstein, et al. (Eds.), Clinical interpretation of the WAIS-III and WMS-III (pp. 183-219). San Diego, CA: Academic Press.

Helmstaedter, C. (2004). Neuropsychological aspects of epilepsy surgery. Epilepsy & Behavior 5, Supplement 1, pp. S45-S55.

Horton, A. M. Jr. (2008). The Halstead-Reitan Neuropsychological Test Battery: Past, Present, and Future. In: A. M. Horton, Jr. and D. Wedding (Eds.), The Neuropsychology Handbook, Third Edition (pp. 251-278). New York: Springer Publishing Company.

Irvine, S. H. and Berry, J. W. (Eds.). (1988). Human abilities in cultural context. New York: Cambridge University Press.

Jan M. M. (2005). Clinical Review of Pediatric Epilepsy. Neuroscience, Vol. 10 (4): 255-264

Jones-Gotman, M., Smith, M. L., and Zatorre, R. J. (1993). Neuropsychological testing for localizing and lateralizing the epileptogenic region. In: J. Engel, Jr. (Ed.), Surgical treatment of the epilepsies (pp. 245-261). New York: Raven Press, Led.

Jones-Gotman, M., Harnadek, M. C. S, and Kubu, C. S. (2000). Neuropsychological Assessment for Temporal Lobe Epilepsy Surgery. Canadian Journal of Neurological Sciences, 27, Supplement 1, pp. S39-S43.

Kale, R. (2002). Global Campaign against Epilepsy: the treatment gap. Epilepsia, 43(Suppl. 6): 31-3.

Kwan, P. and Brodie, M. J. (2001). Neuropsychological effects of epilepsy and antiepileptic drugs. The Lancet, Volume 357, Issue 9251, pp. 216-222.

Lecours, A. R., Mehler, J., Parente, M. A, et al., (1988). Illiteracy and brain damage III: A contribution to the study of speech and language disorders in illiterates with unilateral brain damage (initial testing). Neuropsychologia, 26, pp. 575-589.

Lezak, M. (1995). Neuropsychological Assessment (3rd ed.). New York: Oxford University Press.

Loring, D. W. and Bauer, R. M. (2010). Testing the Limits: Cautions and concerns regarding the new Wechsler IQ and Memory scales. Neurology, 74, February 23, 2010; pp. 685-690.

Luria, A. R. (1973). The Working Brain: An introduction to neuropsychology. Basic Books; translation: Penguin Books Ltd.

Malafronte, V. A. (1998). F. Davis Johnson (Ed.). The Complete Book of Frisbee: The History of the Sport & the First Official Price Guide. Rachel Forbes (illus.). Alameda, CA: American Trends Publishing Company.

Mbuba, C. K., Ngugi, A. K., Newton, C. R., Carter, J. A. (2008). The epilepsy treatment gap in developing countries: A systematic review of the magnitude, causes, and intervention strategies. Epilepsy, 49(9): 1491-1503.

Meyer, AC, Dua, T., Ma, J., Saxena, S., and Birbeck, G. (2010). Global disparities in the epilepsy treatment gap: a systematic review. Bull World Health Organ 2010; 88:260-266 [doi: 10.2471/BLT.09.064147].

Miller, R. J. (1973). Cross-cultural research in the perception of pictorial materials. Psychological Bulletin, 80, pp. 135-150.

Meador, K. J. (2006). Cognitive and memory effects of the new antiepileptic drugs. Epilepsy Research, Volume 68, Issue 1, pp. 63-67.

Motamedi, G. K. and Meador, K. J. (2004). Antiepileptic drugs and memory. Epilepsy & Behavior, Volume 5, Issue 4, pp. 435-439.

Novelly, R. A. (1992). The debt of neuropsychology to the epilepsies [Review]. American Psychologist, Volume 47, pp. 1126-1129.

Ortinski, P. and Meador, K. J. (2004). Cognitive side effects of antiepileptic drugs. Epilepsy & Behavior, Volume 5, Supplement 1, pp. 60-65.

Ostrosky, F., Ardila, A., Rosselli, M., Lopez-Arango, G., and Uriel-Mendoza, V. (1998). Neuropsychological test performance in illiterates. Archives of Clinical Neuropsychology, 13, pp. 645-660.

Padilla, A. M. (1999). Hispanic psychology: A 25-year retrospective look. In: D. Dinnel and W. J. Lonner (Eds.), Merging past, present, and future in cross-cultural psychology: Selected papers from the Fourteenth International Congress of the International Association for Cross-Cultural Psychology (pp. 73-81). Netherlands: Swets.

Reitan, R. M. and Wolfson, D. (1986). The Halstead-Reitan Neuropsychology Test Battery. In: D. Wedding, A. M. Horton, Jr., and J. S. Webster (Eds.), The neuropsychology handbook (pp. 134-160). New York: Springer Publishing Company.

Reynolds, E. H. (2000). The ILAE/IBE/WHO Global Campaign against Epilepsy: Bringing Epilepsy "Out of the Shadows". Epilepsy & Behavior, 1: S3-S8.

Rosselli, M., Ardila, A. and Rosas, P. (1990). Neuropsychological assessment in illiterates II: Language and praxic abilities. Brain and Cognition, 12, pp. 281-296.

Rosselli, M. and Ardila, A. (2003). The impact of culture and education on non-verbal neuropsychological measurements: A critical review. Brain and Cognition, 52, pp. 326-333.

Russell, E. W. (1997). Developments in the psychometric foundations of neuropsychological assessment. In: G. Goldstein and T. M. Incagnoli (Eds.), Contemporary approaches to neuropsychological assessment, pp. 15-65. New York: Plenum.

Salazar, G. D., Garcia, M. P., and Puente, A. E. (2007). Clinical Neuropsychology of Spanish Speakers: The Challenge and Pitfalls of a Neuropsychology of a Heterogeneous Population. In: B. P. Uzzell, M. Ponton, and A. Ardila (Eds.), International Handbook of Cross-Cultural Neuropsychology (pp. 283-302). Mahwah, NJ: Lawrence Erlbaum Associates Inc Publishers.

SAS Version 9.2. (2008). Statistical Analysis System, SAS Institute Inc Cary, NC, USA.

Spencer, S. and Huh, L. (2008). Outcomes of epilepsy surgery in adults and children. Lancet Neurology, Volume 7, Issue 6, pp. 525-537.

Tellez-Zenteno, J. F., Dhar, R., Hernandez-Ronquillo, L., and Wiebe, S. (2007). Long-term outcomes in epilepsy surgery: antiepileptic drugs, mortality, cognitive and psychosocial aspects. Brain, Volume 130, pp. 334-345.

Walker, A. J., Batchelor, J., Shores, E. A. and Jones, M. (2010). Effects of cultural background on WAIS-III and WMS-III performances after moderate-severe traumatic brain injury. Australian Psychologist, 45(2): pp. 112-122.

Wechsler, D. (1997a). Wechsler Adult Intelligence Scale - Third Edition. Administration and Scoring Manual. San Antonio, TX: The Psychological Corporation.

Wechsler, D. (1997b). Wechsler Memory Scale - Third Edition. Administration and Scoring Manual. San Antonio, TX: The Psychological Corporation.

Wilde, N., Strauss, E., Chelune, G. J., Loring, D. W., Martin, R. C., Hermann, B. P. and Sherman, M. S. (2001). WMS-III performance in patients with temporal lobe epilepsy: Group differences and individual classification. Journal of the International Neuropsychological Society, 7, pp. 881-891.

World Health Organization (2008). Country Statistics, Global Health Observatory Data Repository. WHO Website: http://apps.who.int/ghodata/?vid=17400&theme=country#

World Health Organization (2010). Epilepsy in the WHO Eastern Mediterranean Region: Bridging the gap. World Health Organization, Regional Office for the Eastern Mediterranean.

9

Psychic Seizures and Their Relevance to Psychosis in Temporal Lobe Epilepsy

Kenjiro Fukao
Department of Neuropsychiatry, Graduate School of Medicine, Kyoto University
Japan

1. Introduction

Psychic seizures are defined as simple partial seizures manifesting themselves as psychic phenomena. Their current classification is neither comprehensive nor rational enough, mainly because psychic seizures are defined as subjective phenomena and not as objective dysfunctions. The high incidence of psychiatric symptoms in patients with temporal lobe epilepsy (TLE) is not only resulted from misdiagnosis but causally linked to psychic seizures. In this article the relationship between psychic seizures and various psychiatric disturbances found in TLE patients is discussed from a clinical viewpoint. Psychosis in TLE patients has historically been of special interest, whose relationship with schizophrenia is still unresolved. In this article the possible effect of psychic seizures on the generation of psychosis in TLE patients is discussed based on the author's own data obtained from magnetoencephalographic localization of psychic seizures and psychosis in TLE patients.

2. Psychic seizures and their classification

Psychic seizures have been classified from a practical viewpoint into cognitive, emotional, mnemonic seizures and miscellaneous. In the following, psychic seizures are described according to the current classification (Fish, 1997), and the limitation of the current classification is discussed.

2.1 Cognitive seizures

Cognitive (or dyscognitive) seizures indicate disturbances of cognitive functions by seizure activities. Although theoretically disturbances of higher cognitive functions cannot be localized only to temporal lobes, as they are realized by collaboration of many circuits all around the brain, clinically cognitive seizures are most often observed in TLE.

2.1.1 Perceptual illusions

Perceptual illusion are defined as various kinds of illusions in perception in any mode of senses (Penfield & Jasper, 1954). Historically various kinds of perceptual illusions have been described as psychic seizures, for example, increase and decrease of the size of objects (macropsia and micropsia, respectively), augmentation and diminution of stereoscopy in visual mode. There are also corresponding auditory phenomena, namely, increase and decrease of the volume of sounds and various alterations of hearing sounds.

From the present point of view, these phenomena can be understood as dysfunctions of higher sensory processing, both visual and auditory modes of which are made in the temporal necortex. In this sense, perceptual illusions could be classified as sensory seizures.

Whereas higher cognitive functions in general tend to be disturbed by seizure activities, facilitation is also seen among lower cognitive functions. Thus micropsia, the most often seen phenomenon of this kind, could be either augmentation of distant sight or diminution of near sight.

2.1.2 Aphasic seizures

Sensory and motor aphasias can appear as simple partial seizures, although these disturbances would appear more frequently as postictal symptoms which are observed after complex partial seizures involving the speech-dominant hemisphere. On the other hand, occasionally abundant but empty speech is observed after complex partial seizures involving the non-dominant hemisphere. This is regarded as a release phenomenon rather than a stimulated symptom, whereas aphasias are defective symptoms.

2.1.3 Higher cognitive seizures

Higher cognitive functions in general tend to be disturbed by epileptic seizures, and combinations of disturbances of various cognitive functions bring about the states called complex partial seizures. It is worth remembering that concept of complex partial seizures, characterized by "impairment of consciousness", had implied originally combinations of elementary partial seizures, each of which interferes each of cognitive functions. While "impairment of consciousness" is defined by lacks of responsiveness and memory registration, these two lacks can be caused by disturbances of elementary cognitive functions. For example, combination of sensory aphasia, apraxia and amnesia could hardly be differentiated from impairment of consciousness.

Although it is hard to suppose that any of higher cognitive functions might be facilitated, instead of disturbed, by seizure activities, as higher cognitive functions are thought to be realized by exquisite collaborations of many elementary circuits, there remains the possibility of epileptic facilitation. At least, some patients report subjective sense of transient facilitation of general cognition and it sounds like a mystic experience.

2.1.4 Depersonalization and derealization

Depersonalization and derealization can be caused by seizure activities. These phenomena could be interpreted as weak forms of complex partial seizure or impairment of consciousness. Some patients complain of "absurdity" about the surroundings during the seizure (Penfield & Jasper, 1954), which can be interpreted as some peculiar kind of alteration of perception of the surroundings, presumably similar to derealization.

2.2 Emotional seizures

Emotional seizures can basically be epileptic induction of any kind of emotion. In reality, however, this category is almost solely used for ictal fear, namely, fearful or anxious emotion induced by seizure activities. Although there are actually cases showing pleasant emotions as seizure symptoms, their incidence is decisively lower than that of fear, anxiety, sadness or loneliness. The reason why those negative emotions are much more often induced by epileptic seizures than positive ones might be explained that the neural structure relevant to emotions, amygdala, is tend to respond as negative when stimulated grossly, corresponding to an evolutionally lower hierarchy in emotional system.

Incidentally, embarrassment has also been suggested to be a possible type of emotional seizure (Devinsky et al., 1982), which is a social emotion and thought to reside in a rather high hierarchy, presumably anteromedial frontal lobes, although occurring with much lower incidence.

Regarding the possibility that seizure activities inhibit, instead of induce, any of the emotions, there has been no consideration as far as the author knows. Although it is theoretically possible, thought to be very difficult to detect in practice.

2.3 Mnemonic seizures

Mnemonic or dysmnesic seizures, which mean seizures causing dysfunction of memory, include amnesia, recollection and reminiscence, déjà vu, jamais vu and experiential hallucination induced by seizure activities. As it is known that temporal lobe is the most important part of the brain for memory function, it is reasonable that seizure activities residing in temporal lobes cause various derangements of memory function.

2.3.1 Amnesia

The simplest effect of seizure activities on memory functions is a negative one, namely, amnesia. Although clinically it is difficult to differentiate from complex partial seizures with impairment of consciousness, there are actually "pure amnesic seizures" in some TLE patients with seizure activities confined within hippocampus (Palmini et al., 1992).

2.3.2 Experiential hallucination

Experiential hallucination is defined as hallucination of any mode of sense which is related to the patient's personal experiences and can be regarded as reappearance and projection of memorized images (Penfield & Jasper, 1954). Each mode of sense is involved either solely or in combination resulting in appearance of audiovisual images just like fragments of movies. The mode most frequently involved is auditory. Patients often report the voices calling their name at the time of the first attack in their childhood.

Experiential hallucination of only olfactory or gustatory mode is hard to be discerned, because hallucination of these primitive modes of sense easily evoke associated memory recall with multiple sensory images, and associated images could hardly be separated from the initial ictal hallucination.

2.3.3 Recollection and reminiscence

Ictal recollection is autonomous remembrance of past experiences induced by seizure activities. Contrasted to experiential hallucination, sensuous property of recollection is not strong, and its content is confined within the inner space, not projected to the outer. Emotions accompanying the past experience are often revived. The content of recollection is ordinarily limited to the same every time, although it may change in course of time, in the time scale of years. The content is occasionally the scene that the patient experienced the first attack, apparently resembling traumatic recollection. This resemblance is not only coincidental, because the patient's impression of the first attack is thought to be very strong and therefore memorized particularly strongly, that is comparable to traumatic memory.

Reminiscence is a weaker form of recollection, which does not make any concrete image within the inner space and is experienced only as a feeling of remembering something.

2.3.4 Déjà vu

Déjà vu (or déjà vécu) is a paradoxical feeling of knowing things that is never known by the subject. It is not rare in healthy people and therefore not necessarily a pathological phenomenon. Ictal déjà vu tends to be, however, stronger and longer than healthy ones. Emotion accompanying ictal déjà vu is often nostalgic and pleasant, but some patients feel uncanny and uneasy. Penfield suggested that ictal déjà vu should be caused by incomplete recollection and mixed perception of the present environment and the past one (Penfield & Jasper 1954). Whereas, Gloor suggested that déjà vu should be caused by misidentification resulted from deterioration of matching mechanism by seizure activities (Gloor 1990). Recent studies suggested that déjà vu is appearance of a special sense of familiarity separable from either recollection or identification, whose neural substrate is supposed to be parahippocampal gyri (Spatt 2002).

2.3.5 Jamais vu

Jamais vu (or jamais vécu) is thought as the opposite of déjà vu, namely, a feeling of not knowing things that is already known by the subject. It is, however, much more difficult to understand the mechanism than déjà vu. Ictal jamais vu is much rarer than ictal déjà vu and seems hard to induce by electrical stimulation. In addition, practically jamais vu is hard to be differentiated from derealization. Emotion accompanying ictal jamais vu is usually anxiety or uneasiness, with rare exception of pleasant freshness.

2.3.6 Prescience

There is a special kind of psychic seizure called "prescience", which indicates a sense of knowing what is going to happen from now on (Sadler et al., 2004). While it is theoretically difficult to take prescience as a disorder of memory, which is bound only to the past and never to the future, practically prescience is frequently seen in combination with déjà vu. It seems necessary to clarify what is the common between déjà vu and prescience, which may be the clue to understand the mechanism of déjà vu.

2.4 Miscellaneous

There are various kinds of psychic phenomenon evocated by seizure activities, which are difficult to be classified under any rubric of cognitive, emotional and mnemonic. Here are described only some of them.

2.4.1 Forced thinking

Forced thinking is a comparatively often found type of psychic seizure. It is an autonomous appearance of a fixed thought induced by seizure activities. Apparently it resembles obsession, but it does not necessarily evoke anxiety as obsession does. Some patients take the thought as a gift from the god and develop religious delusion.

2.4.2 Changes of time perception

Some TLE patients report changes of time perception, namely, increase or decrease of the speed in which surroundings proceed. They are experienced just like quick motion or slow motion of the movies. While these phenomena are usually reported only in visual mode without corresponding changes in auditory mode, some patients complain of feeling that others' speeches abruptly change into being too quick to understand.

2.4.3 Autoscopy and out-of-body experience

Autoscopic phenomena have been reported as manifestations of temporal, parietal and occipital lobe epilepsies (Devinsky et al., 1989). Autoscopic hallucinations, in which a mirror image of the patient is seen, is supposed to originate from the parieto-occipital lobes preferentially of the right side (Maillard et al., 2004). Dramatic type of autoscopy, in which the patient see himself/herself act independently from his/her will, just as seeing a movie, is reported mainly as hallucinations appearing during postictal confusional states of TLE. Out-of-body experience, in which the patient feels like seeing himself/herself from the outside of his/her body, is reported to originate from temporo-parietal junctions (Blanke et al., 2004). These autoscopic phenomena induced by seizures could be understood as distortion of body scheme, although the displacement of the viewpoint in out-of-body experiences is hard to explain.

2.4.4 Feeling of a presence

Feeling of a presence is a hallucinatory feeling as if someone is nearby. This phenomenon is most often observed in TLE patients and seemingly related to ictal fear. It is also interesting from the viewpoint of the similarity to a type of psychotic hallucination observed in schizophrenia.

2.5 Problems in the classification of psychic seizures

The essential difficulty in the classification of psychic seizures is that they are defined as subjective phenomena and not as objective dysfunctions. Observation of their existence in the patients depends solely on the patients' report, which is often not reliable because of insufficiency of the patients' ability to express the phenomena in his/her words.

From a clinical point of view, current classification of psychic seizures is regarded as the minimum but still usable. On the other hand, from a scientific point of view, replacing the term by "limbic seizures", which has been done in a standard textbook (Engel & Williamson, 2008) seems immature because the neural substrates of psychic seizures have not been sufficiently determined, although a large part of them seems to be localized to limbic structures. Furthermore, our data has suggested the conflicting possibility, as shown in the following.

2.6 Co-existence of psychic seizures and other simple partial seizures in TLE

Aiming at objectifying psychic seizures, the author performed an investigation on existence and co-existence of psychic seizures and other simple partial seizures in 38 TLE patients. The patients with TLE were investigated on existence and co-existence of autonomic, auditory, olfactory and psychic seizures, which were subclassified into fear/anxiety, déjà vécu, jamais vécu, reminiscence, forced thinking, visual alteration and others.

The sides of the focus of the patients are left in 16, right in 17 and bilateral or not determined in 5 patients.

Fifteen patients had autonomic seizures (5 left, 8 right and 2 undetermined), 23 had psychic seizures (10 left, 10 right and 3 undetermined), 8 had auditory seizures (3 left, 4 right and 1 undetermined), 2 had olfactory seizures (1 left and 1 right) and none had gustatory seizures. Subclassified psychic seizures were déjà vécu in 11 (6 left, 4 right and 1 undetermined), reminiscence in 10 (4 left, 4 right and 2 undetermined), fear/anxiety in 5 (2 left, 3 right), jamais vécu in 2 (0 left and 2 right), forced thinking in 2 (0 left and 2 right) and visual alteration in 2 (1 left and 1 right) (Table 1).

	Left	Right	Bilateral or undetermined
Autonomic seizures	5	8	2
Auditory seizures	3	4	1
Olfactory seizures	1	1	0
Gustatory seizures	0	0	0
Psychic seizures	10	10	3
déjà vécu	6	4	1
reminiscence	4	4	2
fear/anxiety	2	3	0
jamais vécu	0	2	0
forced thinking	0	2	0
visual alteration	1	1	0
Total	16	17	5

Table 1. The type of simple partial seizure and the side of the focus in the subject group.

Co-existence among each of the subcategories of psychic seizure was as follows: fear/anxiety and reminiscence in 2, déjà vécu and reminiscence in 2, fear/anxiety and déjà vécu in 1, etc.

Co-existence between the subcategories of psychic seizure and each type of simple partial seizure was as follows: 3 out of the 5 patients with fear/anxiety, 2 out of the 11 with déjà vécu, 4 out of the 10 with reminiscence and 1 out of the 2 with forced thinking had also autonomic seizures; 4 out of the 11 with déjà vécu, 1 out of the 2 with jamais vécu, 2 out of the 10 with reminiscence and both the 2 patients with visual alteration had also auditory seizures; 1 out of the 11 with déjà vécu, 1 out of the 2 with jamais vécu, 1 out of the 10 with reminiscence had also olfactory seizures.

The results are schematically visualized in Fig. 1 where simple partial seizures are connected each other by the bonds with breadth corresponding to the number of patients. It is noteworthy that the connections between autonomic seizures and the subcategories of psychic seizure are not significantly stronger than those between auditory seizures and them. This result suggests that psychic seizures should not necessarily be based solely on temporolimbic dysfunctions but also on dysfunction of temporal neocortex.

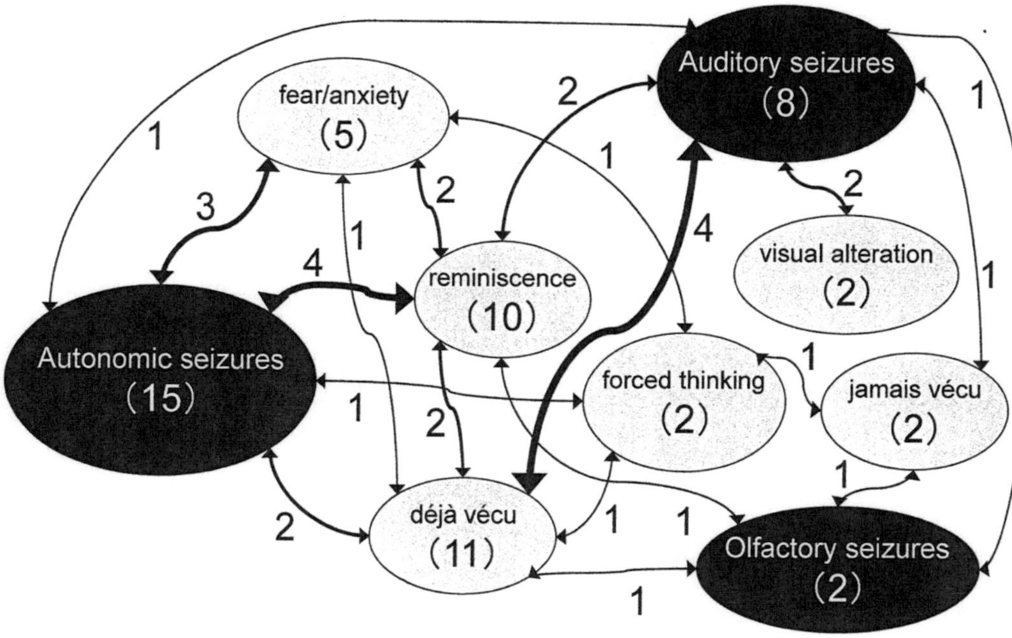

Fig. 1. Co-existence of simple partial seizures and different types of psychic seizure

In summary, co-existence was not conformed between subcategories of psychic seizure and other types of simple partial seizure, and it was fairly rare among subcategories of psychic seizure. Consequently, it seems adequate to assume considerably independent neural networks for the neural basis of different psychic seizures.

3. Various psychiatric disturbances accompanying TLE

TLE patients manifest psychiatric disturbances with significantly higher incidence than normal population. The explanation of this fact is not satisfactory that generally patients with chronic medical problem are inclined to have psychiatric disturbances. It is supposed that in some proportion of patients epileptic activities by themselves are the cause of the psychiatric symptoms. In this context, amygdalar dysfunction in TLE is the most important.

3.1 Depression and dysphoria

Epilepsy patients show a higher tendency of having depression than normal population (Barry et al., 2007). As it is well-known that depression is frequently seen in the patients with chronic diseases in general, the frequency of depression in epilepsy patients could partly be ascribed to the fact that epilepsy is one of them.

Whereas, it has long been told that epilepsy patients have a special kind of mood disorder, which is not simply depressive or manic, but of periodical character. It is called "interictal dysphoric disorder" (Blumer et al., 2004). The mechanism of this syndrome is not well clarified, but it is thought to have some causal relationship with seizure activities, because it seems that the mood change depends on the appearance of seizures. Namely, there is a tendency that depressive mood is strengthened before the appearance of seizures and manic mood appears after it. Although interictal dysphoric disorder is not confined to TLE, it is most often seen in patients with TLE and its mechanism is supposed to be related to amygdalar dysfunction.

3.2 Anxiety disorders

Epilepsy patients are inclined to develop anxiety disorders. One reason is purely psychological, namely, they are naturally anxious about presenting seizures in public and this anxiety can become pathologically strong. Especially those who have auras perceived only by themselves can get psychologically conditioned and present panic attacks triggered by perceiving auras. Furthermore, there is possible enhancement of conditioning in TLE patients based on amygdalar dysfunction. Ictal fear can by itself condition panic attack, because panic attack and ictal fear are in common supposed to be the results of paroxysmal dysfunction of amygdala. Moreover, amygdalar dysfunction may cause anxiety symptoms also in a persistent form, represented as generalized anxiety disorder.

3.3 Psychosomatic inclination

As amygdala contains also the center of autonomic system, its dysfunction may cause various autonomic disorders. For example, epilepsy patients are often sensitive to the moisture and have seizures at the time of the passing of a rain front. Another example is the sensitivity to the lights which are not flashing or flickering lights inducing photosensitivity, but glimmering lights from the cloudy sky. Psychosomatic inclination among epilepsy patients may be ascribed not only to the pure psychological complication but also to the amygdalar dysfunction directly linked to the epileptic activities.

3.4 Aggressivenss or "explosiveness"

Historically epilepsy patients have been believed to have aggressive tendency in general, and aggressiveness or "explosiveness" has been regarded as a component of so-called epileptic personality change, which is most remarkable in TLE patients. In reality, it is not so often to find extraordinary aggressiveness in epilepsy patients at present, probably due to the propagation of anticonvulsant medication from early stages of the disease. Aggressiveness observed in some TLE patients may be understood by amygdalar dysfunction, because stimulation of amygdala is known to cause not only fear and anxiety but also aggression.

3.5 Obsessive-compulsive disorder

It has long been known that obsessive-compulsive disorder (OCD) is frequently seen in TLE patients. The mechanism of the comorbidity of OCD and TLE is supposed to be plural, one of which being comorbidity of autism spectrum disorder and TLE. However, it seems certain that there are some cases in which confined epileptic activities cause obsessive-compulsive behavior. To date, laterality and subdivision within the temporal lobe of the focus causing OCD is not determined.

3.6 Epileptic personality change

"Epileptic personality change" or a special kind of personality disorder caused by suffering epilepsy has long been discussed and remains under dispute. Several elements such as viscosity, circumstantiality, explosiveness, and religiosity have been believed to characterize this type of personality change. Although believed to be caused by epilepsy, because it becomes increasingly manifest as the duration of the illness increases, it is regarded as non-specific "organic personality change" by some researchers (Slater et al., 1963). According to Bear and Fedio (1977), the elements included in epileptic personality change are: 1. humorlessness or sobriety, 2. sadness or depression, 3. emotionality, 4. circumstantiality, 5. philosophical interest, 6. sense of personal destiny, 7. viscosity or interpersonal stickiness, 8. dependence, 9. aggression, 10. obsessionality, 11. paranoia or suspiciousness, 12. sense of guilt, 13. hypergraphia or excessive writing, 14. changes to or diminution of sexual drive, 15. hypermorality, 16. religiosity, 17. elation or mood change, and 18. anger or irritability.

Gastaut and Waxman and Geschwind attempted to treat these features in neuropsychological style (Gastaut, 1956; Waxman & Geschwind, 1975). In particular, the personality change characteristic of TLE (limbic epilepsy personality syndrome; LEPS) was interpreted as the opposite of Klüver-Bucy syndrome, which is a group of symptoms seen in monkeys and humans with bilateral anterior temporal lesions. Whereas Klüver-Bucy syndrome comprises (1) oral tendency or cognitive deficit, (2) indifference or diminution of emotionality, (3) hypermetamorphosis or difficulty in attention fixation and (4) increase in sexual activity, LEPS comprises (1) strengthened cognition, (2) increased emotionality, (3) viscosity or difficulty in attention shift and (4) diminution or alteration of sexual activity. This apparent opposition has been explained by interictal epileptic activities within the temporolimbic structures (Gastaut, 1956), or "sensory-limbic hyperconnection" established by a kindling effect (Bear, 1979). In any case, the explanation is based on the positive or excessive function of the temporolimbic structures.

The author has shown a case which suggests presumed persistent seizure activities confined around the right amygdala had caused acute psychosis and rapid development of epileptic personality change (Fukao, 2010).

3.7 Psychosis in TLE

Psychosis in TLE patients is a rather independent problem, which has a long history of research and dispute. The question whether psychosis in epilepsy patients or epileptic psychosis is peculiar and different from schizophrenia and other psychoses is still unresolved (Gibbs, 1951; Slater et al., 1963). Although some studies suggested that epileptic psychosis is nothing but enhanced occurrence of schizophrenia in people with epilepsy and that there is no significant difference in the symptomatology between them (Adachi et al., 2011). However, it is frequently seen that symptoms of psychic seizures appear in the episodes of acute psychosis in TLE patients.

Importance has been attached to the distinction between acute and chronic forms of psychosis in epilepsy patients, because the acute form is clinically hard to discriminate from postictal confusional state that is not genuinely psychotic. However, since the recognition of postictal psychosis diffentiated from postictal confusional state (Logsdail & Toone, 1988), different mechanism is supposed for acute psychosis in TLE patients from confusion or impairment of consciousness, for example, "hyperarousal" after the cluster of seizures (Wolf, 1991). On the other hand, progression from postictal acute psychosis into chronic psychosis was confirmed in substantial part of TLE patients (Tarulli et al., 2001). It is, therefore, uncertain whether the distinction between acute and chronic forms of psychosis in TLE patients is really important or not.

Furthermore, some part of psychosis occurring among TLE patients are apparently induced by anticonvulsant drugs. Some anticonvulsant drugs like topiramate and zonisamide are evidently more prone to induce psychiatric disturbances than others, while there has been a general concept of "forced normalization" that means the reciprocal appearance of psychotic disorders when the patient's seizures were inhibited by any drugs (Krishnamoorthy & Trimble, 1999).

One point to be clarified in the problem of psychosis in TLE is the relationship between psychic seizures and psychosis, which has not been confirmed by clinical evidence. The author addressed this point by magnetoencephalographic study on TLE patients as described below (Fukao et al., 2009; Fukao et al., 2010).

4. Magnetoencephalographic localization of psychic seizures and psychosis in TLE

In this section the author' s data obtained from magnetoencephalographic (MEG)

localization of epileptic activities in TLE patients is described briefly and its implication on the relationship between psychic seizures and psychosis is discussed.

4.1 Methods

4.1.1 Patients

The subject of this study comprises 57 patients who had been diagnosed as having TLE based on EEG findings and clinical symptoms. Out of the 57 patients, all having complex partial seizures, those who had autonomic seizures were 25 in numbers, those with auditory seizures were 10 and those with psychic seizures were 16. Six patients had both autonomic and psychic seizures, and 5 patients had both auditory and psychic seizures. One patient had all of the three types of aura. One patient had psychic, olfactory and gustatory seizures, and another patient had only olfactory seizures. Those who had no aura were 16.

Sixteen out of the 57 patients had history of psychosis, whether chronic or episodic. Age of the patients with psychosis ranged from 20 to 46 (mean = 29) years old and the mean duration of the illness was 19 years. Eight out of the 16 patients had chronic psychosis and the remaining 8 had episodic psychosis.

4.1.2 Magnetoencephalographic measurements and spike-dipole typing

MEG measurements were performed using Magnetic Source Imager (a dual sensor system containing 37 channels of first order gradiometer in each sensor; Biomagnetic Technologies Inc., San Diego, CA, USA), whose sensors were positioned symmetrically on both temporal regions of the patients, with simultaneous recordings of scalp EEG which helped the examiner find epileptiform discharges. MEG was recorded as epochs each of which consisted of a 6 seconds segment including 5 seconds before and 1 second after the trigger by manual button press. The sampling rate was 200 Hz. The duration of the recording session was 1.5-2 hours including the time needed to place EEG electrodes on the scalp. After filtering the all MEG epochs by 3-30 Hz digital band-pass filter, equivalent current dipoles of magnetic spikes collected (spike-dipoles) were calculated off-line by the attached software, and overlaid on MR images.

"Clustering" of spike-dipoles was approved only when more than 10 dipoles show concentration within a cube with edges of 3 cm. Then spike-dipole clusters found in the patients were classified according to anatomical localization and orientation of the dipoles.

As a preparation of the study, we classified the patients according to the type of "clustering" of magnetic spike-dipoles. We found two major types of the pattern of "clustering" of spike-dipoles in the patients. One of which is positioned on the lower part of the temporal lobe and oriented from anterior toward posterior as shown in Fig. 2, and we name this "inferotemporal-horizontal (IH) type". The other is positioned on the upper part of the temporal lobe and oriented from superior toward inferior as shown in Fig. 3, and we name this "superotemporal-vertical (SV) type". Of the 57 patients showing spike-dipole clusters, 33 patients showed IH type of spike-dipole and 27 showed SV type. Among them 10 patients showed both types, ipsilaterally in 8 and contralaterally in 2. Three patients had bilateral SV type, and one had bilateral IH type. None had more than two spike-dipole clusters. Thus 88% of the patients had either IH type or SV type of spike-dipole, and the presence of the two types were statistically exclusive (Fisher's exact test: $p = 0.027$). In the following studies, we investigated only these two, or four, when each divided into two according to the side, types of spike-dipole.

4.1.3 Correlation analyses

Correlation analyses were performed using Fisher's exact test between these four types and the presence of autonomic, auditory and psychic seizures and history of psychosis.

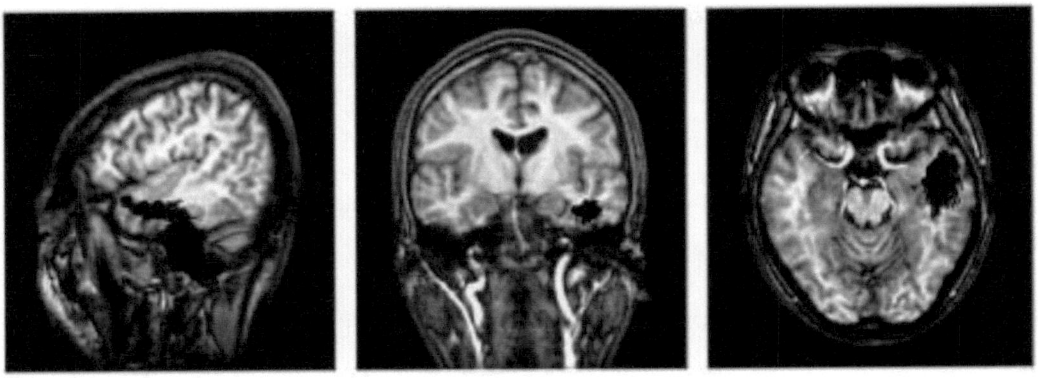

Fig. 2. An example of IH (inferotemporal-horizontal) type of spike-dipole distribution.

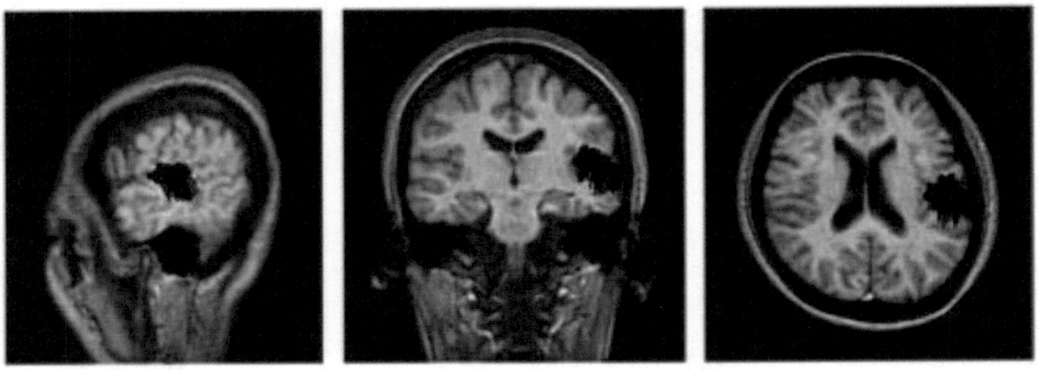

Fig. 3. An example of SV (superotemporal-vertical) type of spike-dipole distribution.

4.2 Results

The results of the correlation analyses are summarized in Table 2. Left IH type positively correlated to autonomic seizures. Right SV type positively correlated to both auditory and psychic seizures. Left SV type positively correlated to history of psychosis.

	Left	Right
IH type	Autonomic seizures	
SV type	Psychosis	Auditory seizures Psychic seizures

Table 2. The positive correlations between left and right IH and SV types of spike-dipoles and autonomic, auditory and psychic seizures and psychosis.

4.3 Interpretation

As IH type correlated not only to autonomic seizures on the left side but also correlated to hippocampal sclerosis on both sides, it seems very certain that IH type of spike-dipole should represent hippocampal or temporolimbic discharges. On the other hand, the correlation of SV type to auditory seizures suggests that SV type should represent temporal neocortical discharges.

In view of the discussion in section 2, the neural substrates of psychic seizures are supposed to be interconnected networks within and around the temporal lobes. Therefore, the correlation of SV type to psychic seizures suggests that SV type should represent also interconnected networks. The correlation of SV type to psychosis should be interpreted in the same way as psychic seizures, not as simple as auditory seizures.

It is, nevertheless, suggested by our results that temporolimbic discharges should not be essential to psychic seizures and psychosis, compared to temporal neocortical discharges. This is a remarkable finding in view of the other researches' suggestion that temporolimbic dysfunction should be essential to psychic seizures and psychosis.

The results described above imply that psychosis in TLE should be causally related to epileptic activities occurring in the superior portion of the left temporal lobe. From the neuropsychological viewpoint, the superior portion of the left temporal lobe is related to the function of awareness. Therefore, while seizures occurring on the right side are perceived by the patient as various psychic phenomena, those occurring on the left side might not be perceived as such phenomena due to the disturbance of awareness. Furthermore, prolonged seizure activities on the area might bring about the disturbance of self-consciousness resulting in psychosis characterized as an ego disorder. Thus psychosis in TLE patients might be caused by epileptic activities on the superior portion of the left temporal lobe.

4.4 Magnetoencephalography's advantage in this study

More than a half century ago, Penfield wrote, "The relationship of epileptic discharge, in temporal cortex and in the gray matter that is hidden deep in the Sylvian fissure, to psychotic states deserves exhaustive study" (Penfield & Jasper, 1954). Our findings seem to confirm his prediction. However, as neural activities detected by EEG and MEG are same in origin, the reason why these findings became available only by MEG has to be explained.

The key point is thought the orientation of the current of the epileptic discharges. As magnetic field is produced around the electric current on the plane tangential to the current, MEG can detect magnetic fields tangential to the corresponding neural currents. This condition makes possibility that MEG could detect spike activities that are not detectable by EEG. To be concrete, if spike generators existed deep within the Sylvian fissure, whether frontal, parietal or temporal operculum, and those spikes would elude detection by scalp EEG. In contrast, MEG can detect spike activities in considerably deep part of the cortex within the fissure, as far as the originating current is tangential to the scalp, accordingly, vertical to the fissure (Iwasaki et al., 2003). As can be seen, our SV type of spike-dipoles represents such deep current generators. Moreover, the elusiveness of deep Sylvian discharges is also the case with intracranial recording using subdural electrodes, because usually the electrodes are not inserted into the Sylvian fissure but laid striding over it. In those cases with deep spike foci, therefore, MEG would be more powerful in spike detection than subdural electrodes.

5. Conclusion

There are various kinds of psychic seizures, or psychic phenomena induced by seizure activities in the brain and the classification is fundamentally difficult because of their subjective nature. Although the neural substrates of psychic seizures are supposed to reside mostly in temporal lobes, they are not necessarily localized and possibly some of them are extensive networks. Various psychiatric disturbances observed in TLE patients should be understood in relation to psychic seizures. Furthermore, because seizure activities may inhibit psychic functions or induce negative phenomena, the concept of psychic seizures could theoretically be extended to cognitive impairments induced by seizure activities, which are clinically elusive. Our results from MEG studies suggest that cognitive impairments induced by seizure activities on the left temporal lobe should lead to the generation of psychosis in TLE patients. This is a novel finding and interpretation that MEG made possible.

6. References

Adachi N, OnumaT, Kato M, Ito M, Akanuma N, Hara T, Okubo Y & Matsuura M. (2011). Analogy between psychosis antedating epilepsy and epilepsy antedating psychosis. *Epilpsia* 52, pp. 1239-1244

Barry JJ, Lembke A, Gisbert PA & Gilliam F (2007). Affective disorders in epilepsy. In: Ettinger AB, Kanner AM (Eds.) *Psychiatric Issues in Epilepsy: A Practical Guide to Diagnosis and Treatment, 2nd edition,* pp. 203-247, Lippincott Williams & Wilkins, Philadelphia

Bear DM & Fedio P. (1977). Quantitative analysis of interictal behavior in temporal lobe epilepsy. *Arch Neurol* 34, pp. 454-467

Bear DM. (1979). Temporal lobe epilepsy: A syndrome of sensory-limbic hyperconnection. *Cortex* 15, pp. 357-384

Blanke O, Landis T, Spinelli M & Seeck M. (2004). Out-of-body expeience and autoscopy of neurological origin. *Brain* 127, pp. 243-258

Blumer D, Montouris G & Davies K. (2004). The interictal dysphoric disorder: recognition, pathogenesis, and treatment of the major psychiatric disorder of epilepsy. *Epilepsy Behav* 5, pp. 826-840

Devinsky O, Hafler DA & Victor J. (1982). Embarrassment as the aura of complex partial seizure. *Neurology* 32, pp. 1284-1285

Devinsky O, Feldmann E, Burrowes K & Bromfield E. (1989). Autoscopic phenomena with seizures. *Arch Neurol* 46, pp. 1080-1088

Engel JJr & Williamson RD. (2008). Limbic seizures. In: Engel JJr, Pedley TA (Eds.) *Epilepsy: A Comprehensive Textbook, 2nd edition,* pp.541-552, Lippincott-Raven Publishers, Philadelphia

Fish DR. (1997). Psychic seizures. In: Engel JJr, Pedley TA (Eds.) *Epilepsy: A Comprehensive Textbook,* pp.543-548, Lippincott-Raven Publishers, Philadelphia

Fukao K, Inoue Y & Yagi K. (2009). Magnetoencephalographic characteristics of psychosis in temporal lobe epilepsy. *J Neuropsychiatry Clin Neurosci* 21, pp. 455-462

Fukao K. (2009). A study on the relationship between psychic seizures and cognitive functions in epilepsy patients. *Ann Jpn Epi Res Found* 20, pp. 95-100

Fukao K, Inoue Y & Yagi K. (2010). Magnetoencephalographic correlates of different types of aura in temporal lobe epilepsy. *Epilepsia* 51, pp. 1846-1851

Fukao K. (2010). Psychopathology of epileptic psychosis. In: Matsuura M, Inoue Y. (Eds.) *Neuropsychiatric Issues in Epilepsy.* pp. 35-44, John Libbey Eurotext, Montrouge

Gastaut H. (1956). Etude électroclinique des épisodes psychotiques survenant en dehors des crises cliniques chez les épileptiques. *Rev Neurol* 94 , pp. 587-594

Gibbs FA. (1951). Ictal and non-ictal psychiatric disorders in temporal lobe epilepsy. *J Nerv Ment Dis* 109, pp. 522-528

Gloor P. (1990). Experiential phenomena of temporal lobe epilepsy: Facts and hypotheses. *Brain* 113, pp. 1673-1694

Iwasaki M, Nakasato N, Shamoto H & Yoshimoto T. (2003). Focal magnetoencephalographic spikes in the superior temporal plane undetected by scalp EEG. *J Clin Neurosci* 10, pp. 236-238

Krishnamoorthy ES & Trimble MR. (1999). Forced normalization: clinical and therapeutic relevance. *Epilepsia* 40(Suppl 10), pp. 57-64

Logsdail SJ & Toone BK. (1988). Post-ictal psychoses: A clinical and phenomenological description. *Br J Psychiatry* 152, pp. 246-252

Maillard L, Vignal JP, Anxionnat R & TaillandierVespignani L. (2004). Semiologic value of ictal autoscopy. *Epilepsia* 45, pp. 391-394

Palmini AL, Gloor P & Jones-Gotman M. (1992). Pure amnestic seizures in temporal lobe epilepsy. Definition, clinical symptomatology and functional anatomical considerations. *Brain* 115, pp. 749-769

Penfield W & Jasper H. (1954). *Epilepsy and the Functional Anatomy of the Human Brain.* Little Brown, Boston

Sadler RM & Rashey S. (2004). Prescience as an aura of temporal lobe epilepsy. *Epilepsia* 45, 8, pp. 982-984

Slater E, Beard AW & Glithero E. (1963). The schizophrenia-like psychoses of epilepsy. *Br J Psychiatry* 109, pp. 95-150

Spatt J. (2002). Déjà vu: possible parahippocampal mechanisms. *J Neuropsychiatry Clin Neurosci* 14, pp. 6-10

Tarulli A, Devinsky O & Alper K. (2001). Progression of postictal to interictal psychosis. *Epilepsia* 42, pp. 1468-1471

Waxman SG & Geschwind N. (1975). The interictal behavior syndrome of temporal lobe epilepsy. *Arch Gen Psychiatry* 32, pp. 1580-1586

Wolf P. (1991). Acute behavioral symptomatology at disappearance of epileptiform EEG abnormality. Paradoxical or "forced normalization"." In: Smith D, Treiman D, Trimble M. (Eds.) *Advances in Neurology*, Volume 55, pp. 127-142, Raven Press, New York

10

Personality Profiles of Patients with Psychogenic Nonepileptic Seizures

Krzysztof Owczarek[1,*] and Joanna Jędrzejczak[2]
[1]*Department of Medical Psychology, Medical University, Warsaw*
[2]*Department of Neurology and Epileptology,*
Medical Centre for Postgraduate Education, Warsaw
Poland

1. Introduction

Despite the substantial prevalence of psychogenic disorders (from 50% to 70% of patients who report to physicians of various specialities are treated for psychogenic, functional disorders), these disorders continue to be relatively poorly understood (Hamilton et al., 1996; Mace & Trimble, 1996). Psychogenic nonepileptic seizures (PNES) are one variety of psychogenic disorders. PNES are sudden changes in behaviour, usually of limited duration, which imitate an epileptic attack, but are not accompanied by EEG changes occurring during a genuine epileptic attack.

PNES are a diagnostic and therapeutic challenge. Some patients referred to epilepsy centres because of drug-resistant epileptic seizures have PNES. The consequences of a false diagnosis of epilepsy have profound effects like changes in antiepileptic drugs (AEDs), changes of doctors, frequent medical staff interventions, and numerous hospitalizations. Swift correct diagnosis and implementation of correct therapeutic intervention may protect this group from many adverse psychological and social effects and save society costs of unnecessary pharmacological treatment and disability pension. At present, from 7 to 16 years often elapse between the first dissociative seizure and correct diagnosis. This leads to symptom chronicity, making treatment difficult (De Tinary et al., 2002; Reuber et al., 2002). Some studies consistently report that up to one third of patients become chronically ill (Bodde et al., 2009). Economic concerns are also by no means trivial. The costs of inaccurate diagnosis in terms of public money are really colossal: patients with a diagnosis of epilepsy receive disability pension, are unsuccessfully (and unnecessarily) treated with antiepileptic drugs, move in and out of hospital, and wander from doctor to doctor in a never-ending quest for help. Researchers in the USA who studied the cost of treatment of patients with PNES found that the average cost of medical treatment dropped by 84% within six months of a correct diagnosis (Martin et al., 1998).

Dissociative disorders cause significant diagnostic problems. Prolonged dysfunction of this type, particularly paresis and dysaesthesia, may be related to an unresolved

* Corresponding Author

personality disorder. The most frequent forms of psychogenic disorders are limb paresis, headaches, backaches and psychogenic nonepileptic seizures. They often lead to incorrect medical diagnosis and may seriously jeopardise the implementation of appropriate and effective treatment. According to The International Statistical Classification of Diseases and Related Problems, the common theme of all conversion or dissociative disorders is partial or complete loss of normal integration between memories of the past, sense of identity, sensory sensitivity, and body movement control (ICD-10, 1992). Whatever their specific type, all dissociative disorders may subside within a few weeks or months, particularly if their onset coincided with a traumatic life event. More persistent dissociative disorders such as paralysis or dysaesthesia may be related to insoluble or interpersonal problems. These disorders used to be classified as various "hysterical conversions". Nowadays their origins are believed to be psychogenic and can be temporally related to traumatic events, insoluble or "unbearable" situations, or dysfunctional relations with the social environment. The symptoms often reflect the patient's ideas of how a somatic disease would manifest itself. The symptomatology is often very complex and confusing and may involve several body systems or functions. The patient may present with symptoms mimicking cardiologic, gastric, muscular-skeletal, urogenital, or neurological symptoms or may complain of pain or fatigue. All these forms of dysfunction have one common origin, i.e., somatisation. Somatisation means the individual propensity to present somatic symptoms and attract the attention of health care providers. At first glance, the symptom picture does not resemble psychiatric symptoms. The anomalies are rooted in the patient's social situation or are related to occupational responsibilities. The DSM-IV–TR diagnostic system relates somatisation to vegetative disorders, which present themselves in the form of somatic complaints, conversion disorders, hypochondriacal disorders, somatisation disorders, persistent psychogenic pain, or somatoform disorders. Incorrect diagnosis of conversion disorders as organic diseases is a serious clinical concern. It may lead to many misunderstandings in various medical specialities.

Psychogenic nonepileptic seizures are one variety of dissociative disorders. From 5 to 33% of all patients referred for epilepsy assessment actually suffer from PNES. In Europe, thousands of patients, most of them young, suffer from this type of seizures. Psychological criteria for the differential diagnosis of seizures are lacking, both in the literature and clinical practice. Frequently patients with PNES may be submitted to unnecessary intense treatment with antiepileptic drugs. The future of these patients largely depends on the accuracy of their diagnosis. Approximately 22% of drug-resistant epilepsies are in fact pseudo-drug-resistant. One frequent reason for drug-resistance is the psychogenic nature of some of the seizures. Such episodes are wrongly assumed to be epileptic seizures. Because of this, the true picture of epilepsy is blurred and this interferes with the proper treatment of the patient's true epilepsy. Wrong diagnosis leads to wrong treatment and the consequences for the patient may be dramatic. Prolonged inadequate and ineffective treatment is also a problem for physicians because it undermines their sense of competence and their confidence in contemporary medical expertise and the effectiveness of medication. On top of this, there are the social aspects of the problem. Instead of getting better and returning to normal life and work, wrongly treated patients remain on disability pension, convinced that they are seriously, organically ill.

2. Personality profiles of patients with psychogenic nonepileptic seizures: Our findings

Although the methods used to diagnose psychogenic non-epileptic seizures are expanding rapidly, it is still very difficult to identify proper causes. Psychological evaluation of patients with PNES using personality profile indicators (levels of anxiety and somatisation) may help us gain a better understanding of the etiology of psychogenic non-epileptic seizures. In our research, we utilize the Minnesota Multiphasic Personality Inventory (MMPI).

The MMPI is one of the most widely used psychological instruments. The first version of the test was constructed by Starke R. Hathaway and J. Charnley McKinley. The first theoretical and clinical publications on this test appeared in 1940. Since the original version of the MMPI was constructed many decades ago, it has been modified and amended, normalized, standardized and submitted to other procedures to improve its reliability and validity. The MMPI has 566 self-report items that respondents answer in a True/False format. The items cover a wide array of contents including general health, behaviours, social adjustment, marital problems, family problems, attitudes toward other people, attitudes toward generally accepted normative systems, tradition, religion, etc. The test is scored using a scoring template and raw scores are transformed into standardized scores based on available norms. The standardized scores are presented on a standard ten scale that can theoretically range from 0 to 100, with a mean score of 50 and a standard deviation of 10. A routine psychological interpretation of the MMPI is based on the respondent's normative profile (psychograph) which has three control scales used to assess the profile's validity and 10 clinical (personality) scales (Table 1). In order to obtain a truly informative assessment one must analyze the scale profile and their configuration.

Control scales
Lie scale (L)
Low frequency (F)
Correction (K)
Clinical scales
Hypochondriasis (Hs)
Depression (D)
Hysteria (Hy)
Psychopathic Deviate (Pd)
Masculinity-Femininity (Mf)
Paranoia (Pa)
Psychasthenia (Pt)
Schizophrenia (Sc)
Mania (Ma)
Social Introversion (Si)

Table 1. Normative MMPI profile. Scales and scale abbreviations

MMPI results are used to plan patient treatment and interventions whereas repeated assessments can be used to assess therapeutic outcome. The MMPI is used in differential diagnostics on psychiatric wards, in psychological assessments, medical clinics specializing in various disorders, and institutions such as penitentiaries, police, military etc. The MMPI comes in several forms: paper-and-pencil booklets, audiocassettes, cards. The most useful form is a computerized version that greatly reduces administration time, simplifies scoring and offers a greater variety of interpretations based on the scale interactions and empirical indices.

Thanks to computer technology, it is possible to make rapid computations that have considerably widened our diagnostic possibilities and improved the original version of the MMPI. Computerized versions enable about 200 scales, diagnostic indices, and configurative indices to be analyzed, greatly enhancing our interpretative capacity. Today, a psychological diagnosis with the help of the MMPI not only takes advantage of these new possibilities, but also forces us to select our data more carefully and interpret the information on the different dimensions of personality more accurately. This requires competence in personality and clinical psychology and clinical experience.

3. Participants

The study was conducted at the Department of Neurology and Epileptology, Medical Centre for Postgraduate Education in Warsaw (Poland). Based on long-term video-EEG monitoring data the patients were divided into two groups: group I consisted of 70 patients (58 F and 12 M) with PNES and group II - 42 patients (30 F and 12 M) had epileptic seizures. The majority of the PNES (group I) were of the following three types: episodes imitating tonic-clonic seizures (35 patients), episodes imitating simple partial seizures, partial complex seizures, mioclonic seizures with dominating sensory or vegetative sensations accompanied by limited response to external stimulation (28 patients), and more than one form of psychogenic seizure (7 patients). In group II, 19 of the 42 epileptic patients presenting partial complex seizures had secondary generalised tonic-clonic episodes. Mean age was 24.5 and 26.3 respectively. Upon completion of the selection procedure, the MMPI was administrated to all participants.

4. The normative MMPI psychological profile

The MMPI scores were first submitted to a procedure which enabled the construction of normative personality profiles and differences in means obtained by the two groups were analyzed (Fig. 1).

The psychological profiles of our groups differed significantly, both with respect to the shape and values of the hypochondriasis (Hs) and hysteria (Hy) scores ($p \leq 0.001$). In patients with PNES, the mean Hs and Hy scores were higher ($p \leq 0.001$) than the D score. In patients with epileptic seizures, the Hs, D and Hy was reversed - D was significantly higher than Hs and Hy ($p \leq 0.01$). Elevated Hs and Hy scores and lower D scores are typical for individuals with a powerful need to interpret their problems in a way which is at once rational and socially acceptable. Such patients have a sense of entitlement (Jędzejczak & Owczarek, 1999; Owczarek & Jędzejczak, 2001). In the present study, higher Hs and Hy scores compared with D scores suggest the presence of a conversion mechanism (the so-called conversion dip). Analysis of the subscale data additionally suggests that the existence of psychogenic nonepileptic seizures or the predisposition to such seizures is reflected in a personality profile.

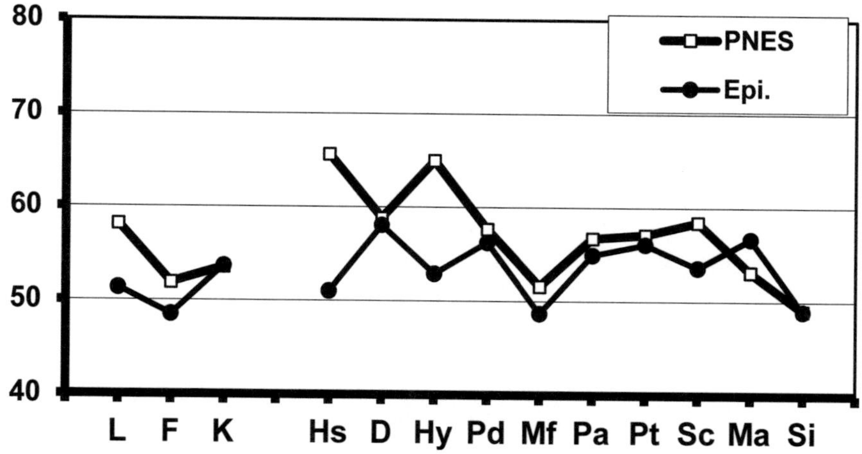

Fig. 1. Averaged MMPI profiles for the two groups. Symbols on the horizontal axis signify control scales (L - Lie scale, F - Low frequency, K - Correction) and clinical scales (Hs - Hypochondriasis, D - Depression, Hy - Hysteria, Pd - Psychopathic Deviate, Mf - Masculinity-Femininity, Pa - Paranoia, Pt -Psychasthenia, Sc - Schizophrenia, Ma - Mania, Si - Social Introversion); PNES - patients with psychogenic nonepileptic seizures; Epi.- patients with epilepsy.

4.1 Differences in Hysteria (Hy) scores

The next thing we did was to analyze the differences between the two groups on the Hysteria subscales. The differences between the mean scores of participants with PNES and

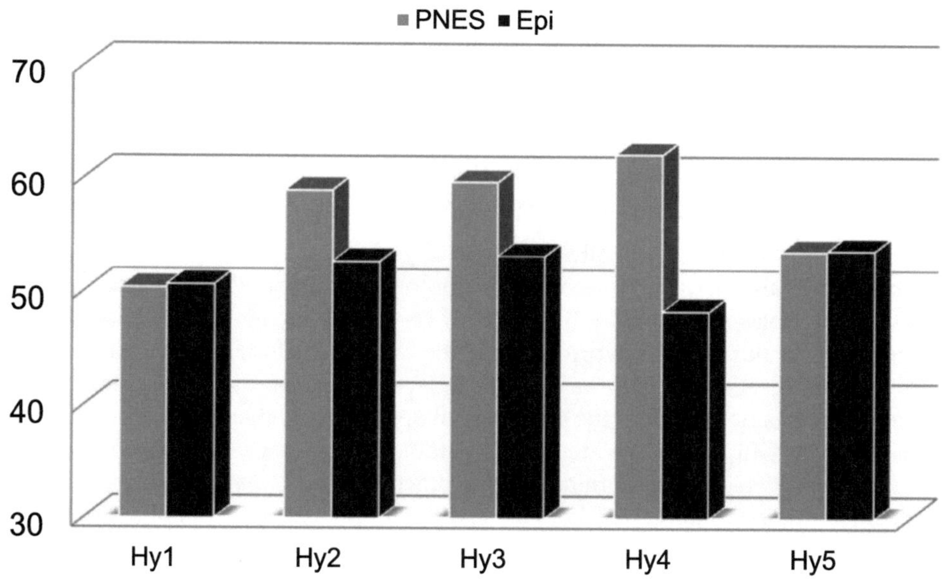

Fig. 2. Minnesota Multiphasic Personality Inventory (MMPI) - hysteria subscales. Hy1- denial of social anxiety; Hy2-need for affection; Hy3-lassitude-malaise; Hy4-somatic complaints; Hy5-inhibition of aggression; PNES-patients with psychogenic nonepileptic seizures; Epi.-patients with epilepsy.

participants with epilepsy on all five subscales, Hy1 - denial of social anxiety; Hy2 - need for affection; Hy3 - lassitude-malaise; Hy4 - somatic complaints; Hy5 - inhibition of aggression, are presented in Figure 2.

The greatest differences were found for subscale Hy4 ($p \leq 0.001$). Participants with psychogenic nonepileptic seizures reported significantly more nonspecific somatic complaints. The higher scores for subscale Hy3 in the PNES group compared with the group with epilepsy ($p \leq 0.01$) is indicative of greater weakness and fatigue. The significant difference for the subscale Hy2 ($p \leq 0.01$) indicates that participants with PNES have an excessive need of affection and were also very trusting and optimistic.

4.2 Discussion

Analysis of the Hysteria subscales sheds more light on the underlying mechanisms of PNES. The high Hy2 scores in participants with PNES suggest an excessive need of emotional contact. These people are extremely "emotionally adhesive" and they relate to people willingly and often indiscriminately. We noticed that when these people reported to an epilepsy clinic they often brought various cuddly toys, teddy bears, frogs, rabbits, etc. They demonstrate a great need to remain in the limelight, are extremely trusting and often naïve but on the other hand, they fear competition, confrontation and criticism. They tend to view other people as sensible, honest, and compassionate and this may be why they are frequently disappointed. They often experience internal tension and conflict between what they expect and what they get. If these internal conflicts intensify and if they lack socially acceptable forms of discharge of their mounting tension, they may resort to conversion as a form of adjustment.

Elevated Hy3 (lassitude-malaise) and Hy4 (somatic complaints) scores are the consequence of excessive Hy2 (need for affection). These people often tend to present as people suffering from a serious somatic disease. They arouse interest and sympathy, which help to reduce the emotional deficits caused by their excessive need for affection. This factor adds to a conversion mechanism. Psychogenic nonepileptic seizures are a source of primary gain in the form of energetic discharges of internal conflicts and tensions and secondary gain in the form of other people's interest and care (Devinsky, 1998). Our study confirmed this hypothetical mechanism (high Hy2, Hy3 and Hy4 scores).

The presence or predisposition to PNES was generally confirmed by the personality profiles. A relationship may also exist between the symptoms of nonepileptic seizures and the psychological variables measured by the MMPI. These problems need to be analyzed further at a deeper level. In our study, interpretation of the findings included the mean values of the variables in both groups, i.e. we based our interpretation on the most clear-cut central tendencies. This does not preclude the unequivocal operation of other mechanisms and causes of PNES, however. Only when we identify all pathological personality mechanisms of PNES will we be able to develop precise guidelines for prevention and treatment of these behaviour disorders. Such findings should also help to classify psychogenic nonepileptic seizures taking into consideration the etiology of personality disorders.

We know for certain that anxiety and somatisation contribute to PNES (Szaflarski et al., 2000; Owczarek, 2003a, b; Griffith et al., 2007). High levels of both these factors increase the likelihood of occurrence and recurrence of these behaviour disorders. We shall discuss this in the remaining sections of this article.

5. Anxiety dimensions

According to the literature, the factors frequently reported to lead directly or indirectly to psychogenic nonepileptic seizures include anxiety, difficulty controlling internal tensions and needs, and attention disorders (Devinsky, 1998; Donofrio et al., 2000; Swanson et al., 2000; Mökleby et al., 2002; Owczarek, 2002). Patients with PNES who have blatant anxiety symptoms can be classified as having one of the following four anxiety disorders: anxiety disorder without agoraphobia, anxiety disorder with agoraphobia, post-traumatic stress disorder (PTSD), or acute post-traumatic stress. Most symptomatic for anxiety disorders, with or without agoraphobia, are panic attacks with other symptoms which accompany these episodes or emerge in consequence of these episodes, or changes in behaviour. Panic attacks manifest with palpitations, sweating, chest pain, depersonalization, derealisation, loss of sense of control, the feeling that one is dying, etc. These symptoms may be classified as pseudoepileptic attacks or erroneously classified as epileptic seizures (like epileptic aura, partial simple seizures, and others). PTSD-related anxiety disorders and acute PTSD symptoms manifest as more general anxiety symptoms with more pronounced dissociation (Alper et al., 1997; Donofrio et al., 2000; Prueter et al., 2002). The basic difference between these two anxiety disorders is that the former one lasts over a month whereas acute PTSD lasts from 2 days to 4 weeks.

Roy and Barris (1993) compared patients with PTSD and patients with epilepsy on the Salkind Morbid Anxiety Inventory. Patients with PNES had significantly higher anxiety ($p \leq 0.001$) and significantly higher affective responses ($p \leq 0.001$). These results confirm the observations, made elsewhere, that patients with PNES are more anxious and more prone to affective reactions.

5.1 MMPI anxiety scales

Originally, the MMPI had two scales to measure anxiety and defence mechanisms: Anxiety (A) and Repression (R). These were identified by G. S. Walsh in 1956. High scores on scale A indicate general pessimism, apprehension and psychological discomfort, low self-confidence, and excessive focus on oneself and one's problems. High scorers' social attitude is pervaded with excessive docility, uncritical obedience to authority, submissiveness and shyness. High scorers on scale R typically resort excessively to defence mechanisms such as repression and rationalization. They constantly feel threatened and their excessively controlling ego helps them to defend themselves cognitively, affectively, and volitionally. They are slow to act and have great difficulty making decisions. One of their most dreaded fears is the fear of making a fool of themselves and the fear of social failure.

These scales greatly contributed to the normative description of personality and W. G. Dahlstrom, G. S. Welsh and L. E. Dahlstrom conducted an analysis of the MMPI scale configurations (Dahlstrom et. al.,1986). Their factor rotations yielded two factors, anxiety and somatisation. Therefore, it was now possible to obtain MMPI measures of these personality parameters on the basis of the scale configurations which these American researchers identified.

The anxiety configurations are as follows:

Anxiety Scale (AxS)
AxS = (L+Hs+Pa) - (D+Pt)
Expression-Repression Scale (ERI)
ERI = (L+K+Hy) - (Pd+Ma)
Neuroticism Index (NS)
Ns = (Hs+D+Hy)
Triad Elevation Index (TI)
TI = (Hs+D+Hy)/3 - (Pa+Pt+Sc)/3
Frustration Tolerance Index (FT)
FT = (Ma+Pd) (Hy+D)

If it is true that a tendency toward affective reactivity underlies PNES then this effect should be observed in emotional parameters of personality profiles of patients with PNES. These patients also differ in these respects from patients with epilepsy who do not have PNES. The MMPI enables us to assess these parameters vis-à-vis the population norm. This is extremely important when we want to identify the factors that contribute to the pathogenesis and consolidation of behaviour disorders. According to Dahlstrom and coworkers (1986), we can use the MMPI to control the levels of somatisation and anxiety as well as other personality parameters that may contribute to PNES. We will now analyze the anxiety indices. We will analyze the role of somatisation in the next paragraph.

5.2 Results

Significant between-group differences were obtained for the anxiety scales. Patients with psychogenic nonepileptic seizures scored significantly higher than the epileptic group on the Anxiety Scale (AxS) and the Expression-Repression Scale (ERI) .

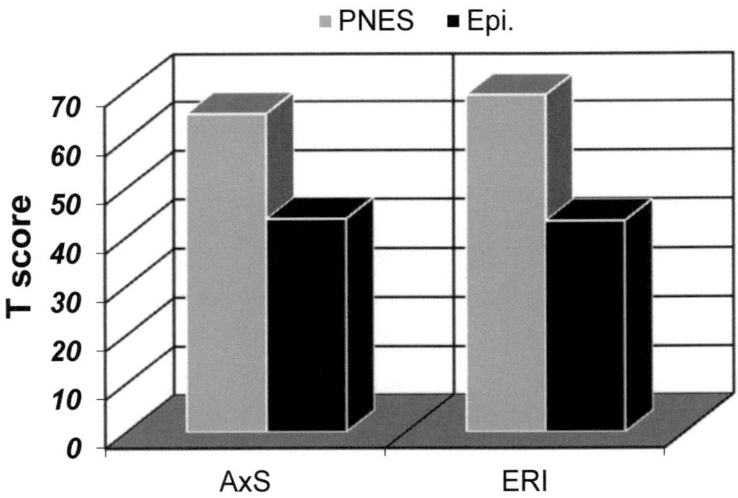

Fig. 3. Mean values of clinical parameters for the MMPI Anxiety Scale (AxS) and Expressive-Repressive Index (ERI) in the studied groups; PNES - patients with psychogenic nonepileptic seizures; Epi. - patients with epilepsy.

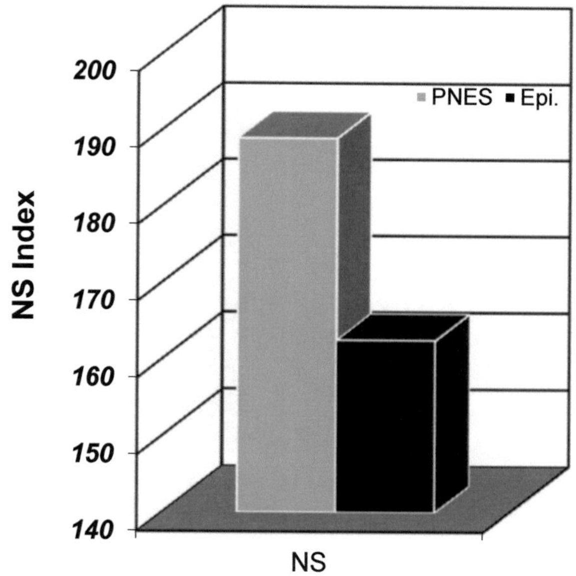

Fig. 4. Mean Neuroticism Index (NS) in the studied groups. PNES - patients with psychogenic nonepileptic seizures; Epi.-patients with epilepsy.

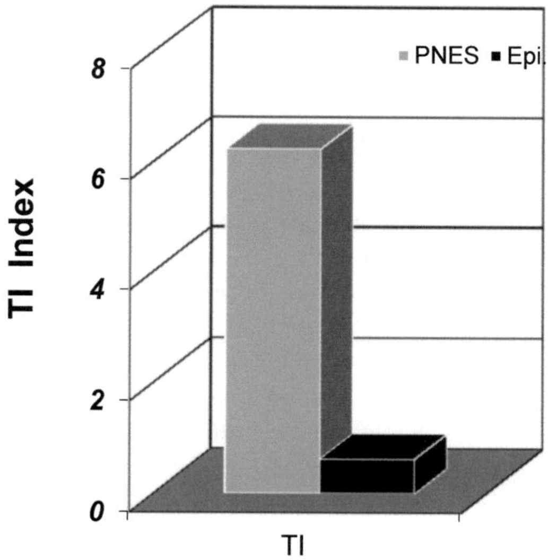

Fig. 5. Mean Triad Elevation Index (TI) scores. PNES - patients with psychogenic nonepileptic seizures; Epi.-patients with epilepsy.

Statistically significant differences between the two groups were also found for the Neuroticism Index (NS), Triade Elevation Index (TI), and Frustration Tolerance Index (FT).

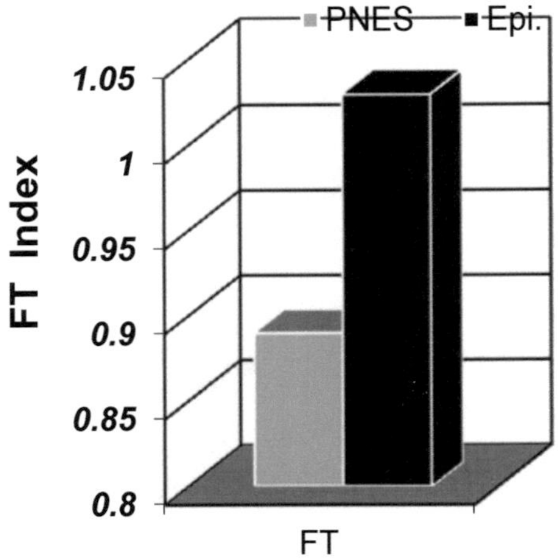

Fig. 6. Mean Frustration Tolerance Index (FT); PNES-patients with psychogenic nonepileptic seizures; Epi.-patients with epilepsy.

These findings suggest that the existence of psychogenic nonepileptic seizures or the predisposition to such seizures is reflected in the anxiety dimensions of a personality profile. Psychological evaluation of anxiety may help us to gain a better understanding of, and discrimination between, patients with psychogenic nonepileptic seizures and epileptic seizures.

5.3 Discussion

The psychological causes of PNES postulated by the researchers whose work was reviewed in the introduction to this chapter are reflected in the MMPI anxiety parameters. Once again, we need to make it clear that the PNES and patients with epilepsy differed with respect to the studied personality dimensions. In PNES participants, the mean AxS and Expression-Repression scores were elevated, suggesting that anxiety and maladaptive defence mechanisms are the mechanisms underlying psychogenic nonepileptic seizures. Defence mechanisms enable rational and socially acceptable need satisfaction. However, when the methods used to reduce mounting tension are inadequate, PNES help to abreact stress (primary gain). PNES also help to attract care and attention (secondary gain). Both types of gain reinforce the tendency toward seizures. In other words, PNES are a "pathological adjustment". They help the patient to ward off other intense negative emotions caused by unsolved and unsolvable life situations.

Reduced frustration tolerance is the direct cause of the inability to cope with psychological discomfort. When a person is unable to cope, he or she resorts to PNES to solve his/her problems. Attention is shifted from the psychological discomfort to "health problems" which are felt to be objective and independent. People who have PNES are often stubborn, determined and "attached" to their presenting somatic complaints. They often reject the doctor's opinion and deny the validity of medical tests, making it very hard to treat psychogenic epileptic seizures.

More generally, elevated anxiety scores suggest that problems of patients with PNES are related to an emotional dysfunction. Elevated emotional indices in the MMPI in their personality profiles unequivocally point to anxiety, psychological tension and increased defence against intense affect. Since these patients' defence mechanisms are inefficient, relations with the proximal environment are dysfunctional and more general social adjustment is poor. Patients are therefore motivated to seek other ways of attracting caring attention.

Elevated Neuroticism and triad elevation confirm the neurotic nature of the symptoms. Because of their emotional deficits, patients with PNES often have disturbed social relations. In the literature (e.g. Roy & Barris, 1993), PNES patients' poor social communication is emphasized. The feeling that one is living in a rejecting social environment and cannot communicate one's grudges, fears, and needs directly causes these patients to feel frustrated and is a source of unresolved and mounting tension. Psychogenic nonepileptic seizures are a way of attracting attention, concern, understanding and compassion. Family, co-workers , and even close friends are seldom aware of the problems these patients are experiencing.

5.4 Somatisation indices

Somatisation disorders consist of various recurrent and often changing somatic symptoms such as gastro-intestinal, heart and lung, neurological, urological, sexual complaints, and others. Patients may report these symptoms wrongly and often exaggerate or dramatize them. They sometimes produce their own original theories of complex, multi-organ disease. PNES may be the only health problem, but more often than not they co-occur with other somatisations. Most patients with somatisation disorders have a history of frequent hospitalizations and medical tests whose results were negative. If there is comorbidity, it does not justify the range or intensity of complaints and depressed mood (ICD-10). Somatisation disorders are usually accompanied by exaggerated concern with one's health. Patients usually address very clear expectations to the medical staff. They want medical diagnosis that sounds professional and confirms their symptoms. When medical examination fails to find any organic foundation for their health complaints, patients usually conceal the fact that they have consulted a doctor or been to hospital. They often behave as if they were manipulating people and they make repeated attempts to take advantage of the medical personnel. Somatisation disorders can co-occur with anxiety, depression, and suicidal ideation or attempts. These patients are reluctant to accept the verdict that the nature of their problems is psychological. Despite the lack of confirmation of organic etiology, they are determined to have even more laboratory tests and demand even more medical consultations and hospitalizations. Somatisation disorders are chronic and are usually accompanied by maladjusted interpersonal, family and social functioning.

According to Dahlstrom and coworkers (1986), the MMPI can be used to assess somatisation factors. These researchers identified the following personality predictors of somatisation disorders using factor analysis: Hypochondriasis (Hs), Somatic Complaints (Hy4), Physical-Somatic complaints.

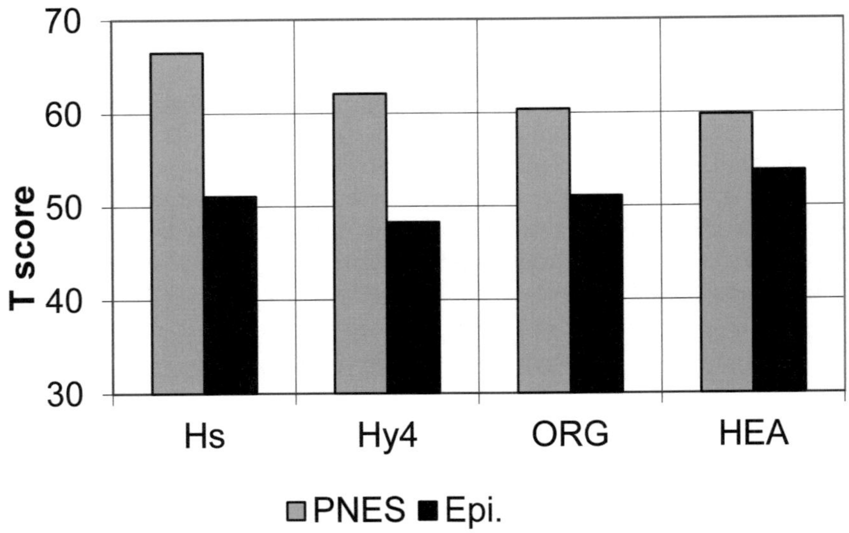

Fig. 7. Differences in somatisation indices in the studied groups: PNES - patients with psychogenic nonepileptic seizures; Epi. - patients with epilepsy; Hypochondriasis (Hs), Somatic Complaints (Hy4), Organic Symptoms (ORG), Poor Health – (HEA). Complaints (Si6), Organic Symptoms (ORG), and Poor Health (HEA).

Significant between-group differences were obtained for the following somatisation indices (mean scores): Hypochondriasis ($p < 0.001$), Somatic Complaints ($p < 0.001$), Organic Symptoms ($p < 0.015$) and Poor Health ($p < 0.05$). No significant differences were found for Physical-Somatic Complaints. These findings suggest that the existence of psychogenic nonepileptic seizures or the predisposition to such seizures is reflected in the values of the somatisation indices in the personality profile.

5.5 Discussion

The differences that we found in the studies reported above enabled us to identify several psychological variables that can be used to make a differential diagnosis between psychogenic nonepileptic seizures and epileptic seizures. The mean somatisation indices in the three studied groups differed greatly and these differences have a repetitive, statistically confirmed pattern. Patients with PNES had higher scores on all measured parameters than patients with epilepsy. The greatest difference was found for Hypochondria. The high Hs scores in the PNES group signify excessive concern with one's body and its functioning and the reporting of many somatic complaints. Paradoxically, patients with real somatic symptoms (epileptic seizures) had much lower Hs scores. As far as the three remaining indices of subjective somatic disorder are concerned, the largest difference was found for Somatic Complaints. The PNES patients scored highest on this scale which means that they report the highest intensity of neurological symptoms such as headaches, fainting, nausea, trembling, distorted vision etc. They also habitually resort to such defence mechanisms as repression and affect conversion. The differences on the two remaining scales, Poor Health (HEA) and Organic Symptoms (ORG) are also significant but they are less pronounced and therefore these scales are not so discriminating. The different groups are less distinct as far as these scales are concerned, probably because they measure symptoms, which are more related to the functioning of the organism as a whole, and to somatic complaints localized in the gastro-intestinal and urogenital systems.

In our study, patients with PNES had higher scores than patients with epilepsy on four out of five somatisation measures. This means that PNES is related to a higher tendency to view one's health problems as organic. The differences between the groups were only statistically nonsignificant for one parameter, Somatic Problems (Si6). This is a subscale of the Social Introversion (Si) scale and signifies concern with one's health and appearance. However, its value depends more heavily on constitutional factors than the values of the remaining indices that are more environmentally determined. This finding is consistent with earlier reports, which have drawn attention to the importance of family and environmental determinants of somatisation (Livingstone et al., 1995; Garralda, 1996; Devinsky, 1998; Owczarek, 2003b).

Wood and coworkers (1998) conducted a psychological study of families of patients with psychogenic nonepileptic seizures and found that these patients' family environments contained several factors inducing and reinforcing somatisation. Families of patients with PNES had significantly higher levels of poor adjustment to life and other psychological problems than families of patients with epilepsy. The psychological profiles of patients with PNES and their families are very similar and their dominant features are excessive criticism, hostility, and focus on health problems. Their preferred method of coping with anxiety, depression, and life failures is focusing on an illness - their own and/or their families'. This is the way of avoiding daily hassles and it is a convenient and socially accepted explanation and justification of disappointment and failure. In addition, other anomalies such as communication problems or unfair systems of reward and punishment in the family are ignored or suspended when a family member falls ill. Somatisation helps to bond family members. However, because its motives are not fully conscious, they are difficult to identify and because somatisation is so persistent and is often cross-generationally transmitted, therapy is extremely difficult.

More generally, elevated anxiety and somatisation indices mean that the problems of patients who demonstrate nonepileptic seizures are definitely neurotic disorders. The elevated personality indices in the MMPI unequivocally confirm that the personality profiles of these patients show signs of anxiety, psychological tension, and heightened defence against intense emotions. Failure of defence mechanisms leads to impaired functioning and overall poor social adjustment. The observed factors may also be responsible for frequent somatisations, including psychogenic nonepileptic seizures, as a way of reducing tension. In our study, we analysed average measures of existing personality dysfunctions in patients with psychogenic nonepileptic seizures. If only we could gain a better understanding of this dysfunction in each particular patient, this would help us to personalize psychological intervention. A therapist should always try to accentuate certain aspects of therapy and focus less on other aspects depending on the diagnosis of each patient's personality problems.

6. Treatment

When dealing with patients in whom PNES is suspected we must pay attention to such issues as adequate rapport, motivation to cooperate and adequate information concerning the need to conduct specific tests and examinations. The importance of the patient's attitude toward the physician/therapist cannot be overemphasized. An atmosphere of suspicion must be avoided at all costs. The patient should feel that he/she is undergoing a routine procedure to identify the cause and nature of his/her problems. The principle of trust building and honesty must be adhered to because this will have a profound effect on the success of the examination and further treatment. Once the diagnostic problem has been unequivocally resolved and the nonepileptic nature of the seizures has been established, we must deal with the problem of how to inform the patient about the nature of his/her disorder, because the way we tackle this problem will have an extremely important effect on the prognosis. Most patients tend to react with intense guilt. This guilt is exacerbated if the patient's environment blames him/her for deceiving it and accuses him/her of malingering. Another potential reaction is denial of the diagnosis and lack of confidence in the physician's competence. In this case the patient will have a natural tendency to try to convince the doctor of the epileptic nature of his/her symptoms and we must expect considerable intensification of premeditation, increased frequency of seizures and increased dramatisation of their consequences. In both cases, the situation may become very serious and the prognosis may be bad. We must therefore take several inevitable precautions and obey several well-tested clinical rules:

- do not blame the patient
- do not make light of his/her symptoms
- do not reinforce his/her tendency toward seizures
- listen rather than giving advice

It is extremely important that we obey all these rules consistently and simultaneously and that we practice patience and insightfulness in both diagnosis and therapy. Patients themselves often want to share their ideas concerning the source of their troubles and tell us about their daily hassles. This way we can gather information about the contingencies of the seizures. We must remember, however, that patients are unaware of the real causes of PNES. Meanwhile, a clinician must identify the mechanism of the disorder in each individual patient and usually needs to pick the relevant information out of the irrelevant background.

According to an American research (Ettinger et al., 1999) in the New York agglomeration, about one-half of patients with PNES are referred to psychiatrists, fewer than 20% to psychologists and the rest to other specialists. Treatment efficacy reports differ depending on the criteria of improvement, observation duration and treatment model.

According to the literature, satisfactory improvement (between 50 and 100% reduction of seizure frequency) is achieved in 34-76% of cases depending on the source (Ettinger et al., 1999; Irwin et al., 2000; Silva et al., 2001). We know from clinical experience that when treating patients with PNES it is important to convince the patient that he/she is capable of gaining control of previously uncontrollable seizures, which left him/her feeling helpless, and of preventing further seizures. In patients with PNES who feel their functioning is aimless and ineffective and their environment is hostile or blaming, one of the goals of therapy is to find positive and socially acceptable forms of emotional abreaction and tension reduction.

Uhlmann et al. (2011) analyzed the effectiveness of PNES therapies in various centres providing such therapies. The most frequent approach is cognitive-behavioural therapy, based on the bio-psycho-social model. According to this model, neurotic symptoms are stable reaction patterns triggered by a variety of life situations, which produce tension and anxiety. The clinician's goal is to extinguish inadequate reactions to stimuli and supplement them with new, adequate reactions and habits. Patients in therapy learn new ways of integrating incoming information and new responses and this enables them to adjust to reality better. Of course, when we explore emotions and behaviour of people with psychogenic nonepileptic seizures we find shortcomings in various areas relating to the development of this pathological form of adaptation. Therefore, common forms of psychotherapy are designed and several different objectives are addressed: to strengthen the personality defence mechanisms and coping strategies, to improve emotional resilience, and others. Therefore, given the variety of forms and methods of treatment guidelines, it is difficult to do well-conducted, prospective, randomized studies and to control psychotherapy outcomes. As the authors very aptly point out, "also it should be stressed that complete and immediate ending of seizures must not always be the first goal of the treatment". First, one must find the primary cause of dissociative seizures and concentrate on its treatment. Both psychodynamic and cognitive-behavioural treatments equip patients with several specific skills. If one does not take care of the primary cause, this disorder will transform into other pseudo-neurological, pseudo-cardiological, pseudo-sexuological, or others (Uhlmann et al., 2011).

The authors accurately present the stage of progress in explaining the phenomenon of psychogenic nonepileptic seizures. Their attention is directed to issues related to the effectiveness of therapy and they provide a comprehensive analysis of work in progress. There is nothing missing in their report. Methodological difficulties in constructing valid studies in this field are correctly assessed. On this basis, we can conclude that we still have a long way to go before we know how to treat dissociative seizures effectively and to achieve long-lasting effects.

7. Concluding remarks

It is now quite clear that individuals with PNES and individuals with epilepsy have different MMPI personality profiles. Nevertheless, the jury is still out on the aetiology, diagnosis, and treatment of PNES (Owczarek et al., 1995; Jędrzejczak et al., 1999; Rowan, 2000; Storzbach et al., 2000; Griffith et al., 2007). True, differences of opinion no longer concern the imponderables. Rather, it is now a question of distribution of accents. If we focus on empirical facts we need to point out the presence of the so-called conversion dip (Hy, Hs > D) in profiles of persons with PNES. These persons do not signal their psychological discomfort directly. They do so indirectly, in ways that are more symbolic. This is one of the reasons why PNES are usually so sudden and dramatic. The symptoms attract attention and compassion and PNES patients are hungry for both. They are unable to get them, abreact their amassed emotions, stresses and tensions at the energy level of behaviour, but manage to avoid other negative reinforcements and failures.

Fig. 8. Evolution of pathological adaptation of the original cause of the excessive need for emotional contact to the effects in the form of psychogenic nonepileptic seizures

An extensively developed need for emotional contacts is probably at the roots of the disorder. We have empirically confirmed this hypothesis: patients with PNES have elevated need for affection (Hy2 scores). They are "emotionally adhesive" for purely instrumental reasons. This superficiality facilitates emotional contacts with the environment in the short run and they seek such contacts consistently and indiscriminately. Time and time again our attention was drawn to the fact that these grown-up people came to the clinic carrying all sorts of cuddly toys, which helped them to express their emotionality. This is another of our findings. It looks as if one of the conditions sine qua non of PNES is poor frustration tolerance. Our PNES participants had low Frustration Tolerance Indices (FT). Low frustration tolerance leads directly to inability to cope effectively with psychological discomfort and the uncontrollable urge to achieve immediate problem resolution, both of which reinforce the tendency toward PNES recurrence. One of the consequences of this dysfunction is that persons with PNES suddenly switch their attention from their psychological problems with which they are unable to cope to "health problems", which they feel to be objective and independent and which a doctor, not they themselves, should resolve. However, patients with PNES do not really want to get better. Their psychogenic disorder is often accompanied by a peculiar "attachment" to their somatic complaints. In order to make therapy difficult and make recovery impossible, they negate the doctor's opinion and the validity of medical tests. They rarely comply with doctor's orders but they are often impatient and complain that treatment is not working. Even if we manage to convince them to comply with a prescribed therapy, their compliance may be short-lived. Meanwhile, they have numerous personality problems that need to be treated and this takes time and effort.

We found that patients with PNES also obtained high scores on somatisation indices. High Hypochondriasis scores mean concentration on one's health and especially in its complications. High scores on Somatic Complaints mean that patients report many different somatic symptoms. Paradoxically, patients with real somatic symptoms (epilepsy) obtained much lower scores on this index. A high Somatic Complaints score also means uncritical and excessive use of such defence mechanisms as affect conversion. The fact that patients with PNES also have depressed mood and actually feel extremely unhappy most of the time makes problems even worse. PNES patients with severe mood problems show similar, low levels of health-related quality of life (HRQOL) to patients with severe mood problems who have epilepsy (Szaflarski et al., 2003; Griffith et al., 2007). In our study patients with PNES had even higher levels of depression in the MMPI than patients with epilepsy.

The only somatisation index which was not elevated in patients with PNES was Physical-Somatic Complaints. This parameter is empirically related to Social Introversion and means the tendency to worry about one's health and appearance. This variable is more constitutionally determined than the other somatisation indices, which are more environmentally determined. This suggests that family and environmental factors are more responsible for somatisation than constitutional factors. Our study confirmed this observation. We found much higher levels of poor adjustment to life, failure and other psychological problems in the families of patients with PNES.

Psychological profiles of patients with PNES and their families were quite similar and dominated by excessive criticism, hostility, and concentration on health problems. Our findings suggest the presence of factors responsible for the development and consolidation of somatisation in the families of patients with PNES disorders. When a family member develops PNES, this provides a convenient and safe explanation for the family's problems. Concentration on an illness - one's own or a family member's - helps to cope with anxiety, depression and failure. PNES provides an escape from daily hassles and is a convenient and socially acceptable "explanation" for disappointment and failure. When a family member is ill, problems with communication, unfair reward and punishment systems, etc. are no longer important. Paradoxically, somatisation helps to bind the family together. Because the motives underlying PNES are unconscious, they are difficult to identify. What is more, they have often been operating for several generations and become a family tradition. Sometimes a doctor is the only one who wants the patient to get better, even the patient does not want to recover. No wonder that treatment of psychogenic nonepileptic seizures is such hard work, takes so long and seldom succeeds.

8. References

Bodde N.M., Bartelet D.C., Ploegmakers M., Lazeron R.H., Aldenkamp A.P., & Boon P.A. MMPI-II personality profiles of patients with psychogenic nonepileptic seizures. Epilepsy & Behavior, 2011, 20: 674-680.

Bodde N.M., Brooks J.L., Baker G.A., Boon P.A., Hendriksen J.G., Mulders O.G., & Aldenkamp A.P. Psychogenic non-epileptic seizures - definition, etiology, treatment and prognostic issues: a critical review. Seizure, 2009, 18: 543-553.

Dahlstrom W.G., Welsh G.S., & Dahlstrom L.E. MMPI Handbook. University of Minnesota Press, Minneapolis, 1986.

De Timary P., Fouchet P., Sylin M., Indriets J.P., de Barsy T., Lefebvre A., & van Rijckevorsel K. Non-epileptic seizures: delayed diagnosis in patients presenting with electroencephalographic (EEG) or clinical signs of epileptic seizures. Seizure, 2002, 11: 193-197.

Devinsky O. Nonepileptic psychogenic seizures: quagmires of pathophysiology, diagnosis, and treatment. Epilepsia, 1998, 39: 458-462.

Donofrio N., Perrine K., Alper K.R., & Devinsky O. Depression and anxiety in patients with non-epileptic seizures. In: J.R. Gates & A.J. Rowan (eds.). Non-epileptic seizures. Butterworth-Heinemann, Boston, 2000: 151-157.

Ettinger A.B., Devinsky O., Weisbrot D.B., Ramakrishna R.K., & Goyal A. A comprehensive profile of clinical, psychiatric, and psychosocial characteristics of patients with psychogenic nonepileptic seizures. Epilepsia, 1999, 40: 1292-1298.

Garralda M.E. Somatisation in children. Journal of Child Psychology & Psychiatry & Allied Disciplines 1996, 37:13-33.

Griffith M.N., Szaflarski J.P., Schefft B.K., Isaradisaikul D., Meckler J.M., McNally K.A., & Privitera M.D. Relationship between semiology of psychogenic nonepileptic seizures and the Minnesota Multiphasic Personality Inventory profile. Epilepsy & Behavior, 2007, 11: 105-111.

Hamilton J., Campos R., & Creed F. Anxiety, depression and management of medically unexplained symptoms in medical clinics. Journal of the Royal College of Physicians of London, 1996, 30(1): 18-20.

ICD-10 - International Statistical Classification of Diseases and Health Related Problems. Tenth Revision. World Health Organization, 1992.

Irwin K., Edwards M., & Robinson R. Psychogenic non-epileptic seizures: management and prognosis. Arch. Dis. Child., 2000, 82: 474-478.

Jędrzejczak J., Owczarek K.: Psychogenic pseudoepileptic seizures: Analysis of the clinical data in 1990-1997. Krankenhauspsychiatrie, 1999, 10 (1): 36-40.

Jędrzejczak J., Owczarek K., & Majkowski J. Psychogenic pseudoepileptic seizures: clinical and electroenephalogram (EEG) video-tape recordings. Seizure, 1999, 6: 473-479.

Livingston R., Witt A., & Smith G.R. Families who somatize. Dev. Behav. Psychiatr., 1995, 16: 42-46.

Mace C.J, Trimble M.R. Ten-year prognosis of conversion disorder. Brit J Psychiatr 1996, 169 (3), 282-8.

Mökleby K., Blomhoff S., Malt U. Fr., Dahlström A., Tauböll E., & Gjerstad L. Psychiatric comorbidity and hostility in patients with psychogenic nonepileptic seizures and healthy controls. Epilepsia, 43 (2): 193-198.

Owczarek K., Majkowski J., & Jędrzejczak J. Evaluation of interpersonal relations between epileptic patients during hospitalization. Epilepsia, 1995, 36 (suppl. 3): 190.

Owczarek K., Jędrzejczak J.: Patients with coexistence psychogenic pseudoepileptic and epileptic seizure: a psychologilal profile. Seizure- Eur. J. Neurol., 2001, 10 (8): 566-569.

Owczarek K. Anxiety as a differential factor in epileptic versus psychogenic pseudoepileptic seizures. Epilepsy Research, 2003a, 52: 227-232.

Owczarek K. Somatisation indexes as differential factors in psychogenic pseudoepileptic and epileptic seizures. Seizure - Eur. J. Neurol., 2003, 12 (3): 178-181.

Prueter Ch, Shultz-Venrath U., & Rimpau W. Dissociative and associated psychological symptoms in patients with epilepsy, pseudoseizures, and both seizure forms. Epilepsia, 2002, 43 (2): 188-192.

Reuber M., Fernandez G., Bauer J., Helmstaedter C., & Elger C.E. Diagnostic delay in psychogenic nonepileptic seizures. Neurology, 2002, 58: 493-495.

Rowan A.J. Diagnosis of non-epileptic seizures. In: J.R. Gates & A.J. Rowan (eds.). Non-epileptic seizures. Butterworth-Heinemann, Boston 2000: 15-30.

Roy A. & Barris M. Psychiatric concept in psychogenic non-epileptic seizures. In: A.J. Rowan & J.R. Gates (eds.). Non-epileptic seizures. Butterworth-Heinemann, Boston 1993, 143-151.

Silva W., Giagante B., Saizar R., et al. Clinical features and prognosis of nonepileptic seizures in a developing country. Epilepsia, 2001, 42: 398-401.

Storzbach D., Binder L.M., Salinsky M.C., Campbell B.R., & Mueller R.M. Improved prediction of nonepileptic seizures with combined MMPI and EEG measures. Epilepsia, 2000, 41: 1452-1456.

Swanson S.J, Springer J.A., Benbadis S.R., & Morris G.L. Cognitive and psychological functioning in patients with non-epileptic seizures. In: J.R Gates & A.J Rowan (eds.). Non-epileptic seizures. Butterworth-Heinemann, Boston 2000: 123-137.

Szaflarski J.P., Ficker D.M., Cahill W.T., & Privitera M.D. Four-year incidence of psychogenic nonepileptic seizures in adults in Hamilton County, OH. Neurology, 2000, 55: 1561-1563.

Szaflarski J.P., Szaflarski M., Hughes C., Ficker D.M., Cahill W.T., & Privitera M.D. Psychopathology and quality of life: psychogenic non-epileptic seizures versus epilepsy. Med. Sci. Monit., 2003, 9(4): 113-118.

Uhlmann C., Rösche J., Tschöke S., & Fröscher W. Empirical evidence for treatment of dissociative seizures: New studies and therapeutic recommendations. Epileptologia, 2011, 19 (2): 87-92.

Wood B.L., McDaniel S., Burchfiel K., & Erba G. Factors distinguishing families of patients with psychogenic seizures from families of patients with epilepsy. Epilepsia, 1983, 9 (4): 432-437.

11

Psychogenic Pseudoepileptic Seizures – From Ancient Time to the Present

Joanna Jędrzejczak[1,*] and Krzysztof Owczarek[2]
[1]Department of Neurology and Epileptology
Medical Centre for Postgraduate Education, Warsaw
[2]Department of Medical Psychology
Medical University, Warsaw
Poland

1. Introduction

Clinicians who work with patients with epilepsy are confronted with many diagnostic and therapeutic challenges when have to differentiate between epileptic and psychogenic nonepileptic seizures (PNES). At the end of the twentieth century, the introduction of electroencephalography (EEG) recording with simultaneous monitoring of patient behaviour helped to correct false positive and false negative diagnoses of the nature of convulsive conditions. This technological advancement sensitized physicians to the high incidence of patients with PNES receiving referrals to clinical centres specializing in the treatment of epilepsy. When PNES is erroneously diagnosed as epilepsy, patients are at risk of prolonged, unnecessary, and above all, ineffective treatment with antiepileptic drugs. These drugs do not reduce the number of psychogenic convulsive incidents. Moreover , ineffective treatment leads to frequent visits to outpatient clinics and hospitalizations. It also leads to frequent change of doctors, strategies and forms of treatments. All this increases the cost of erroneous diagnosis and inadequate treatment.

PNES are defined as "episodes of altered movement, sensation or experience similar to epilepsy, but casued by a psychological process and not associated with abnormal electrical discharges un the brain" (Reuber and Elger, 2003) In current diagnostic schemes PNES are categorized as a manifestation of dissociative or somatoform (conversion) disorder (ICD-10).

This mean that they are caused by unconscious, symbolically expressed psychological processes leading to conversion, i.e. the pressing need to interpret one's problems in ways which are both rationally and socially acceptable. This psychological mechanism has tangible gains, such as reduction of anxiety, and is a specific defence against the experience of other powerful and negative emotions. The external observer often feels that the patient is faking symptoms, unwittingly or even deliberately. He/she responds with irritation or even wants to blame the patient and make him/her feel guilty. The patient is not simulating, however. The misery associated with dysfunction is genuine and tangible. It is true,

* Corresponding Author

however, that patients with PNES do sometimes want to achieve certain goals and benefits which they could not naturally achieve by other means.

The role of cerebral organic factors in PNES is unclear. This type of seizures often coincides with mild brain injury. It has been reported for a long time that patients with nonepileptic seizures have more or less pronounced changes in their EEG recordings between seizures. They may even demonstrate typical epileptic EEG discharges. Changes in the brain structures have also been found using neuroimaging techniques. Both types of changes lead to the false diagnosis of epilepsy, especially when focal or generalized sharp waves, spikes or complexes of these elements and short waves are present. EEG changes have been variously interpreted in terms of cortical-subcortical connections or brain tissue hyper-reactivity, both of which can cause psychogenic disorders directly or indirectly. It therefore seems legitimate to assume that the predisposition to psychogenic seizures may be discussed in psychodynamic or psychosocial terms in lieu of brain dysfunction diagnosed earlier on the basis of EEG and neuroimaging. (Reuber et al., 2002, Jędrzejczak et al., 2004, Reuber, 2009, Auxéméry et al, 2010)

Treatment of PNES is equally controversial. There is no unified method of treatment as there is in epilepsy. Because PNES is so complex and multifaceted and is often accompanied by anxiety, depression or attention deficit, it is highly advisable that both psychiatrists and psychologists cooperate in the treatment. At present the diagnosis and treatment of PNES is largely in the hands of neurologists (epileptologists). It is currently thought that the doctor must first either rule out the epileptological nature of the disorder or diagnose the comorbidity of epilepsy and PNES and determine the proportion of both types of seizures. He/she can then begin to treat the PNES. Rational psychotherapy based on meticulous diagnosis and tailored to the specific patient can be effective.

2. Historical background

Psychogenic nonepileptic seizures have been traditionally associated with the concept of hysteria. However, reports of PNES are as old as reports of epilepsy. In various ancient cultures and civilizations the causes of various types of seizures were found . We also find accounts of various methods of prevention and treatment. Ancient written sources can roughly be divided into two groups: philosophical-medical and religious. The religious literature quotes and reflects the knowledge presented in philosophical and medical texts to various degrees. Historical sources often contain many superstitions and prejudices about epilepsy. Many people suffered not so much from the seizures themselves but from misunderstandings concerning their prevention and treatment. Today we can assert that the attitudes of whole societies toward epilepsy and the way patients who suffered from epileptic incidents were treated attested to the level of their social conventions and the progressiveness of their culture.

2.1 Accounts of epileptic seizures in the religious literature

In the Old Testament, all diseases are viewed as punishments for the sins of individuals or entire societies and as cautions against the violation of divine laws. Epilepsy was no exception except that it was called the "falling disease" in the original text. Yahve was the one who inflicted people with epileptic attacks and only he could alleviate their suffering. In chapter 21 of the First Book of Samuel we find the story of young David escaping from Saul to the Philistines. In order to avoid being taken hostage he simulates epileptic seizures whose symptoms, according to contemporary nomenclature, resembled tonic-clonic seizures: "So he pretended to be insane in their presence: and while he was in their hands he acted like a madman, making marks on the doors of the gate and letting saliva run down his beard" (I Sam 21:13). This is surely the oldest existing account of nonepileptic seizures.

Where there are nonepileptic seizures there must be a model which can be emulated. Without much further searching we find in the same book a description of the peculiar behaviour of Saul, king of Israel, who discards his clothes then falls to the ground and lies there all day and all night. This reminds us of an epileptic state probably preceded by a psychomotor seizure. These seizures were a punishment which God inflicted on Saul for his vile and disobedient acts. Saul repeatedly tried to kill David who was designated to be the next king of Israel. In those days seizures were thought to be symptoms of a mysterious malady but their specific form, course and circumstances were submitted to rational rules and interpretations in terms of cause, negative consequences and advantages.

In the fourth book of the Pentateuch we find the story of Balaam, a Midianite, who prophesized the coming of the Holy Spirit. He saw this prophesy in his mind's eye when he was stricken to the ground and learned the will of the God Most High. Inscriptions discovered in the Middle Jordan Valley and dated at 750 BCE attest to the activity of Balaam, son of Beor, in the days of the Middle Kingdom. Known for his talents, he was summoned by Balak, king of the Moabites, to curse the group of Israelites who settled in Moab upon leaving Egypt. Baalam was promised a handsome reward for his efforts. Three times did Balak make offerings to God in Balak's presence and three times did Yahve bless the Izraelites, enraging the king. In the biblical account of this event we read that Balaam fell to the ground during the blessings. The ancient Hebrew term for this falling is *nophel,* the same term that was used to describe Saul's seizures. The word evidently conveys the action of the spirit, be it odd or impure. This analysis of the stories of Saul, David and Balaam is testimony to the belief, which was widespread in ancient times but is also quite popular today, in the tragic inevitability of poor prognosis in patients, their inferiority, and interpretation of human seizures in terms of God's punishment for their evil deeds and vile life. Seizures are repeatedly described in the Pentateuch. From our contemporary perspective it is hard to say if these were epileptic seizures or conversion dissociative attacks. What is typical, however, is that they always occur in unusual circumstances, as the only solution to a seemingly impossible situation which cannot be resolved happily or even satisfactorily.

2.2 Middle ages

In ancient times Hippocrates and Aretaeus of Cappadocia left extremely accurate accounts of both types of seizures (Veith, 1965; Lennox & Lennox, 1960). Treatment varied in various cultures depending on the level of medical and philosophical knowledge and was the domain not only of medics but often also of shamans, exorcists and other religious hierarchs. Throughout the Middle Ages, but also partly in modern times, both epilepsy and hysteria were believed to be signs of possession by the devil but were thought nevertheless to be two distinct conditions. The concept of hysteria is rooted in the idea, already discernible in Egyptian medicine that functional disorders are caused by the wandering of the uterus (Ebbel, 1937). As it wandered to different parts of the body, the uterus caused a variety of symptoms. The Hippocrates (2005) school attributed hysterical disorders in women to sexual abstinence. Abstinence was thought to make the uterus "anxious" and to move to the upper parts of the body. If, for example, it reached the respiratory system, the patient had difficulty breathing and turned blue. Restricted breathing was a source of "anguish" and made the patient ill. Mainly spinsters and widows were prone to hysterical attacks and prompt marriage was thought to be the best cure. Fumigation of the uterus was another radical remedy. Hippocrates is also believed to have been the first writer to differentiate between genuine epilepsy and hysterical seizures. He recommended the following diagnostic procedure: press the abdomen energetically above the ovaries. If this evokes a seizure, hysteria shall be diagnosed. Like all his contemporaries, Hippocrates believed epilepsy to be a male disease and hysteria to be a typically female disease (Trillat, 1993). This belief is surprisingly consistent with contemporary observations that many more women than men have psychogenic nonepileptic seizures. (Jedrzejczak et al., 1998, Lesser, 1996, Selkirk et al., 2008). Roman medical authorities discarded the wandering uterus hypothesis but continued to associate hysteria with the female reproductive system. In the 17th century English physicians declared that hysterical attacks originated in the brain, not the abdomen, but they continued to endorse the idea that epileptic attacks and hysterical attacks were related and that they shared the same pathophysiology. Some of them argued that hysteria was caused by an excess of passion (Arts, 2001).

2.3 From the Middle Ages to the Enlightenment

In the Middle Ages hysteria, although not so well-understood as epilepsy, was extremely widespread. The supra-naturalistic climate is partly responsible for this. Demons, devils and other evil powers reigned in both theological and medical circles, affecting social ideas concerning hysterical phenomena. Fear of the devil of panic proportions and religious ecstasy were fertile soil for the development of morbid affective states and hysteria. Bourneville, a representative of the Charcot school, threw some light on this problem. In the last decade of the 19th century he published a series of monographs entitled *Collection Bourneville* or *Bibliothèque diabolique* containing many original 16th and 17th century documents. In this excellent work he compared historical examples of alleged witches and demoniacs with contemporary forms of hysteria pointing that in Middle Ages powerful Catholic church forced a spiritual interpretation of disease - servants of Satan tried to confuse the minds of people (Szumowski, 1994).

The term *demonopathia hysterica* was coined in the Middle Ages. According to magic-theologic theory, hysteria is the consequence of possession. It is a malady of the soul which manifests itself via somatic symptoms. Possessed women, stigmatized with behaviour which others could not understand, were submitted to cruel exorcisms and other bestial procedures, often widely condoned, which frequently maimed them or even killed them (Fig. 1).

Evidence of these murky times and the opinions and customs which affected the lame and the sick can be found in the Dominican textbook whose full title is *Malleus Maleficarum. Acts of witchcraft, and how to protect oneself against them, and remedies, all encompassed in two parts* (Sprenger & Instytor, 1992). This book was written by two inquisitors designated by Pope

Innocent VIII and sent to south Germany to purge those lands from heretics and practitioners of Satanism. The book contains unequivocal evidence that it was not difficult in those days to be accused of witchcraft or black magic. It was hard not to admit to such practices when one was being submitted to elaborate and cruel tortures. People who looked odd or were behaving violently were thought to be possessed by the devil or practicing witchcraft and were submitted to the water test. The suspect (usually female) was thrown into the river to see if she would drown or not. If she didn't, this meant that impure forces were keeping her afloat. If she sank, this meant that she had been too hastily and probably incorrectly accused. One way or another, women accused of witchcraft and submitted to such brutal practices lost their lives. Many centuries had to elapse before the Enlightenment dispersed the darkness of superstition, prejudice and stigma engulfing men and women suffering from psychogenic nonepileptic seizures.

Fig. 1. Hundreds of innocent people were burnt at the stake because of the inquisitors' fear of witchcraft and diabolic possession.

Other group incidences of hysterical behaviour were also observed and described in the Middle Ages. Particularly famous were the epidemics of mass dancing, called the dancing plague or tarantism, which overcame even several thousand people at a time (Fig. 2).

A barking epidemic broke out in France in 1609. In a small town 40 people barked like dogs during hysterical attacks (Szumowski, 1994). However, possession epidemics usually broke out in monasteries. The monastery epidemics in Loudon and Louviers in the 15th century were famous. Possession could also be sexually motivated. This was particularly obvious during the epidemic which broke out in a monastery in Auxonne. Mediaeval hysteria can clearly be divided into demonic hysteria and mystical hysteria. Demonic hysteria is probably more akin to what we typically call hysteria. But mystic hysteria often conveys the image of a person capable of sacrifice and suggestible. Many of these phenomena were viewed as supernatural and perhaps we should view them as meta-psycho-physiological phenomena.

Fig. 2. Peasants in the throes of hysteria manifested in mass dancing.

Wier, a distinguished 16th century physician and expert in diabolic phenomena, wrote that "the number of patients who say that they are possessed by the devil is infinite" (Szumowski, 1994). In the Middle Ages, hysterical women prone to suggestion guessed the exorcist's intentions and demonstrated various symptoms. They often had split personalities and gave the impression of being possessed by the devil. Their movements were lewd and lascivious, their language was obscene, and the devil was allegedly to blame. One nun spat out the Holy Communion straight in the priest's face. When in a state of hysterical frenzy, people (usually women) exhibited extraordinary physical strength. "One heroine of the madness epidemic in Auxonne, a woman of slight build, grasped a stone holy water stoup which normally could barely be lifted by two men, lifted it from its pedestal and threw it to the ground with the greatest ease" (Szumowski, 1994). The typical symptoms of hysteria include complete or partial dermal anaesthesia. Patients did not react to pricking with a needle or touching with a hot object. These changes in sensory sensitivity played an important role in mediaeval witchcraft trials because the sites of anaesthesia were thought to signify a pact with the devil.

During the Enlightenment many monographs extracting hysterical behaviour from the darkness of shamanism and superstition were written. Susceptibility to hysteria now began to be associated with personal disposition, temperament, or other genetically determined traits and hysterical attacks were attributed to traumatic life events. In the 17th century Sydenham drew attention to the similarity between high fever states and hysterical behaviour. He found that one-sixth of all patients who consulted a physician were hysterics (Pechy, 1772). In the 18th century Boerhaave coined the term "nervous disease" and it was he who introduced a simple albeit drastic diagnostic method, burning with a hot iron, during the epidemic of hysteria. Seizures which did not subside in response to this cruel treatment were diagnosed as epileptic (Sokołowski, 1950).

Pierre Briquet (1796-1881) published the first systematic review of hysteria, *Traité Clinique et Therapeutique de L'Hysterie,* in 1859. He presented three main types of seizures: spasmodic seizures, *syncopal attaca* and hysterical convulsions. Hysterical convulsions were the most frequent type. Their course and clinical manifestations resembled epileptic seizures most closely. Briquet also found that his patients were most familiar with epileptic seizures. According to his ideas - very modern at the time - hysterical symptoms were closely related to their causes which were largely emotional. Briquet's theory was based on the assumption that every painful and unpleasant event is registered in the brain's emotional centres and it is that part of the brain which reacts, resulting in external manifestations. This could help to differentiate between hysterical and epileptic attacks. In hysterical attacks "convulsions are an expression of feelings, passions or life events". In epilepsy "convulsions are a kind of tetanus contraction, quite unlike the movements which occur in physiological conditions" (Briquet, 1859). From our contemporary point of view this is an extremely simplified idea but it is worth pointing out that, unlike, nonepileptic seizures, epileptic attacks reflect the cortical motor representation. Also, automatisms reproduce everyday movements such as buttoning one's clothes. Today it is hard to accept that hysterical attacks are simply the direct outcomes of cerebral emotional states.

2.4 The 19th century – "grand hystérie"

It was not until the 19th and 20th centuries that hysteria began to be understood better. In 1853 Carter presented an extremely mature monograph of hysteria in which he suggested three etiological factors: disposition, circumstances and undisclosed causes (usually sexual passion) (Carter, 1853).

Most of the data on hysteria in those days came from France where Jean Martin Charcot and his students at La Salpêtriére clinic made hysteria the focus of their attention. The unfortunate notion of hysteroepilepsy was introduced in France at that time, suggesting a link between the two disorders. The term was used to describe the coexistence of epileptic and hysterical attacks in the same patient, epileptic attacks triggered by suggestion (what we now call psychogenic nonepileptic seizures) and finally hysteroform epileptic attacks considered to be an intermediate form between epilepsy and hysteria. According to Charcot, hysteria was a disorder of the nervous system, a "neurosis", like epilepsy and migraine. He believed that hysteria, just like the two other diseases, was the consequence of structural or functional damage. In hysteria, the damage is caused by the coexistence of inherent dispositions and an external trigger such as physical or emotional shock (Charcot & Marie, 1892). Richter, a student of Charcot's, clearly established that hysteria was a defined disease quite unrelated to either supernatural forces or epilepsy (Temkin, 1945). Charcot gave extremely detailed accounts of hysterical symptoms which he described, analyzed and photographed but unlike Freud, who also studied hysteria, he was not interested in his patients' life histories. Charcot distinguished two main forms of hysteria - with and without convulsions. Most varieties of hysteria have features which enable differentiation between hysterical and nonhysterical disorders. In hysterepilepsy the most reliable factor was allegedly the "tubal symptom", hypersensitivity in the vicinity of the fallopian tubes. Application of pressure to this area could trigger an attack of hysteroepilepsy but not an epileptic attack. This way, the history of medicine travelled full circle from Charcot to Hippocrates and back.

In 1880 Charcot and Paul Richer (1849-1933), one of his students and collaborators, began an extensive study of over 100 female patients with hysteroepilepsy. They published the results in the book *Études cliniques sur l'hystéro-épilepsie ou grand hystérie* in 1885. The book was richly illustrated by Paul Richer.

According to Charcot, grand (convulsive) seizures can occur in cases of severe hysteria. The patient has contractions, like the ones experienced by patients with epilepsy, then falls but usually does herself no harm, has facial grimaces and violent movements of the extremities, clenches her fist as if she were angry, screams, whines, laughs or weeps (Figs. 3, 4, 5).

One of the classical symptoms of the grand hysterical attack is the arching forward of the whole trunk (*opisthonus*) so that the body, when lying on the floor or the bed, rests only on the head and the heels (*arc de cercle*) and the abdomen protrudes upward (Fig. 6).

Fig. 3. The initial phase of a hysterical attach, Paul Richer, *Études cliniques sur l'hystéro-épilepsie ou grand hystérie* (Paris, 1885).

Fig. 4. The tonic phase of a hysterical attack, Paul Richer, *Études cliniques sur l'hystéro-épilepsie ou grand hystérie* (Paris, 1885).

Fig. 5. The motor phase of a hysterical attack, Paul Richer, *Études cliniques sur l'hystéro-épilepsie ou grand hystérie* (Paris, 1885).

If these seizures are frequent enough, the two ends of the arch, the head and the heels, can touch. The main characters of the Auxonne monastery madness even "walked" in this pose. The body poses and facial countenance sometimes express feelings such as surprise, religious ecstasy, anger or erotic passions (Charcot's *attitudes passionnelles*), and the body moves in harmony with these feelings (Figs. 7 and 8).

Fig. 6. The circular arc - "arc de cercle", Paul Richer, *Études cliniques sur l'hystéro-épilepsie ou grand hystérie* (Paris, 1885).

Fig. 7. Bodily poses and facial countenances expressing intense emotions - attitudes passionnelles, Paul Richer (*Études cliniques sur l'hystéro-épilepsie ou grand hystérie* (Paris, 1885).

Fig. 8. Bodily poses and facial countenances expressing intense emotions - attitudes passionnelles, Paul Richer (*Études cliniques sur l'hystéro-épilepsie ou grand hystérie* (Paris, 1885).

Richer also suggested that the attack of *grand hystérie* goes through four typical stages. In the first, epileptoid, stage the patient has a tonic seizure, usually preceded by an aura. She then begins to have acrobatic movements (*grands mouvements*) in which she assumes spectacular poses like the aforementioned *arc de cercle*. This second stage is followed by the expression of emotions by means of gestures and words (*attitudes passionnelles*). Finally in the fourth stage the attack ends with a kind of delirium. This fourth stage was often omitted in later publications (Arts, 2001).

The cases from Charcot's clinic reported in *Iconographie photographique se la Salpêtrière* are extremely instructive. The images captured in the clinic match the experiences of mediaeval hysterical women who claimed to have had sexual intercourse with the devil. The patient would lied on the bed in a cataleptic slumber and have, for example, sexual visions accompanied by appropriate words and movements. Following the attack the patient would fall into a deep sleep lasting even several days or weeks. This sleep resembled lethargy where breathing and heart function are so weak that the patient looked as if she were dead. The author also reports a case of authentic hysterical agony. On the other hand, some patients manifest enormous physical strength during a hysterical attack. One male patient from Charcot's clinic, barely seventeen years old, unscrewed the iron bed railings, pulled out an iron stair balustrade, lifted a whole heavy iron bedstead together with the bedclothes and threw it onto the neighbouring bed when in the throes of a hysterical attack (Szumowski, 1994).

Many writers tried to refute Charcot's ideas after his death. The Belgian physician Joseph Delboeuf and the Swedish physician Axel Munthe helped to discredit the *Salpêtrière* school. They argued that the patients who demonstrated hysterical attacks at Charcot's Tuesday lectures were coached by a team of doctors and that, in addition to the pleasure they gleaned from their demonstrations, they were rewarded for their convincing performances.

Babiński, himself once a member of the *Salpêtrière* school, also became a critic but his was a global criticism of the entire concept of hysteria (Arts, 2001). On the other hand, his work should be seen as the beginning of psychological interpretation of traumatic neurosis.

Conversion disorders were initially thought to be neurological because their symptoms resembled the symptoms of somatic diseases. Hysteria was called the great malingerer. Freud and Breuer (1896) argued that conversion disorders were psychogenic and were related to early childhood sexuality (Aleksandrowicz, 1998). Freud had the greatest influence on the understanding of hysterical disorders in the 20th century. His libido concept and his principles of psychoanalysis as a method of understanding and solving unconscious internal conflicts continue to be known and applied, with certain modifications, today. According to psychoanalytic assumptions, conversion disorders were symbolic expressions of repressed experiences. Contemporary psychoanalytic formulations do not limit the range of these experiences to the sphere of incestuous desires, however.

3. Contemporary conceptualizations of conversion and dissosiative disorders

3.1 Paroxysmal disorders

In the 19th and 20th centuries the concept of hysteria evolved within the concepts of conversion and dissociation. Earlier classifications qualified dissociative disorders as a hysterical *neurosis pithiatica* syndrome, i.e. the co-occurrence of conversion symptoms and specific personality traits (Aleksandrowicz, 1998). Modern psychiatry (particularly American) tends to separate the two disorders and view them as two distinct diseases. The DSM-IV places conversion in the somatoform disorder group and dissociation in a quite separate group, the dissociative disorder group (American Psychiatric Association, 1994). The DSM-IV has completely eschewed the category of neurotic disorders and classifies psychogenic non-epileptic seizures as conversion disorders. The ICD-10 is less explicit on this issue. In accordance with the European psychiatric perspective, neurotic disorders are included and classified as a subgroup of the larger class of dissociative (conversion) disorders. This group includes amnesia, fugue, ataxia, apraxia and anaesthesia. Meanwhile, depersonalization and derealisation, which can also imitate epileptic seizures and are usually associated with the concept of dissociation, are classified as "other neurotic disorders". Finally, psychogenic pain and vegetative disorders are classified as "somatoform disorders".

These confusions suggest, on the one hand, that it is difficult to diagnose patients with nonepileptic seizures and, on the other hand, that we need to identify the pathological mechanisms of these disorders in order to determine whether the same mechanism is leading to this or that symptom or whether the variety and abundance of symptoms is a reflection of the heterogeneity of the etiopathogenic context (Jędrzejczak, 2007).

To understand paroxysmal disorders (both epileptic and pseudoepileptic), we must understand both the physical and the psychological factors influencing each individual patient. This is particularly important when we want to come to grips with somatoform disorders which greatly contribute to the etiology of nonepileptic seizures. It is now thought that conversion disorders are psychogenic and that their onset strictly coincides with traumatic events, difficult or "unbearable" situations or dysfunctional interpersonal relations. Symptoms often reflect patients' fantasies concerning the course of somatic illness. Symptoms may be symbolically linked to the psychological cause. For example, the patient who saw terrifying scenes becomes blind. Pseudoepileptic seizures in conversion disorders are probably indirect (symbolic) manifestations of anger, fear, helplessness, or loss of control (Betts, 1997). In men seizures symbolizing helplessness can be found in victims of war trauma.

3.2 Affect conversion

The estimated prevalence of conversion disorders in general practice is from 15% to 30% of all hospital admissions (Folks et al., 1984; Wolańczyk et al., 1994, Jedrzejczak et all., 1999, Szaflarski et al., 2000, Angus -Leppan, 2008). Conversion disorders often give the impression that the patient wants to be ill, "is escaping into illness", and unconsciously or even consciously feigning illness. This impression sometimes causes irritation with the patient and the wish to blame the patient and make him/her feel guilty. This reaction is unjustified, however, because the suffering which accompanies the dysfunction is genuine. It is true, however, that patients do sometimes have their own personal agendas and seek certain benefits which they cannot achieve by other means.

The current classification of conversion and dissociative disorders is rooted in 19th century research on hysteria and defence mechanisms. Early psychological theories conceptualized these disorders as unconscious conversions into physical symptoms in order to resolve conflict or reduce anxiety. Stress caused the symptoms. In cases of acute reactions to stress this is probably correct. The trouble begins when, as is sometimes the case, patients benefit from a particular response to stress. If this behaviour is reinforced or rewarded it may be repeated and become chronic. However, it is not always easy to trace the direct and obvious relations between symptom onset (cause) and its indirect psychological consequences and therefore it is hard to reconstruct the psychological mechanism underlying the repetitive disorder.

3.3 Dissociative disorders

Dissociative disorders are processes leading to various clinical symptoms and these symptoms represent a broad range of problems confronting neurologists and psychiatrists.

ICD-10 (1992) gives the following definition: "what dissociative and conversion disorders have in common is partial or complete loss of normal integration between memories of the past, sense of identity, sensory sensations and control of body movements" which are normally under conscious control. In dissociative disorders conscious, selective control of these functions is reduced. The patient is no longer able to manage his/her behaviour at will or move his/her body in a controlled way. "Conversion" sets in, i.e. emotions are transformed into physical manifestations of unpleasant experiences. Dissociation is thought to be caused by stress and stressful life events with which the patient is unable to cope, and the most important consequence is loss of voluntary control of somatic functions which cannot be attributed to organ or system condition (Aleksandrowicz, 1998). Definitions of this group of neurotic syndromes tend to ignore the sphere of disordered experience, particularly unconscious experience. In the psychodynamic approach to neurotic systems, the natural functioning of the human psyche includes both spheres which are accessible to awareness and those which are unconscious and conceal unbearable problems, experiences and memories (Kokoszka, 1998). This approach gave rise to the concept of defence mechanisms, of which repression is the most important. Defence mechanisms help to keep these contents unconscious. If repression is not quite successful, other defence mechanisms such as intellectualisation, isolation or rationalisation are resorted to (Meissner, 1985). If they too are unsuccessful, neurotic symptoms develop. These symptoms serve defensive functions. They reduce conscious anxiety or other unbearable emotions ("primary gain") and the symptom is easier to accept than its cause (Kokoszka, 1988). The symptom symbolizes the concealed repressed content. Patients also glean secondary gains from their neurotic symptoms when the environment responds sympathetically and reduces the misery caused by the symptoms.

3.4 Difficulty classifying paroxysmal disorders

The DSM-IV also includes dissociative disorders in its classification system but distinguishes them from conversion disorders and defines them (e.g. fugue, depersonalization) as rupture of the normally integrated functions of consciousness, identity and environmental perception. It defines conversion disorders as symptoms or deficits concerning any motor functions and sensory functions, suggesting neurological or other health disorders. Conversion is understood as a symptom, not a disease, and it can have various causes.

Most of the dissociative disorders specified in separate ICD-10 categories are syndromes whose occurrence is associated with a single psychological trauma and in which there is one dominant symptom, often with sudden onset, which usually recedes within several weeks or months. These disorders become chronic when there are underlying insolvable problems. The most important dissociative (conversion) disorders are:

- dissociative amnesia
- dissociative fugue
- dissociative stupor
- dissociative motor disorders
- dissociative convulsions
- dissociative anaesthesia and loss (disturbance) of sensation.

Psychogenic nonepileptic seizures are behaviour disorders which resemble epileptic attacks in which the focal or generalized neuronal discharges which are typical for epilepsy are absent. According to the ICD-10 classification system, dissociative (conversion) disorders also include dissociative stupor and dissociative convulsions whose most frequent manifestations can be erroneously confused with epileptic attacks. In dissociative stupor the main symptom is psychomotor inhibition - loss or serious reduction of voluntary movements, lack of response to external stimuli such as noise, voice or touch, loss of contact. Patients usually recline, immobile and speechless. In dissociative convulsions consciousness is fully retained or altered like in stupor. Paroxysmal attacks may resemble the "hysterical attacks" so often described in the 19th century. Dissociative stupor and dissociative convulsions are the forms most frequently confused with epileptic attacks and that is why they are called psychogenic pseudoepileptic seizures.

A review of the literature shows quite clearly that the physician's behaviour and expectations usually modified or intensified the phenomena they described. Charcot and Freud's work illustrates this very well. One extremely interesting question is why non-epileptic seizures were so popular in times when epilepsy was much more stigmatised than it is today and there was much greater risk of detention in the uncomfortable psychiatric hospital. From our contemporary perspective, perhaps non-epileptic seizures symbolized problems with which patients could not possibly cope?

It is still unclear what causes various types of disorders. Many symptoms found in psychogenic nonepileptic seizures can also be found in the course of dissociative (conversion) disorders. These disorders do not have a homogenous clinical picture. On the contrary, their symptomatology is typically heterogeneous (Jędrzejczak, 2007).

Many questions and doubts come to mind.

Why do some patients with conversion disorders have fugues or dissociative amnesia, for example, whereas others have PNES? And pursuing this query further, what is the difference (if there is any difference) between patients who develop dissociative stupor and patients who develop dissociative convulsions? The observation that conversion disorders assumed many different forms throughout the ages further confirms that this is a mutable and heterogeneous phenomenon. Also, not everyone who is stressed develops conversion disorders. Whether they do or not probably depends on many factors such as culture, socioeconomic circumstances, level of education, life experience, sex, etc. Other important factors are personality and habitual coping mechanisms. Finally we must also consider the contribution of the history of mental and somatic illness. It is worth pointing out that culture does not have a major effect on the expression of psychogenic pseudoepileptic seizures. Although the course of PNES varies greatly, similar patterns can be found worldwide. The semiology of PNES is similar in western countries, Taiwan, Lebanon and India (De Paola, 2003; Lai and Yu, 2003; Riachi, 2003; Kriszanmorrthy et al., 2003).

PNES occur with a variety of psychiatric diagnoses, with a variety of psychic risk factors but without a uniform psychological/psychiatric profile. Reuber at al (2004) found distinct types of maladaptive personality disorders as the psychiatric basis for PNES. However, most experts assume that no single mechanisms or even contributing factors has been identified that is sufficient to explain PNES in all patients. PNES occur in a heterogeneous patient population and in contrast to epileptic seizures no commonly accepted classification of PNES exists. It is important to find PNES subtypes that differ with respect to etiology. The first step is to use objective instruments to study clinical semiology of PNES. Using quantitative cluster analysis based on 125 patients with PNES (Jędrzejczak 2007) developed a classification system with four different types of PNES. All types have a characteristic pattern, i.e., motor symptoms intensify progressively starting with immobility (PS4), through simple motor symptomatology (PS2), trembling of the body (PS3) and ending with complex, sudden motor symptoms (PS1). Gröppel et al. (2000) presented such an analysis but it was based on a very small sample (29 patients). They identified three clusters: seizures with a clearly expressed motor component, seizures with minor motor symptoms and trembling and a third cluster of atonic seizures, which they defined as falling down without motor manifestations (only one patient in Gröppel et al.'s sample had this symptom). As far as diagnostics are concerned, this distinction helps to identify differential criteria for psychogenic pseudoepileptic seizures and epileptic seizures. It also provides a new perspective on prognosis. If we manage to distinguish homogeneous groups of patients with psychogenic pseudoepileptic seizures, this may be a milestone on the road to developing a more detailed description of such patients, particularly from the point of view of etiology and the pathophysiology of psychogenic seizures.

Models explaining the pathogenic mechanisms leading to these events are limited. In paper of Baslet (2011) presented is a hypothetical pathophysiological model suggesting an alteration in the influence and connection of brain areas involved in emotion processing onto other brain areas responsible for sensorimotor and cognitive processes. Integrating this information, PNES are conceptualized as brief episodes facilitated by an unstable cognitive-emotional attention system. (Baslet 2011). According to Reuber (2009) PNES should be understood based on multifactorial etiologic model. He differentiates predisposing factors (abuse, neglect, interpersonal problems) means increase vulnerability to development of PNES. The next group covers precipitating factors (stressful experience, physical or mental health symptoms) which can occur days or months before onset of PNES and casue PNES to start. Perpetuating factors that inhibit ability to gain control over seizures. Anxiety was identified as relevant perpetuating factor in more than 50% of patients with PNES (Reuber, 2009).

What we want to know is whether the observed heterogeneity of clinical symptoms is caused by one and the same etiology or whether each clinical manifestation has its own specific pathophysiology. Our present state of knowledge does not allow us to answer this question unequivocally.

4. References

Aleksandrowicz J.: Zaburzenia nerwicowe [Neurotic disorders]. PZWL, Warszawa 1998.

American Psychiatric Association. Diagnostic and Statistical Manual of Mental Disorders. Fourth edition. American Psychiatric Association, Washington, DC, 1994.

Angus-Leppan H.: Diagnosing epilepsy in neurologic clinics: a prospective study. Seizure, 2008, 17: 431-436.

Arts N.: after Jackson A.: The borderland of epilepsy. In: Arts N. (ed.), Epilepsy through the ages. An anthology of classic writings on epilepsy. Van Ziuden Communication B.V., Alphen van den Rijn, 2001, 289-293.

Auxéméry Y, Hubsch C, Fidelle G. : Psychogenic non epileptic seizures: a review.Encephale. 2011, 2:153-158.

Baslet G.; Psychogenic non-epileptic seizures: a model of their pathogenic mechanism. Seizure. 2011, 20:1-13.

Betts T.: Conversion disorder. W. Engel J. Jr. & Pedley T.A. (eds.): Epilepsy: A comprehensive textbook. Lippincot-Raven Publishers, Philadelphia 1997: 2757-2766.

Briquet P.: Traité clinique et thérapeutique de l'hystérie. Paris 1859.

Carter R.B.: On the pathology and treatment of hysteria. John Churchill, London 1853.

Charcot J.M. & Marie P.: Hysteria mainly hystero-epilepsy. W: Tuke D.H.: A Dictionary of Psychological Medicine, 1892.

De Paola L.: Non epileptic seizures in Brazil: Commonalities and Unique cultural expressions. Epilepsia, 2003, 44 suppl. 8: 16

Ebbel B.: The Papyrus Ebers, the greatest Egyptian medical document. Levin i Munksgaaard, Copenhagen 1937.

Folks D., Ford C., Regan W.: Conversion symptoms in a general hospital. Psychosomatics, 1984, 25: 285-295.

Gröppel, G., Kapitany, T, Baumgartner C. Cluster analysis of clinical seizure semiology of psychogenic nonepileptic seizures. Epilepsia 2000, 41: 610-614.

Hippocrates: The Works of Hippocrates. Department of Neurology. University of Athens. Hellas Publications INC, Illinois 2005.

ICD-10 - International Statistical Classifcation of Diseases and Health Related Problems. Tenth Revision. Word Health Organization, 1992.

Jędrzejczak J., Owczarek K., and majkowski J.: Psychogenic pseudoepileptic seizures: clinical and electroencephalogram (EEG) video-tape recordings. Europ. J.Neurol., 1999, 6: 473-479.

Jędrzejczak J, Grabowska - Grzyb A, Królicki L.: Value of SPECT In differential diagnosis of psychogenic pseudoepileptic and epileptic seizures: preliminary results. 6 th European Congress on Epileptology, Vienna 30 May-3 June 2004, Eplepsia 45, suppl.3: 165.

Jędrzejczak Joanna:. Cluster analysis of clinical semiology of psychogenic pseudoepileptic seizures.Epileptology, 2007,15:13-28.

Kokoszka A.: Zaburzenia nerwicowe [Neurotic disorders]. Medycyna Praktyczna, Kraków 1998.

Krishnamorrthy ES., Srinivasan TS.: Non epileptic seizures in India: Commonalities and Unique cultural expressions. Epilepsia, 2003, 44 suppl. 8: 17.

Lai CW., Yu H.Y.: Non epileptic seizures in Taiwan: Commonalities and unique cultural expressions. Epilepsia, 2003, 44 suppl. 8: 16.

Lennox W.G., Lennox M.M.A.: Epilepsy and related disorders. Vol. 1. Little Brown & Co, Boston, 1960.

Lesser RP.: Psychogenic seizures. Neurology, 1996, 46: 1499-1507

Meissner W.W.: Theories of personality and psychopathology: classical psychoanalysis. In: Kaplan H.I. & Sadock B.J. (eds.): Comprehensive Textbook of Psychiatry. Williams and Wilkins, Baltiomore, 1985: 337-418.

Pechy J.: The Whole Works of Dr Thomas Sydenham. Wellington, London 1722.

Riachi N.: Non epileptic seizures in Lebanon: Commonalities and unique cultural expressions. Epilepsia, 2003, 44 suppl. 8: 16.

Reuber M., Fernández G., Helmstaedter C. et al. Evidence of brain abnormality in patients with psychogenic nonepileptic seizures. Epilepsy Behav. 2002, 3:246-248.

Reuber M., Fernández G., Bauer J. et al.: Interictal EEG abnormalities in patients with psychogenic non-epileptic seizures. Epilepsia 2002,43: 1013-1020.

Reuber M., and Elger CE.:Psychogenic nonepileptic seizures:review and update. Epilepsy Bahav. 2003, R: 205-216

Reuber M.:The etiology of psychogenic non-epileptic seizures :toward a biopsychosocial model. Neurol. Clin 2009, 27:84-86.

Selkrik M., Duncan R., Oto M., et al.: Clinical differences between patients with nonepileptic seizures who report antecedent sexual abuse and those who do not. Epilepsia 2008, 49: 1446-1450.

Sokołowski S. Święta choroba, czyli historia padaczki [The sacred disease or the history of epilepsy]. Problemy, 1950, 7:473-478.

Sprenger J, Instytor H.: Malleus Maleficarum. Acts of witchcraft, and how to protect oneself against them, and remedies, all encompassed in two parts. Szymon Kempini, Kraków 1614.

Szaflarski JP., Ficker DM., Cahill WT. et al.: Four -year incidence of psychogenic nonepileptic seizures in adults in Hamilton County, OH. Neurology, 2000, 55: 1561-1563.

Szumowski W.: Historia medycyny filozoficznie ujęta [A history of medicine from a philosophical perspective]. Sanmedia, Warszawa 1994.

Temkin O.: The falling sickness. The Johns Hopkins Press, Baltimore 1945.

Trillat E.: Historia histerii [The history of histeria]. Zakład Narod. Im. Ossolińskich, Wrocław-Warszawa-Kraków 1993.

Veith I.: Hysteria: The history of a disease, University of Chicago Press, Chicago 1965.

Wolańczyk T, Jędrzejczak J, Owczarek K.: Patients with psychogenic pseudoepileptic seizures. Epileptologia, 1994, 2:11-24.

12

Children with Cerebral Palsy and Epilepsy

Emira Švraka
University of Sarajevo, Faculty of Health Studies
Bosnia and Herzegovina

1. Introduction

The mental health reform in Bosnia and Herzegovina came into reality after the war (1992-1995) and the Dayton agreement in 1996. An important part has been the creation of a network of community mental health centers (CMHCs) all over the two entities of Bosnia and Herzegovina. They are all organized as part of the primary health care system. There was a lack of clear legislative and policy framework, job descriptions and funding as well as management. The personnel had just started to work and had many ideas how to improve the service. The users were basically satisfied but not quite aware of what service they could expect. Co-operation with other service units was poor (Lagerkvist, Maglajlić, Puratić, Suši & Jacobsson, 2003).

Families who have children with disabilities and live in poverty are truly in a double-bind. The same poverty-related factors that place their children at higher risk for disabilities also serve as barriers to assess services for their children and themselves. Early childhood practitioners can play a critical role in supporting families by providing services to overcome these obstacles and by working in partnership with specialized early intervention programs to assure that families and children receive needed services (Peterson, Mayer, Summers & Luze, 2010).

Poverty-stricken families, as well as regression in the structure and services caused by war in Bosnia and Herzegovina, and education system which anticipates "special schools" for children with intellectual disabilities, contributed to isolation of children with disabilities into anonymity of their homes, and especially anonymity of the rural areas. Those children are in the families which often lack financial resources for transportation to special schools or rehabilitation institutions.

Before the 1992, medical rehabilitation in Bosnia and Herzegovina had been provided at the level of institutions, usually after the hospital or ambulant treatments. As a concept of rehabilitation, community-based rehabilitation (CBR) was included in the strategic plan of reform of health care in Bosnia and Herzegovina. CBR is strategy for rehabilitation, equal possibilities and social integration of all persons with disabilities.

Priority problems for the families of children and adolescents with intellectual disabilities in Bosnia and Herzegovina are:

- lack of Register for developmental and intellectual disabilities,
- lack of continuous preventive measures for disability manifestation,

- lack of continuous education for the professionals in multiprofessional team for work with families of children with intellectual and development disabilities,
- lack of continuous education of teachers in general schools for work with children in program of educational inclusion,
- lack of adequate and continuous multidisciplinary community support for all education levels for children and adolescents with intellectual disabilities,
- lack of adequate programs for profession selection and employment, as well as accompanying legislation, and,
- high cost of habilitation-education programs for children and adolescents with intellectual disabilities and their insufficient financing.

The aims of the study "The influence of prenatal etiological factors on learning disabilities of children and adolescents with cerebral palsy" were:

- to identify the prenatal etiological factors of cerebral palsy in children and adolescents aged from 6 to 20 years in the Canton of Sarajevo, Bosnia and Herzegovina;
- to determine the learning disabilities among children with the cerebral palsy, and
- to determine relationship between prenatal etiological factors of cerebral palsy and learning disabilities of children and adolescents with cerebral palsy.

Importance of this study is reflected in the fact that this is the first time in Bosnia and Herzegovina that the influence of the prenatal etiological factors on learning difficulties of the children and adolescents with the cerebral palsy has been determined with use of a scientific method.

Knowing the cause of the condition is fundamental to understanding its prevention, its potential complications, and prospective treatment strategies. Thus, there is a great need for better communication and cooperation between all professional dedicated to the care of individuals with intellectual and developmental disabilities (Percy, 2007).

1.1 The influence of prenatal etiological factors on learning disabilities of children and adolescents with cerebral palsy

The study of the influence of prenatal etiological factors on learning disabilities of children and adolescents with cerebral palsy in the Canton of Sarajevo was conducted as a cohort, retrospective study. The research was conducted in homes of 67 (83,75%) participants, and in Center for education and re/habilitation of children with intellectual disabilities, cerebral palsy and autism, "Vladimir Nazor", for 13 (16,25%) participants. It was necessary to make home visits, especially because of the children who were staying at home, without an institution based re/habilitation or any other form of education.

Participants were members of the Association of persons with cerebral palsy in the Canton of Sarajevo. The Association includes 315 members. Of that number, 123 (39,05%) are children and adolescents, age 4 up to 20 years, and 192 (60,95%) are adults. There are only three Associations of persons with cerebral palsy In Federation of Bosnia and Herzegovina, (towns: Sarajevo, Goražde and Zenica). Cerebral Palsy Association of Federation of Bosnia and Herzegovina is established at 17. October of 2011. That day was announced for Day of persons with cerebral palsy of Federation of Bosnia and Herzegovina.

The sample was consisted of 80 participants, children and adolescents with cerebral palsy in the Canton of Sarajevo, age from 6 up to 20 years; 25 children (age 6-11), and 75 adolescents (age 12-20). Mean age was 13,94 years, 47 male (58,75%) and 33 (41,25%) female. The sample was divided in two subgroups, first includes 30 participants whose mothers had problems during the pregnancy, and second includes 50 participants whose mothers didn't have problems during the pregnancy.

The research data were collected using a Questionnaire for the parents of children and adolescents with cerebral palsy, which was developed by the investigator, based on professional and scientific literature, and personal experience. The Questionnaire consists of 69 questions. The first 8 questions are general sociodemographic characteristics of the family. Questions 9-19 are about pregnancy control and problems during pregnancy; 20-23 questions are directed at determining if there were still-born children. Fourth group of questions (25-35) are about delivery. 36-42 questions are directed at determining type of cerebral palsy, motor development and physical and surgical therapy. 43-56 questions are about intellectual and sensor disabilities; 57-67 questions are directed at determining education and academic success. The last two questions are about family's membership in Association of person with cerebral palsy in the Canton of Sarajevo.

Diagnosis of cerebral palsy has been made before this Study by pediatric neurologist at the Pediatric Clinic in Sarajevo, for each child. Specialist of physical medicine and rehabilitation, the author of the Questionnaire, examined all children (neurological status and observation at home or Center "Vladimir Nazor") depending on the history and clinical findings. After that, almost all children had individual home physical therapy: individual program of exercises and manual massage, with occupational therapy. Cognitive development was assessed by a cognitive test of psychologist.

The study was approved by director of the Center "Vladimir Nazor", and president of the Association of persons with cerebral palsy in the Canton of Sarajevo. Before starting the data collection, the research aim and Questionnaire were explained to parents and they agree to participate by signing consent.

Of total sample, 67 (83,75%) mothers were interweaved, 5 (6,25%) fathers, one (1,25%) aunt, and, in 7 cases (8,75%) both parents participated. Of 80 families, 64 (80%) were complete, and 16 (20%) were one parent families: 15 mothers and one father.

2. Epilepsy in cerebral palsy

Epilepsy is one of the most common neurological disorders in childhood. The risk of epilepsy is highest in patients with an associated brain abnormality, such as intellectual disability and cerebral palsy (symptomatic epilepsy). The average annual rate of new cases per year (incidence) of epilepsy is approximately 5-7 cases per 10000 children from birth to age 15 year. Prevalence studies in childhood epilepsy have been carried out in different geographical areas, age groups, and ethnic groups, with different design and methods. Despite these differences, it is possible to rate the prevalence of epilepsy in children as 4-5/1000. According to several population-based studies, this prevalence tends to increase from 2-3/1000 at age 7 to 4-6/1000 at age 11-15. A wide variety of seizure disorders are included under the term epilepsy. Approximately two thirds of cases were considered idiopathic. Children with additional health problems were more likely to continue to have seizures in early adult life than those with epilepsy alone (Beghi M, Cornaggia, Frigeni & Beghi E, 2006).

If the seizures are responsive to medication – as they are in approximately 60% of cases – the individual may never have another seizure. He or she will have to take medication regularly, but epilepsy will note prevent a full productive life. If the seizures are only partly responsive to medications - as they are in approximately 20% of cases – the individual will continue to have some attacks. If the attacks occur in public, he or she may experience negative consequences. Moreover, if an individual has even one seizure per year, he or she may never be able to legally drive again. Partially resistant seizures, therefore, have a clear, negative effect on life. Still, many people with partially resistant seizures also live productive, successful lives. If seizures are fully drug resistant – as they are in approximately 20% of cases – the individual will continue to experience seizures, which may be frequent, despite use of the best medications available. Drug-resistant seizures are called intractable seizures or refractory seizures. Intractable epilepsy is clearly a disability, sometimes called the "invisible disability" (Burnham, 2007.)

Younger age at onset is strongly associated with symptomatic causes and epileptic encephalopathy, both associated with intellectual disability (Berg, Langfitt, Testa, Levy, DiMario, Westerveld & Kulas, 2007).

Epilepsy, and particularly intractable epilepsy, is often associated with cognitive impairment. As noted, severe impairments are seen in children with West and Lennox-Gastaut syndromes; however, cognitive impairment may be associated with many forms of intractable seizures. In children, this often becomes evident during the school years. Children with intractable seizures have lower IQ scores, often in the "low-normal" range of 80-85. Studies have also found a significant correlation between low IQ scores and the longer duration of the child's seizure disorder (Burnham, 2007).

Epilepsy is common and the frequency increases as IQ decreases. Seizures are more frequent and status epilepticus occurs more often. When epilepsy and learning disabilities coexist, there is a much higher chance of challenging behavior, psychiatric disorders and cerebral palsy (Courtman and Mumby, 2008).

One cause of learning problems is frequent absence seizures. These mild, non-convulsive attacks consist only of brief lapses of consciousness. Some children have hundreds per day, however, and clusters of dozen may occur within a few minutes. During these periods, children cannot follow what is going on around them. Children with absence epilepsy, therefore, may give the appearance of being "slow learners" when their intelligence may very well be normal or even higher (Burnham, 2007).

Severe seizures, such as tonic-clonic attacks, cause a major perturbation in the brain. The after effects of such seizures last for hours. Children who have had one or more seizures during the night may show excessive fatigue during the following day and may appear to have forgotten things learned the day before (Burnham, 2007).

Some children, particularly those with complex partial seizures, have isolated epileptic spikes in their EEG between seizures. These are called interictal spikes. Interictal spikes produce no outward manifestation, but they slow the children's ability to process and retrieve information, causing transient cognitive impairments (Burnham, 2007).

Children with an epileptic focus in particular parts of the brain may show selective deficits related to that area. Children with a focus in the left hemisphere (dominant for language), for example, often have trouble with finding and remembering words. Children with a focus in the right hemisphere may have problems with visual memory (Burnham, 2007).

In addition to the cognitive impairments associated with the sedative (sleep-inducing) side effects of anti-epileptic drugs (AED). These side effects are most serious at the start of the therapy. They improve as tolerance develops, but they do not entirely disappear. They are worst with the older drugs, such as Phenobarbital, but they may be seen with almost any of the AEDs. They are more of a problem in children taking multiple drugs (Burnham, 2007).

Depending on the study, up to 75% of young patients do not take their AEDs as prescribed, and are at increased risk of the adverse effects of seizures, including physical injury, learning disabilities, psychosocial problems, poor self-esteem, damage to future life prospects, and death (Wilmont-Lee, 2008).

Cerebral palsy (CP) occurs at present in about 2,2 per 1000 live born children in Sweden. Epilepsy occurs in 15% to more than 60% of children with CP, depending on the type of CP and the origin of the series, compared with 0,5% in the general population. Epileptic seizures associated with brain damage are generally difficult to control. About half the individuals with epilepsy and a neurodeficit can be successfully treated with AEDs in the long term. A good outcome (seizure free ≥ 1 year) has been reported in 38% to 67% of children with CP and epilepsy. Even if the prognosis in this group is relatively poor, as far as complete remission is concerned, it is noteworthy that in our study 38% (21 of 55) had been seizure-free for one year or more. Nine of these children were without AEDs and were still seizure-free after epilepsy surgery and has remained so for 5 years 6 months. Children with CP due to CNS malformation, CNS infection, and gray matter damage had significantly less chance of a good seizure outcome than those with CP due to white matter damage or with unknown etiology. For all groups the chance of becoming seizure-free increased over time. Further studies are required to describe the long-term follow-up in children with CP and epilepsy (Carlsson, Hagberg & Olsson, 2003).

Prevalence of epilepsy in persons with CP varies with the type of motor impairment. It is most common among persons with hemi-and quadriplegic CP. Children with quadriplegic CP tend to have an earlier onset with other types of CP. Epilepsy is present in 79,5% children with severe disability. Of children with quadriplegic CP and severe intellectual disability, 94% have epilepsy. All types of epilepsy occur; but generalized and partial epilepsy are the predominant types. Children with quadriplegic CP are more likely to have generalized epilepsy, and more than half of them require two or more anti-epileptic drugs. In children with hemiplegic CP the predominant type is localization related epilepsy (83%). The frequency of seizures often decreases after age 16 (Odding, Roebroeck & Stam, 2006).

The proportion of patients with spastic quadriplegic CP and epilepsy ranges from 38 to 56,5%. Children with CP and epilepsy also had higher rates of other comorbidities including gross motor dysfunction and intellectual disability. Epilepsy frequently begins in the first year of life in patients with spastic quadriplegic CP (Venkateswaran & Shevell, 2008).

Epilepsy is a common disorder among children with CP. Of all children (n= 127) included in the Dutch population based study, 18,9 had active epilepsy at the time of examination, and a further 21,3% had a history of epilepsy. Of the children with quadriplegic CP 44,8% never had epilepsy, compared with 66,7% of the children with spastic diplegia, triplegia and hemiplegia, and 37,5% of the children with ataxia and dyskinesia (Wichers, Odding, Stam & Nieuwenhuizen, 2005).

CP type	With epilepsy		Without epilepsy		Total
	Male	Female	Male	Female	
Spastic Quadriplegic CP	8	5	9	10	32
Spastic Quadriplegic CP mixta	3	0	0	0	3
Triplegia	1	0	2	2	5
Spastic diplegia	3	2	6	5	16
Spastic Hemiplegic CP (right)	4	1	2	3	10
Spastic Hemiplegic CP (left)	2	3	3	1	9
Dyscinetic CP	0	0	2	0	2
Ataxia	0	1	2	0	3
	21	12	26	21	
Total	33 (41,25%)		47 (58,75%)		80

Table 1. Frequency of epilepsy in 80 children with CP by gender

Of 33 children with cerebral palsy and epilepsy, 21 (63,6%) are male, and 12 (36,4%) are female.

Of 47 children with cerebral palsy, and without epilepsy, 26 (55,32%) are male and 21 (46,68%) are female.

As children grow, the comorbidities of epilepsy grow with them. The focus of trouble simply shifts from the school to the workplace. Cognitive deficits, once acquired, are likely to remain. The after effects of seizures still occur in adults, as do the sedative effects of AEDs (Burnham, 2007).

2.1 Cerebral palsy

Cerebral palsy (CP) is an umbrella term that defines a group of non-progressive, but often changing, syndromes of motor impairment secondary to lesions or anomalies of the brain arising in the early stages of its development. The characteristic clinical feature that is common to all CP syndromes is the presence of pyramidal or extrapyramidal signs. CP is neither a specific disease nor a pathological or etiological entity, and importantly the term CP does not- and should not-necessarily imply or identify a specific cause (Gupta & Appleton, 2001).

The most recent consensus definition recognizes that: "Cerebral palsy describes a group of permanent disorders of the development of movement and posture, causing activity limitation, that are attributed to non-progressive disturbances that occurred in the developing fetal or infant brain. The motor disorders of cerebral palsy are often accompanied by disturbances of sensation, cognition, communication, and behavior, by epilepsy, and by secondary musculoskeletal problems (Venkateswaran and Shevell, 2008).

CP is traditionally classified according to the type of motor symptoms (spastic, dyscinetic, or ataxic) and the location of impairment (hemiplegic, diplegia, or quadriplegic). The spastic subtype accounts for 66%-82% of CP cases, which makes it the most common type (Bottcher, 2010).

A child is often first suspected of having CP in the first year of life when his or her motor milestones are delayed. For example, the child might be late in sitting, crawling, and walking. Parents may also notice that their baby has an atypical way of moving, such as commando crawling (crawling by pulling the body forward with the arms and dragging the legs behind) or that their child always stands or walks on his or her toes. The child may also appear to have "advanced" development, for example, being left-handed or "wanting to stand all the time". Clear evidence of handedness before 2 years of age is unusual and actually suggests impairment in the limb that is not used, whereas a child who seems "too strong" is probably showing evidence of increased stiffness in their muscles due to spasticity (Fehlings, Hunt & Rosenbaum, 2007).

The transitions from childhood to adolescence and from adolescence to adulthood are difficult for all families, but this stage of development can become particularly challenging when a child has a disability. The first challenge begins with learning to cope independently from one's parents. Other important issues include development of friendships, career planning, emotional and sexual relationships, and marriage. It is known that individuals with disabilities such as CP have more difficulty finding satisfying relationships and making career plans. Only 51% of youth with disabilities have plans for postsecondary education, compared with 74% of same-age peers (Fehlings, Hunt & Rosenbaum, 2007).

2.1.1 Prevalence and causes

Despite the huge progress of medicine in general and neonatology (or better: perinatology) in particular, the prevalence of cerebral palsy hasn't dropped in the recent decades and remains stable around 2 per 1000 live births (De Cock, 2009).

In 1998, fourteen centers in eight European countries started a network called Surveillance of Cerebral Palsy in Europe (SCPE). After reached consensus about the criteria to classify CP, they presented the prevalence rates in six countries, and more detailed prevalence estimated of 13 areas. The prevalence of CP rises in time from well below 2,0 per 1000 live births in the 1970s to well above 2,0 in the 1990s, boys form a small majority (58%). It seems fair to assume that these European data are not very different from findings in other parts of the world. For example the prevalence of CP in China is reported to be 1,6 per 1000 children under age 7. In Mississippi (USA) 2,12 per 1000 inhabitants were diagnosed with CP with a higher prevalence for males, and a, non-significant, higher prevalence in black people. The prevalence of CP in Australia is 2,0 to 2,5 per 1000 live births (Odding, Roebroeck & Stam, 2006).

In the last 15 years, intensive medical care performed in Neonatal Intensive Care Units (NICU) has allowed an increase in the survival of very low birth-weight (VLBW) and extremely premature newborns. New risk factors have appeared among infants who previously would have died, and the incidence of neurodevelopmental impairments in survivors of NICU is higher than in normal birth-weight newborns. In particular, due to the high risk of interventricular haemorrhage and periventricular leukomalacia, an increasing prevalence of cerebral palsy has occurred in premature, low birth-weight newborns and children born with asphyxia (Romeo D, Cioni, Scoto, Mazzone, Palermo & Romeo M, 2008).

According to the time of influence, causes of cerebral palsy can be divided to prenatal (from conception until beginning of the delivery), perinatal (beginning of the delivery until age of 28 days) and postnatal (from 29th day of age until two years of age). The majority of international studies indicates that the prevalence of the cerebral palsy is about 2-2,5 cases per 1000 born, although there are some reports about lower and higher prevalence rates (Nordmark, Hagglund & Lagergren, 2001).

Majority of previous research in the world was focused on the prevalence, determination of the motor abilities, and perinatal etiological factors of the cerebral palsy. Evidences indicated that 70-80 % of cerebral palsy is caused by the prenatal factors and that the birth asphyxia has a relatively minor role with the less than 10 % (Jacobsson & Hagberg, 2004).

Prenatal factors
Hereditary factors
Congenital anomalies of the CNS: hidrocephalus, microcephalus, cranistenosis, vascular malformations
Hormonal disorders
AIDS embriopathy
TORCHS infections
Psychological trauma
Intoxication of the mother
X-ray and other forms of radiation
Uncontrolled use of medications
Abortion attempt and previous bad abortion
Bleeding
Use of contraceptives
Fetal alcohol syndrom
Smoking
Insufficient nutrition
Cardiovascular disorders (severe hearth decompensations, shock) and severe anemia
Multifetal pregnancies
Mother age
Other factors

Table 2. Prenatal etiological factors of cerebral palsy

In our study, of the 35 participants with prenatal factors, 30 were participants whose mothers had problems during the pregnancy, 3 children had congenital anomalies, and 2 children had hereditary factors.

CP etiology	With epilepsy	Without epilepsy	Total
Prenatal	15	20	35 (43,75%)
Perinatal	14	23	37 (46,25%)
Postnatal	2	3	5 (6,25%)
Unknown factors	2	1	3 (3,75%)
Total	33 (41,25%)	47 (58,75%)	80 (100%)

X2=1.018 p=0.7969

Table 3. Frequency of epilepsy in 80 children with CP by CP etiology

Children with prenatal etiology for CP have 1,2 times greater chance of developing epilepsy.

In our study with prenatal etiological factors were 35 (43,75%) participants: 15 with epilepsy and 20 without; with perinatal 37 (46,25%) participants: 14 with epilepsy and 23 without epilepsy; with postnatal 5 (6,25%) participants: 2 with epilepsy and 3 without epilepsy, and with unknown etiological factors 3 (3,75%) participants: 2 with epilepsy and 1 without epilepsy.

Time of birth before introduction of modern neonatal intensive care in Bosnia and Herzegovina and war time (1992-1995) produced more representation of perinatal etiological factors.

Cerebral palsy	Etiological factors				Total
	Prenatal	Perinatal	Postnatal	Unknown	
Bilateral spastic CP					
Spastic Quadriplegic CP	5	7	1		13
Spastic Quadriplegic CP mixta	2	1			3
Triparesis	1				1
Paraparesis	3	2		1	6
Unilateral spastic CP					
Spastic Hemiplegic CP (right)	3	2			5
Spastic Hemiplegic CP (left)	1	2	1	1	5
Total	15	14	2	2	33

X2=9,697 p=0.8384

Table 4. Structure of children with cerebral palsy and epilepsy according to the etiological factors

All children, in group with epilepsy have spastic cerebral palsy: 23 bilateral and 10 unilateral spastic CP.

It is often assumed that children who develop CP have had a difficult birth. As more studies are done, however, it is becoming evident that only a small percentage of children with CP have difficulties related to their delivery. Birth asphyxia is a condition in which the newborn is felt to have a low oxygen level during delivery (sometimes resulting in delayed crying and poor respiratory effort at birth), and it is associated with the subsequent development of CP. One study of children with CP found that 78% did not have birth asphyxia, all had prenatal risk factors as well, which might have been the reason they had difficulty during the delivery. In fact, due to this type of research, the term birth asphyxia has been replaced with the term neonatal encephalopaty because this latter term does not imply a causal relationship (Fehlings, Hunt & Rosenbaum, 2007).

Birth weight less than 2500 g			Up to 3500 g		Over 3500 g		Unknown	Total
< 1500	1501-2000	2001-2500	2501-3000	3001-3500	3501-4000	>4001		
7	13	15	9	17	12	5	2	80
35 (43,75%)			26 (32,5%)		17 (21,25%)		2 (2,5%)	80 (100%)

Table 5. Structure of the sample according to the birth weight

Birth weight less than 2500 g had 35 children (43,75%), 2501-3500 g 26 children (32,5%), more than 3501 g 17 children (21,25%) and for 2 children (2,5%) birth weight was unknown.

Very-low-birth weight (VLBW) children are at risk of developing psychiatric symptoms and disorders, especially symptoms of attention deficit/hyperactivity disorder, anxiety and reduced social skills, and also possibly depression and thought problems. Cognitive and learning disadvantages are documented, as well as increased prevalence of neurodevelopmental disabilities, such as cerebral palsy, and minor motor and visuomotor problems (Indredavik, Vik, Heyerdahl, Romundstad & Brubakk, 2005).

2.2 Physical disability

With the rising incidence of CP in time, the distribution over the subtypes changed: fewer cases with diplegia and more with hemiplegic. The motor impairments of CP, in especially the spastic types, lead to other impairments of the musculoskeletal system; for example; among children and adolescents with quadriplegic CP, 75% have hip luxations, 73% contractures, and 72% scoliosis (Odding, Roebroeck & Stam, 2006).

Orthopedic (bone) complications are frequently seen in CP. They can require surgical intervention by an orthopedic surgeon. The extra stiffness in the muscle decreases the muscle growth. This can lead to joint contractures that can affect an individual's gait (walking pattern). The surgeon can do an operation that lengthens the tendon of the muscle to help to increase flexibility. For instance, with a heel cord lengthening, a child can go from always standing on his or her toes to being able to bring the heel down to the floor. A significant complication of the extra stiffness can be subluxation (a partial sliding out). The surgeon can lengthen the muscles around the hip and do a procedure to reshape the pelvis and try to prevent a complete dislocation of the hip. Children with CP are also at risk for osteopenia (low bone density) if they are nonambulatory. This in turn increases the risk of fractures of "thin" bones (Fehlings, Hunt & Rosenbaum, 2007).

In our study 16 (20%) children had operations of the tendons of the muscles (foot, knee and hip).

Children with CP have increased energy consumption by walking compared with appropriately developing peers. Within the spectrum of the condition, increasing energy consumption is associated with an increase in the severity of functional involvement. Functional and community ambulation issues associated with the increased energy consumption of walking may have a direct effect on participation and social integration of the child at home, at school, and in the community setting (Kerr, Parkes, Stevenson, Cosgrove & McDowell, 2008).

Gait abnormalities increase submaximal walking energy expenditure almost 3-fold compared with healthy children. Children with diplegia have a higher fat percentage and are hypoactive compared with healthy children. Wheelchair-dependent adolescent with CP are hypoactive, which is not the case for ambulatory adolescents with CP. Physical, but not mental, fatigue is more common in adults with CP, than in the general population. The strongest predictors for fatigue are bodily pain, deterioration of functional skills, limitations in emotional and physical role function and life satisfaction (Odding, Roebroeck & Stam, 2006).

In the Study "Cerebral palsy in Norway: Prevalence, subtypes and severity", a total of 374 children with CP were identified with a prevalence of 2,1 per 1000 live births .

Cerebral palsy	Walking ability				Total
	Walks without restrictions	Holding a hand	Walker	Wheelchair	
Bilateral spastic CP					
Spastic Quadriplegic CP	2	/	1	10	13
Spastic Quadriplegic CP mixta	/	/	/	3	3
Triplegia	/	/	/	1	1
Paraplegia	3	1	1	1	6
Unilateral spastic CP					
Spastic Hemiplegic CP (right)	5	/	/	/	5
Spastic Hemiplegic CP (left)	4	/	/	/1	5
Total	14	1	2	16	33

Table 6. Structure of the sample of children with CP and epilepsy according to walking ability

Of 33 children with cerebral palsy and epilepsy, 14 (42,4%) were able to walk independently, 1 (3%) child needs to hold a mother's or friend's hand, 2 (6%) children walks with assistive device (walker), and 16 (48,5%) children were unable to walk, in need of wheelchair.

Of total sample of 80 participants, 34 (42,5%) were in need of wheelchair, and 46 (57,5%) were not.

Of total sample of 80 children, 42 (52,5%) were able to walk independently.

In the group of 30 participants, with illnesses during pregnancy, 13 (43,3%) were in need of wheelchair, and 17 (56,7%) were not. In the group of 50 participants, without illnesses during pregnancy, 21 (42%) were in need of wheelchair, and 29 (58%) were not.

Of 30 participants with illnesses during pregnancy, 17 (56,7%) were able to walk independently, and 13 (43,3%) were not. Of 50 participants without illnesses during pregnancy, 25 (50%), were able to walk independently, and 25 (50%) were not.

2.3 Learning disabilities

The term *mental retardation* has been used in the United States and some other countries recently but is being used with decreasing frequency worldwide. In 2006, the American Association on Mental Retardation (AAMR) voted to change its name to the American Association on Intellectual and Developmental Disabilities (AAID). This name change became official January 1, 2007. The meanings of terms are the blend of their literal, definitional, and social meanings. The literal meaning of *intellectual disability* (ID) - a person's skill or power to know and understand, as the thing that deprives him or her of performing or accomplishing specific things - is tempered by widely accepted definitions of *intellectual disability* and similar terms and by specific social contexts within which the term is used. Similarly the literal meaning of *developmental disability* - something in the way a person grows and changes over time that deprives him or her of developing the abilities to perform or accomplish specific things - takes on specific meaning by those who use it in the jurisdictions where they live and work (Brown, 2007).

A child is described as having a *learning disability* or *mental retardation* (preferred by the World Health Organization) if there is firstly an overall intelligence significantly lower than the general population i.e. IQ less than 70. Secondly, the cognitive impairment must have occurred during the developmental period (less than 18 years), and is accompanied by declining social functioning (Courtman & Mumby, 2008).

A register of all persons with ID in a certain area, with detailed information from many different perspectives, would help in estimating the need for services. There are, however, obstacles in keeping such a register. In many countries, including Finland, a conscious and determined mainstreaming policy is underway, and maintaining a register of people with ID is viewed as a step backward. These registers may also be regarded as unethical and feared as a potential means of discrimination. Moreover, our experience has shown that keeping such a register up to date is demanding and the reliability is difficult to guarantee in normal clinical practice (Westerinen, Kaski, Virta, Almquist & Iivanainen, 2007).

About 40% of children with hemiplegic CP have normal cognitive abilities, while children and adolescents with tetraplegic CP are generally severely intellectually impaired (Odding, Roebroeck & Stam, 2006).

CP associated with epilepsy is far more frequently accompanied by intellectual disability than CP without epilepsy. Similarly, the combination of CP and intellectual disability is reported to be associated with a high risk of developing epilepsy (Carlsson, Hagberg & Olsson, 2003).

Intellectual disability	With epilepsy	Without epilepsy	Total
Normal range	4	11	15 (18,75%)
Borderline	1	4	5 (6,25%)
Mild ID*	7	17	24 (30%)
Moderate ID	9	7	16 (20%)
Severe ID	12	8	20 (25%)
Total	33	47	80 (100%)

*ID = Intellectual disability
Intellectual disability was defined as IQ below 70.
Normal range > 91; Border line IQ = 71-90; Mild ID IQ = 51-70; Moderate ID IQ = 35-50;
Severe ID IQ = 21-34
X2=8,081 p=0.0887

Table 7. Frequency of epilepsy in 80 children with CP by intellectual disability

Of 60 children with cerebral palsy and intellectual disability, 28 (46,7%) have epilepsy. Of 20 children with cerebral palsy and normal intellectual status, 5 (25%) children have epilepsy.

Among all children (n= 127) included in the Dutch population based study, only the hemiplegic subtype showed a majority with normal mental capacity; severe learning difficulties were most common in children with quadriplegic CP (Wichers, Odding, Stam & Nieuwenhuizen, 2005).

Cerebral palsy	Intellectual disability					Total
	Normal range	Borderline	Mild ID	Moderate ID	Severe ID	
Bilateral spastic cerebral palsy						
Spastic Quadriplegic CP	/	/	3	2	8	13
Spastic Quadriplegic CP mixta	/	/	/	/	3	3
Triplegia	/	1	/	/	/	1
Paraplegia	1	/	1	3	1	6
Unilateral spastic cerebral palsy						
Spastic Hemiplegic CP (right)	2	/	1	2	/	5
Spastic Hemiplegic CP (left)	1	/	2	2	/	5
Total	4	1	7	9	12	33

Table 8. Structure of children with cerebral palsy and epilepsy according to the intellectual disability

There is a significant difference between the groups: X2=53,130 p=0.0001

Of 33 children with cerebral palsy and epilepsy, 4 (12,1%) have normal mental capacity, 1 (3%) border, 7 (21,2%) have mild intellectual disability, 9 (27,3%) have moderate intellectual disability, and 12 (36,4%) have severe intellectual disability.

Department of Education (USA) report notes that 4,7% of public school age children are diagnosed as learning disabled. Many feel that the figure represents, at best, half of the children who should be so identified. Public Law 94-142, the Education for All Handicapped Children Act, established a definition for learning disabilities. Since that time, several professional and governmental groups have proposed modifications of this initial definition. In practical terms, the major defining characteristic of a child or adolescent with learning disabilities has become the discrepancy between his or her current academic achievement and the child's intellectual ability as measured by standardized tests (Silver, 1989).

In Sweden, practically all children with motor disabilities are brought up by their parents or sometimes foster parents. Furthermore, almost all children with CP and with normal cognitive development are integrated in mainstream school education: special schools for the group being exceptions. Children with learning disabilities are referred to special schools, but parents have the option to choose integration in a mainstream school for their child. That is not unusual in the lower school grades. Children with severe learning disability with IQs at the upper and attend special schools for those with learning disability and children with profound learning disability have some degree of educational activity on a daily basis, i.e. in communication and music therapy. Forty-three of the 105 children with CP and average IQ attended a mainstream school although they used mobility devices or wheeled mobility and 12 of the 26 children with mild learning disability attended a mainstream school with extra support. Only five children with severe learning disability had no form of educational activity (Beckung & Hagberg, 2002).

Education	With epilepsy	Without epilepsy	Total
Home	18	16	34 (42,5%)
School with special program	9	18	27 (33,75%)
Mainstream school	6	13	19 (23,75%)
Total	33	47	80 (100%)

Table 9. Frequency of epilepsy in 80 children with CP by education

There is a significant difference between the groups: X2=3,349 p=0.0187

More children with CP and epilepsy stay at home without any institutional re/habilitation, and more children without epilepsy attend schools with special programs and mainstream schools.

In our Study, of 80 children and adolescents with cerebral palsy, 34 (42,5%) stay at home without any service of education: 15 female and 19 male; 18 with CP and epilepsy, and 16 with CP and without epilepsy.

Of 80 children and adolescents with CP, 27 (33,75%) children attend schools with specially adapted programs.

Of whole sample of children and adolescents with CP, 19 (23,75%) children attend mainstream schools: 9 female, and 10 male; 6 with cerebral palsy and epilepsy, and 13 with CP and without epilepsy.

Education	Intellectual disability					Total
	Normal range	Border line	Mild ID	Moderate ID	Severe ID	
Home	/	/	1	5	12	18
Elementary school with specially adapted programs	/	/	5	1	/	6
Vocational programs in special elementary school	/	/	/	3	/	3
Mainstream elementary school	3	1	/	/	/	4
Mainstream secondary school	1	/	1	/	/	
Total	4	1	7	9	12	33

Table 10. Structure of the sample of children with CP and epilepsy according to the type of education

Nonverbal learning disabilities refer to developmental disorders of motor function (developmental coordination disorder), visuo-spatial processing, mathematics (dyscalculia), memory, prefrontal executive function, and social-emotional cognition and behavior (Ramanujapuram, 2007).

In a study of 149 children with hemiplegic, one third of children met the criteria for having a specific learning disability in reading, spelling, or math, and nearly half of the children with learning disabilities had problems in two or three areas despite an average verbal IQ. Learning disabilities have been shown to predict lower participation in children with CP (Bottcher, 2010).

Learning mathematics involves not one but a cluster of academic activities, each of which places different cognitive and social demands on the student. It is not enough to give students a test of arithmetic computation and feel satisfied that their mastery of mathematical learning is understood. Assessment of calculation is ordinarily carried out with pencil-and-paper tests, but for learning children with written language disability (e.g. reading or handwriting) assessment should also include oral presentations and responses of problems based on pictures or objects. Care must be taken to determine whether students understand the purpose or meaning of the calculation they are performing (Silver, 1989).

Children born prematurely, particularly those with birth weights below 1500 g, considered as a group, do less wll at school than their fully-developed contemporaries. This applies specially to the ability to solve mathematical problems, but it is also affects behaviour in the form of hyperactivity and ampared fine motor ability. Particularly exposed are those children who have been so ill during their neonatal period that they had to be treated in respirators, or who suffered cerebral haemorrhage (Janson S, 2001).

2.4 Communication problems

Many people with CP are able to speak fluently and clearly. Some have difficulty with articulation, which can make speech difficult for a listener to understand. Others may be unable to speak because of motor problems and require alternate strategies, such as picture or symbol displays to communicate. This is more likely to be the case for people with extrapyramidal CP. Some people with CP have language learning disabilities (Fehlings, Hunt & Rosenbaum, 2007).

Communication disorders are developmental speech and language disorders which include expressive language disorders, mixed receptive-expressive language disorders, phonological disorder, stuttering and other unspecified communication disorders. Prevalence rates for communication disorders range from 1-13% children (Ramanujapuram, 2007).

Many children with more severe spastic CP experience communication problems due to disturbed neuromuscular control of speech mechanism, i. e, dysarthria, that diminish the ability of the child to speak intelligible. However, substantial dysarthria are most often seen in children with severe CP and intellectual disability, while most children with mild or moderate CP and average cognitive level of functioning have normal or near-normal expressive language and articulation skills (Bottcher, 2010).

In our Study, of all sample, 31 (38,75%) children with CP used nonverbal and sign communication, and 49 (61,25%) children used verbal communication (i.e. speech).

Illnesses during pregnancy	Speech impairment		
	Yes	No	Total
With illnesses during pregnancy	22 (73,3%)	8 (26,7%9	30 (100%)
Without illnesses during pregnancy	34 (68%)	16 (32%)	50 (100%)
Total	56 (70%)	24 (30%)	80 (100%)

X2=0,063 p=>0,05 RR=1,07

Table 11. Structure of children with CP and speech impairment according to the illnesses during pregnancy

Of 80 participants, 56 (70%) children have speech impairment.

In subgroup with illnesses during pregnancy, of 30 participants, 22 have speech impairment. In subgroup without illnesses during pregnancy, of 50 participants, 34 have speech impairment.

Of 60 participants with CP and intellectual disability, 50 children have speech impairment. Of 20 participants with CP and normal IQ, 6 children have speech impairment.

Epilepsy	Speech impairment		
	Yes	No	Total
With epilepsy	27	6	33
Without epilepsy	29	18	47
Total	56 (70%)	24 (30%)	80 (100%)

X2=7,186 p=0.0001
RR=2,89 CI=0,9978-8,3877

Table 12. Structure of children with CP and speech impairment according to the epilepsy

Children with epilepsy have 2,89 greater chance of having speech impairment.

Of 33 children with CP and epilepsy, 27 have speech impairment.

Of 47 children with CP and without epilepsy, 29 have speech impairment.

2.5 Visual impairments

The objective of the study "Prevalence and selected characteristics of childhood vision impairment was to examine the descriptive epidemiology of vision impairment among 6-10 years old children in metropolitan Atlanta, Georgia, USA. Children with vision impairment (n=310; 42% black, 56% white; 57% male, 43% female), defined as a best corrected visual acuity in the better eye of 20/70 or worse, were identified through the Metropolitan Atlanta Developmental Disabilities Surveillance Program (MADDSP). The MADDSP conducts active population-based surveillance of five developmental disabilities: vision impairment, intellectual disability, cerebral palsy, hearing impairment, and autism among 3-10 year old children in the five-county metropolitan Atlanta area. The severity of the child's vision impairment was categorized according to World Health Organization guidelines (20/70 through 20/400), blindness (worse than 20/400), or unknown. The overall vision impairment prevalence rate was 10,7 per 10000 children. Prevalence did not differ significantly by either sex or race. 59% had low vision; 64% of the children had one or more coexisting developmental disability. The most common combination was intellectual disability, CP, and epilepsy, which was found in 85 (27%) of the children with vision impairment in the study (Mervis, Boyle & Yeargin-Allsopp, 2002).

Over the last decade the term cerebral visual impairment (CVI) has come to indicate the clinical picture of the child who present with visual impairment due to cerebral pathology, generally in the retrogeniculate visual pathway. Perinatal hypoxia is probably the most frequent cause of CVI and leads to severe neurological sequel that worsen the clinical picture. The visual deficit in CVI involves three different aspects: the perceptual (the sheer visual impairment with subnormal visual acuity and visual field defects), the neuropsychological (attention, recognition, and spatial orientation), and the oculomotor aspects. The latter are so severe that not only does it bring about strabismus and nystagmus but other oculomotor mechanism, such as saccadic movements, pursuit, and vergence, are affected (Salati, Borgatti, Giammari & Jacobson, 2002).

Visual perception is the complex processes that enable us to perceive a wide array of visual qualities such as movement, depth, spatial relations, facial expressions, and, eventually, the identity of objects. The normal functioning of the visual system is thought to hinge on both the integrity of the areas subserving visual-perceptual processes and early visual experience. Several studies have associated visuoperceptual impairment with reduction in the white matter in the parietal and occipital lobes in groups of children with spastic CP. The visuoperceptual impairments of children with spastic CP appear to be unrelated to general intelligence, nonverbal intelligence, or the presence of epilepsy (Bottcher, 2010).

Depending of the study, the prevalence of visuomotor and perceptual problems among children with spastic CP varies from 39% to 100% (Stiers & Vanderkelen, 2002)

In the Canadian Study "Comorbidities and clinical determinants of outcome in children with spastic quadriplegic (SQ) cerebral palsy", were included 92 patients (55 males and 37 females). This study demonstrated that the comorbidities of visual impairment, hearing loss, epilepsy, and the need for assisted feeding occur in a high proportion of children with SQ CP ranging from 33% (assisted feeding) to 80% (visual difficulty). Whether the child was born preterm or at term did not appear to significantly affect any of these eventual outcomes. It is demonstrated that visual impairment is the most common comorbidity in group of children with SQ. Studies have demonstrated that visual abnormalities vary from 10 to 39%. Periventricular leukomalacia (PVL) appears in prior studies to have correlation with observed visual impairment depending on the extent of white matter injury (Venkateswaran & Shevell, 2008).

In a representative series of 176 children with cerebral palsy, aged 5 to 8 years, associations were studied between additional neurodisabilities, activity limitation, and participation restrictions in the domains of mobility, education and social relations as proposed in the International Classification of Functioning Disability and Health (ICF). Visual impairment occurred in 20%, and infantile hydrocephalus in 9% of the children (Beckung & Hagberg, 2002).

Problems during pregnancy	Visual impairment		
	Yes	No	Total
With problems during pregnancy	18 (60%)	12 (40%)	30 (100%)
Without problems during pregnancy	22 (44%)	28 (56%)	50 (100%)
Total	40 (50%)	40 (50%)	80 (100 %)

X2=1,333 p>0,05

Table 13. Structure of children with visual impairment according to the subgroup

Of total sample (80 participants), 40 (50%) have visual impairmants.

Of 30 participants whose mothers had problems during pregnancy, 18 (60%) have visual impairments.

Of 50 participants whose mothers didn't have problems during pregnancy, 22 (44%) have visual impairments.

It is well known that impaired vision early in life from strabismus or ocular disorders can lead to permanent amblyopia because of reorganization of central visual pathways, and early hearing impairment can lead to impaired central auditory perception (Johnston, 2009).

In our research, of total sample, 40 (50%) participants have strabismus.

Of 30 participants whose mothers had problems in pregnancy, 17 (56,7%) have strabismus.

Of 50 participants whose mothers didn't have problems in pregnancy, 23 (46%) have strabismus.

Children whose mothers had illness during pregnancy have 1,2 times greater chance of strabismus.

Epilepsy	Visual impairment		
	Yes	No	Total
With epilepsy	15	18	33
Without epilepsy	25	22	47
Total	40	40	80

Table 14. Structure of children with CP and visual impairment according to the epilepsy

Of whole sample, 4 children (5%) were blind in both eyes, 2 children with epilepsy, and 2 without. One girl without epilepsy, had myopia gravis (- 10) and nystagmus.

2.6 Hearing impairment

Hearing loss is known to be prominent in patients with kernicterus, congenital infections, very low birthweight (VLBW) or severe hypoxic ishemic injury. Hearing impairment in groups of patients with CP ranged from 4 to 15% (Venkateswaran and Shevell, 2008).

In the Study of neuroimpairments, activity limitations, and participation restrictions in children with cerebral palsy, in Sweden, of 176 children with CP, severe hearing impairment have 2 children (Beckung & Hagberg, 2002).

In the study of the health status of 408 school-aged children with cerebral palsy, range 5 to 13 years, hearing ability among the children was very good; overall 97% could hear perfectly, while 2% (n=8) were deaf. There was suggestive evidence that a higher percentage of children with hearing difficulties was found in the Gross Motor Function Classification System (GMFCS) level V, but the numbers were small. In the whole sample, only 12 children (3% of the total) had a hearing ranking other than 1 (best); however, in GMFCS level V, 9 of 84 children (almost 11%) were classified by their parents as fitting into a hearing ranking other than 1 (Kennes, Rosenbaum, Hanna, Walter, Russell, Raina, Bartlett & Galuppi, 2002).

In our study, of total sample of 80 children and adolescents with CP, hearing impairment have 3 children (3,75%). All three children had intellectual disability, and two of them had epilepsy.

2.7 Behavioral and emotional problems

Professionals and parents need to be aware that children with cerebral palsy are at higher risk of psychological problems than their non-disabled peers and this may be attributable to problems in adjustment to their adverse circumstances as well as having an organic basis. Attention should be paid to the effective management of pain, particularly in children unable to self-report for whom a reliable instrument for assessing pain now exists. The difficulties most commonly reported here were peer problems; as these may have implications for later psychological adjustment, follow up work into adolescence and beyond will be important. It may be that for many children with cerebral palsy and their families, chronic psychological problems will have a greater impact than the physical impairments and this possibility also needs to be investigated in longitudinal studies (Parkes, White-Koning, Dickinson, Thyen, Arnaud, Beckung & all, 2008).

Emotional and psychosocial problems are disproportionately high in people with epilepsy, particularly in people with intractable epilepsy. In one large study, approximately 50% of the children with intractable epilepsy were identified as having serious psychosocial problems. In another study, clear-cut psychiatric disorders were identified in 33% of children with epilepsy, as compared with 7% in the general population and 2% in children with other chronic illnesses. Some of the most problems include anxiety, depression, irritability, aggression, and irrational periods of rage. In children at risk for suicide, there is a fifteen-fold overrepresentation of children with epilepsy (Burnham, 2007).

Difficulties in psychosocial adjustment appear to be the major manifestation of learning disabilities. Children with learning disabilities experience less acceptance, lower popularity, more peer rejection, and increased neglect by peers than do normally achieving children or low-achieving peers. In addition to low self-esteem, social skills deficits and general psychosocial adjustment difficulties, many children with learning disabilities experience more serious psychopathology or seek psychiatric help. It has been estimated that 30-70% of children with learning disabilities will experience ongoing comorbid symptoms of attention-deficit/hyperactivity disorder (Ramanujapuram, 2007).

Children with epilepsy are at increased risk for depression and anxiety, and the effect of these conditions on their performance in the classroom can be significant. This is especially true for children who recently experienced their first seizure or may have been recently diagnosed with epilepsy. The condition of epilepsy carries with it a high degree of stigma and it can be confusing and scary for children, resulting in feelings of learned helplessness and loss of control. Adjustment disorders with behavioral or emotional features are not uncommon after the diagnosis and should always be considered as a potential etiology of learning problems (Titus & Thio, 2009).

3. Conclusion

There are large number of health and educational institutions in the Canton of Sarajevo which are working on re/habilitation and education of children and adolescents with intellectual and developmental disab**ilities,** but there isn't a unique database about the people with disabilities, as well as with the cerebral palsy. Lack of unique database indicates poor network among these institutions and Associations in the Canton of Sarajevo.

It is necessary to initiate a Project to open the Registers of neurology developmental disorders in Bosnia and Herzegovina as necessary factor for contemporary exchange of data and link toward European Register, as well as in order to improve organization of health care for children and adults with disability.

Children with prenatal etiology for CP have 1,2 times greater chance of developing epilepsy.

In our study with prenatal etiological factors were 35 (43,75%) participants: 15 with epilepsy and 20 without; with perinatal 37 (46,25%) participants: 14 with epilepsy and 23 without epilepsy; with postnatal 5 (6,25%) participants: 2 with epilepsy and 3 without epilepsy, and with unknown etiological factors 3 (3,75%) participants: 2 with epilepsy and 1 without epilepsy.

It is necessary to initiate more projects of Family Counseling, for education of young couples about etiological factors of learning disabilities, especially prenatal.

More children with CP and epilepsy stay at home without any institutional re/habilitation, and more children without epilepsy attend schools with special programs and mainstream schools.

In our Study, of 80 children and adolescents with cerebral palsy, 34 (42,5%) stay at home without any service of education: 15 female and 19 male; 18 with CP and epilepsy, and 16 with CP and without epilepsy.

Children with epilepsy have 2,89 greater chance of having speech impairment; X2=7,186 p=0.0001

RR=2,89 CI=0,9978-8,3877

Of 33 children with CP and epilepsy, 27 have speech impairment.

Of 47 children with CP and without epilepsy, 29 have speech impairment.

It is also necessary to conduct continuous education of the teaching staff at schools with general curriculum and parents of the children without intellectual disability, as well as health professionals.

In recent years, the probability of long-term survival has increased even among children with a severe level of disability. This means that appropriate services will need to be provided for children with CP through adolescence and into adulthood (Beckung & Hagberg, 2002).

Identification and heightened awareness of comorbidities of CP may assist physicians in guiding caregivers, delivering appropriate counseling, and helping families access suitable resources for their child. Such efforts will lessen the burdens of disability, minimizing secondary complications, and hopefully improve overall quality of life for both the child and family (Venkateswaran & Shevell, 2008).

It is necessary to implement contemporary principles in the management of the cerebral palsy in order to make improvements at all levels of health care, and not only the cure seeking treatment.

Preventions should be primary, secondary, and tertiary, with the guidelines given in this study.

Finally, it is important that all children with a diagnosis of CP should be followed up, and when no obvious cause has been identified or there is any evidence of regression, children should be referred to a pediatric neurologist (Gupta & Appleton).

4. Summary

The aims of the study "The influence of prenatal etiological factors on learning disabilities of children and adolescents with cerebral palsy" were:

- to identify the prenatal etiological factors of cerebral palsy in children and adolescents aged from 6 to 20 years in the Canton of Sarajevo, Bosnia and Herzegovina;
- to determine the learning disabilities among children with the cerebral palsy, and
- to determine relationship between prenatal etiological factors of cerebral palsy and learning disabilities of children and adolescents with cerebral palsy.

Importance of this study is reflected in the fact that this is the first time in Bosnia and Herzegovina that the influence of the prenatal etiological factors on learning difficulties of the children and adolescents with the cerebral palsy has been determined with use of a scientific method.

The study of the influence of prenatal etiological factors on learning disabilities of children and adolescents with cerebral palsy in the Canton of Sarajevo was conducted as a cohort, retrospective study. The sample was consisted of 80 participants, children and adolescents with cerebral palsy in the Canton of Sarajevo, age from 6 up to 20 years; mean age was 13,94 years, 47 male (58,75%) and 33 (41,25%) female. The sample was divided in two subgroups, first includes 30 participants whose mothers had problems during the pregnancy, and second includes 50 participants whose mothers didn't have problems during the pregnancy. The research data were collected using a structural interview for the parents of children and adolescents with cerebral palsy, which was developed by the investigator, based on professional and scientific literature, and personal experience. The Interview consists of 69 questions.

Of 33 children with cerebral palsy and epilepsy, 21 (63,6%) are male, and 12 (36,4%) are female. Of 47 children with cerebral palsy, and without epilepsy, 26 (55,32%) are male and 21 (46,68%) are female.

According to the time of influence, causes of cerebral palsy can be divided to prenatal (from conception until beginning of the delivery), perinatal (beginning of the delivery until age of 28 days) and postnatal (from 29th day of age until two years of age). In our study with prenatal etiological factors were 35 (43,75%) participants: 15 with epilepsy and 20 without; with perinatal 37 (46,25%) participants: 14 with epilepsy and 23 without epilepsy; with postnatal 5 (6,25%) participants: 2 with epilepsy and 3 without epilepsy, and with unknown etiological factors 3 (3,75%) participants: 2 with epilepsy and 1 without epilepsy. Of 33 children with cerebral palsy and epilepsy, 14 (42,4%) were able to walk independently, 1 (3%) child needs to hold a mother's or friend's hand, 2 (6%) children walks with assistive device (walker), and 16 (48,5%) children were unable to walk, in need of wheelchair. Of total sample of 80 participants, 34 (42,5%) were in need of wheelchair, and 46 (57,5%) were not. Of total sample of 80 children, 42 (52,5%) were able to walk independently.

Of 33 children with cerebral palsy and epilepsy, 4 (12,1%) have normal mental capacity, 1 (3%) border, 7 (21,2%) have mild intellectual disability, 9 (27,3%) have moderate intellectual disability, and 12 (36,4%) have severe intellectual disability. In our Study, of 80 children and adolescents with cerebral palsy, 34 (42,5%) stay at home without any service of education: 15 female and 19 male; 18 with CP and epilepsy, and 16 with CP and without epilepsy.

Of 80 participants, 56 (70%) children have speech impairment. In subgroup with illnesses during pregnancy, of 30 participants, 22 have speech impairment. In subgroup without illnesses during pregnancy, of 50 participants, 34 have speech impairment.

Of total sample (80 participants), 40 (50%) have visual impairmants. Of 30 participants whose mothers had problems during pregnancy, 18 (60%) have visual impairments. Of 50 participants whose mothers didn't have problems during pregnancy, 22 (44%) have visual impairments.

In our study, of total sample of 80 children and adolescents with CP, hearing impairment have 3 children (3,75%). All three children had intellectual disability, and two of them had epilepsy.

There are large number of health and educational institutions in the Canton of Sarajevo which are working on re/habilitation and education of children and adolescents with intellectual and developmental disab**ilities,** but there isn't a unique database about the people with disabilities, as well as with the cerebral palsy. Lack of unique database indicates poor network among these institutions and Associations in the Canton of Sarajevo.

Finally, it is important that all children with a diagnosis of CP should be followed up, and when no obvious cause has been identified or there is any evidence of regression, children should be referred to a pediatric neurologist (Gupta & Appleton).

5. Acknowledgments

I thank the children and families of the Canton of Sarajevo, Bosnia and Herzegovina, who participated in the Study "Learning difficulties in children with cerebral palsy", 2004, and the Study "Quality of life in families of the school age children with intellectual disabilities", 2008.

I wish to thank to Professor Slobodan Loga, MD, PhD, psychiatrist, Academy of Sciences and Arts of Bosnia and Herzegovina, for his professional help with the both researches, as mentor.

I am grateful to professor Bengt Lagerkvist, MD, PhD, pediatrician, from a joint master project on child and adolescent psychiatry and psychology between Sarajevo University and Umeå University, Sweden.

I am grateful to professor Ivan Brown, PhD, Faculty of Social Work, University of Toronto, Canada, for his professional help with the second research in Family Quality of Life.

6. References

Andersen, G. L, Irgens, L. M, Haagaas, I, Skranes, J. S, Meberg, A. E. & Vik, T. (2008). Cerebral palsy in Norway: Prevalence, subtypes and severity. *European Journal of Paediatric Neurology* 12: 4-13.

Beckung, E. & Hagberg, G. (2002). Neuroimpairments, activity limitations, and participation restrictions in children with cerebral palsy. *Developmental Medicine & Child Neurology* 44: 309-316.

Beghi, M, Cornaggia, C. M, Frigeni, B, & Beghi E. (2006). Learning Disorders in Epilepsy. *Epilepsy*, 47 (Suppl. 2): 14-18.

Berg, A. T, Langfitt, J. T, Testa, F. M, Levy, S. R, DiMario, F, Westerveld, M. & Kulas, J. (2008). Global cognitive function in children with epilepsy: A community-based study. *Epilepsy*, 49 (4): 608-614.

Bottcher, L. (2010). Children with spastic cerebral palsy, their cognitive functioning, and social participation: a review. *Child Neuropsychology*, 16: 209-228.

Brown, I. (2007). What Is Meant by Intellectual and Developmental Disabilities? In: *A Comprehensive Guide to Intellectual & Developmental Disabilities*. Brown I. & Percy M, 3- 17 ISBN-13: 978-1-55766-700-7 Brookes Publishing Co. Baltimore

Burnham, W. M. (2007) Epilepsy. In: *A Comprehensive Guide to Intellectual & Developmental Disabilities*. Brown, I. & Percy, M. 287- 295 ISBN-13: 978-1-55766-700-7 Brookes Publishing Co. Baltimore.

Carlsson, M; Hagberg, G. & Olsson, I. (2003). Clinical etiological aspects of epilepsy in children with cerebral palsy. *Developmental Medicine & Child Neurology*, 45: 371-376.

Courtman, S. P. & Mumby, D. (2008). Children with learning disabilities. *Pediatric Anesthesia*, 18: 198-207.

De Cock, P. (2009). The cerebral palsies: A changing panorama –the Leuven experience. In: *Current aspects of cerebral palsy therapy: Scientific meeting with international participation*. Novi Sad. ISBN 978-86-87837-00-3. COBISS.SR-ID 238249735.

Fehlings, D; Hunt C. & Rosenbaum P. (2007). Cerebral Palsy. In: *A Comprehensive Guide to Intellectual & Developmental Disabilities*. Brown, I. & Percy, M. 287- 295 ISBN-13: 978-1-55766-700-7 Brookes Publishing Co. Baltimore.

Gupta, R. & Appleton, R. E. (2001). Cerebral palsy: not always what it seems. *Archives of Disease in Childhood*. 85: 356-360.

Indredavik, M. S, Vik, T, Heyerdahl, S, Romundstad P, Brubakk A. M. (2005). Low-birthweight adolescents: Quality of life and parent child relations. *Acta Paediatrica*: 94: 1295-1302.

Jacobsson, B. & Hagberg, G. (2004). Antenatal riscs factors for cerebral palsy. *Best Pract Clinic Obstetric Gynaecol*, 18 (3), 425-436

Janson, S. (2001). Children's and young people's health. *Scandinavian J Public Health* 29 (Suppl 58) p. 103- 116

Johnston, M. V. (2009). Plasticity in the Developing Brain: Implications for Rehabilitation. *Developmental disabilities Research Reviews* 15: 94-101.

Kennes, J.; Rosenbaum, P.; Hanna, S. E.; Walter, S.; Russell, D.; Raina, P.; Bartlett, D, & Galuppi, B. (2002). Health status of school-aged children with cerebral palsy:information from a population-based sample. *Developmental Medicine & Child Neurology*, 44: 240-247.

Kerr, C.; Parkes, J.; Stevenson, M.; Cosgrove, A. P, & McDowell, B. C. (2008). Energy efficiency in gait, activity participation, and health status in children with cerebral palsy. *Developmenta Medicine & Child Neurology* 50: 204-210.

Lagerkvist, B.; Maglajlić, RA.; Puratić, V.; Suši, A, & Jacobsson, L. (2003). Assessment of community mental health centres in Bosnia and Herzegovina as part of the ongoing mental health reform. Sarajevo: *The SweBiH Mental Health Conference.* p. 95-107.

Mervis, A. C; Boyle, A. C. & Yeargin-Allsopp M. Prevalence and selected characteristics ooof childhood vision impairment. *Developmental Medicine & Child Neurology, 44: 538-541.*

Nordmark. E.; Hagglund, G. & Lagergren, J. (2001). Cerebral Palsy in south Sweden. Prevalence and clinical features. *Acta Pediatrica* 90: 1271-1276

Odding, E.; Roebroeck, M. E. & Stam, H. J. (2006). The epidemiology of cerebral palsy: Incidence, impairments and risk factors. *Disability and Rehabilitation*, 28(4): 183-191.

Parkes, J.; White-Koning, M.; O Dickinson, H.; Thyen, U.; Arnaud, C.; Beckung, E. & all. (2008). Psychological problems in children with cerebral palsy: a cross-sectional European study. *The Journal of Child Psychology and Psychiatry* 49: 4, p. 405-413.

Percy, M. (2007). Factors that Cause or Contribute to Intellectual and Developmental Disabilities. In: *A Comprehensive Guide to Intellectual & Developmental Disabilities.* Brown I. & Percy M, editors. Paul H. Brookes Publishing Co. Baltimore, 125-148

Peterson, C. A.; Mayer, L. M.; Summers, J. A, & Luze, G. J. (2010). Meeting Needs of Young Children at Risk for or Having a Disability. Early Childhood Education Journal. 37: 509-517.

Ramanujapuram, A. (2007). Neuropsychiatry of Learning Disabilities. *Internet Journal of Neurology*;,Vol. 6 Issue 1. Database: Academic Search Premier. ISSN: 1531295X

Romeo, D.; Cioni, M.; Scoto, M.; Mazzone, L.; Palermo, F. & Romeo, M. (2008). Neuromotor development in infants with cerebral palsy investigated by the Hammersmith Infant Neurological Examination during the first year of age. *European Journal of Paediatric Neurology* 12: 24-31.

Salati, R; Borgatti, R; Giammari G. & Jacobson L. (2002). Oculomotor dysfunction in cerebral visual impairment following perinatal hypoxia. *Developmental Medicine & Child Neurology*, 44: 542-550.

Silver, B- L. (1991). The Assessment of Learning Disabilities: Preschool Through Adulthood. 8700 Shoal Creek Boulevard. Austin, Texas 78757. ISBN 0-89079-393-X

Stiers, P. & Vanderkelen, R. (2002). Visual-perceptual impairment in a random sample of children with cerebral palsy. *Developmental Medicine & Child Neurology*, 44: 370-382.

Titus, J. B, & Thio, L.L. (2009). The effects of antiepileptic drugs on classroom performance. *Psychology in the Schools*, Vol. 46 (9). P. 885-891.

Westerinen, H.; Kaski, M.; Virta, L.; Almquist, F, & Iivanainen, M. (2007). Prevalence of intellectual disability: a comprehensive study based on national registers. *Journal of Intellectual Disability Research.* Volume 51,part 9. p. 715-725.

Wichers, M. J.; Odding, E.; Stam, H. J, & Nieuwenhuizen, O. V. (2005). Clinical presentation, associated disorders and aetiological moments in Cerebral Palsy: A Dutch population-based study. *Disability and Rehabilitation*, 27(10): 583-589.

Wilmot-Lee, B. (2008). Improving medicine taking in epilepsy. *Paediatric Nursing*. 20, 9, 37-43.

Venkateswaran, S. & Shevell, M. I. (2008). Comorbidities and clinical determinantsofoutcome in children with spastic quadriplegic cerebral palsy. *Developmental Medicine & Child Neurology*. 50: 216-222.